T0235689

IFIP Advances in Information and Communication Technology

654

Editor-in-Chief

Kai Rannenberg, Goethe University Frankfurt, Germany

Editorial Board Members

TC 1 – Foundations of Computer Science
  Luís Soares Barbosa, University of Minho, Braga, Portugal

TC 2 – Software: Theory and Practice
  Michael Goedicke, University of Duisburg-Essen, Germany

TC 3 – Education
  Arthur Tatnall, Victoria University, Melbourne, Australia

TC 5 – Information Technology Applications
  Erich J. Neuhold, University of Vienna, Austria

TC 6 – Communication Systems
  Burkhard Stiller, University of Zurich, Zürich, Switzerland

TC 7 – System Modeling and Optimization
  Fredi Tröltzsch, TU Berlin, Germany

TC 8 – Information Systems
  Jan Pries-Heje, Roskilde University, Denmark

TC 9 – ICT and Society
  David Kreps, National University of Ireland, Galway, Ireland

TC 10 – Computer Systems Technology
  Ricardo Reis, Federal University of Rio Grande do Sul, Porto Alegre, Brazil

TC 11 – Security and Privacy Protection in Information Processing Systems
  Steven Furnell, Plymouth University, UK

TC 12 – Artificial Intelligence
  Eunika Mercier-Laurent, University of Reims Champagne-Ardenne, Reims, France

TC 13 – Human-Computer Interaction
  Marco Winckler, University of Nice Sophia Antipolis, France

TC 14 – Entertainment Computing
  Rainer Malaka, University of Bremen, Germany

IFIP – The International Federation for Information Processing

IFIP was founded in 1960 under the auspices of UNESCO, following the first World Computer Congress held in Paris the previous year. A federation for societies working in information processing, IFIP's aim is two-fold: to support information processing in the countries of its members and to encourage technology transfer to developing nations. As its mission statement clearly states:

> *IFIP is the global non-profit federation of societies of ICT professionals that aims at achieving a worldwide professional and socially responsible development and application of information and communication technologies.*

IFIP is a non-profit-making organization, run almost solely by 2500 volunteers. It operates through a number of technical committees and working groups, which organize events and publications. IFIP's events range from large international open conferences to working conferences and local seminars.

The flagship event is the IFIP World Computer Congress, at which both invited and contributed papers are presented. Contributed papers are rigorously refereed and the rejection rate is high.

As with the Congress, participation in the open conferences is open to all and papers may be invited or submitted. Again, submitted papers are stringently refereed.

The working conferences are structured differently. They are usually run by a working group and attendance is generally smaller and occasionally by invitation only. Their purpose is to create an atmosphere conducive to innovation and development. Refereeing is also rigorous and papers are subjected to extensive group discussion.

Publications arising from IFIP events vary. The papers presented at the IFIP World Computer Congress and at open conferences are published as conference proceedings, while the results of the working conferences are often published as collections of selected and edited papers.

IFIP distinguishes three types of institutional membership: Country Representative Members, Members at Large, and Associate Members. The type of organization that can apply for membership is a wide variety and includes national or international societies of individual computer scientists/ICT professionals, associations or federations of such societies, government institutions/government related organizations, national or international research institutes or consortia, universities, academies of sciences, companies, national or international associations or federations of companies.

More information about this series at https://link.springer.com/bookseries/6102

Lekshmi Kalinathan · Priyadharsini R. ·
Madheswari Kanmani · Manisha S. (Eds.)

Computational Intelligence in Data Science

5th IFIP TC 12 International Conference, ICCIDS 2022
Virtual Event, March 24–26, 2022
Revised Selected Papers

Springer

Editors
Lekshmi Kalinathan 🆔
Sri Sivasubramaniya Nadar College
of Engineering
Chennai, India

Priyadharsini R.
Sri Sivasubramaniya Nadar College
of Engineering
Chennai, India

Madheswari Kanmani
Sri Sivasubramaniya Nadar College
of Engineering
Chennai, India

Manisha S.
Sri Sivasubramaniya Nadar College
of Engineering
Chennai, India

ISSN 1868-4238   ISSN 1868-422X (electronic)
IFIP Advances in Information and Communication Technology
ISBN 978-3-031-16366-1   ISBN 978-3-031-16364-7 (eBook)
https://doi.org/10.1007/978-3-031-16364-7

© IFIP International Federation for Information Processing 2022
This work is subject to copyright. All rights are reserved by the Publisher, whether the whole or part of the material is concerned, specifically the rights of translation, reprinting, reuse of illustrations, recitation, broadcasting, reproduction on microfilms or in any other physical way, and transmission or information storage and retrieval, electronic adaptation, computer software, or by similar or dissimilar methodology now known or hereafter developed.
The use of general descriptive names, registered names, trademarks, service marks, etc. in this publication does not imply, even in the absence of a specific statement, that such names are exempt from the relevant protective laws and regulations and therefore free for general use.
The publisher, the authors, and the editors are safe to assume that the advice and information in this book are believed to be true and accurate at the date of publication. Neither the publisher nor the authors or the editors give a warranty, expressed or implied, with respect to the material contained herein or for any errors or omissions that may have been made. The publisher remains neutral with regard to jurisdictional claims in published maps and institutional affiliations.

This Springer imprint is published by the registered company Springer Nature Switzerland AG
The registered company address is: Gewerbestrasse 11, 6330 Cham, Switzerland

Preface

The Department of Computer Science and Engineering at Sri Sivasubramaniya Nadar College of Engineering, Kalavakkam, Chennai, India, proudly hosted the 5th International Conference on Computational Intelligence in Data Science (ICCIDS 2022), as a virtual event during March 24–26, 2022.

The building blocks of computational intelligence involve computational modeling, natural intelligent systems, multi-agent systems, hybrid intelligent systems, etc. The topics covered by this conference include computational intelligence in machine learning, deep learning, the Internet of Things (IoT), and cyber security. All intelligence-based algorithms require data as a fundamental building block, and cyber security systems act as information gatekeepers, protecting networks and the data stored on them. With machine learning, cyber security systems can analyze patterns and learn from them to help in preventing similar attacks and respond to changing behavior. IoT and machine learning deliver insights hidden in data for rapid automated responses and improved decision making. Machine learning for IoT can be used to project future trends, detect anomalies, and augment intelligence by ingesting images, video, and audio.

This year's conference program included two pre-conference workshops with three sessions each followed by prominent keynote talks and paper presentations. The two workshops, "Image and Video Analysis" and "Data Science – CAPSTONE" were conducted in parallel on March 24, 2022. The conference provided an opportunity for students, researchers, engineers, developers, and practitioners from academia and industry to discuss and address the research and methods in solving problems related to data science and to share their experience and exchange their ideas in the field of data science and computational intelligence.

On the first day of the conference, speakers from Niramai Health Analytix, VIT, and SSN Institutions elaborated on the topics of "Medical Image Processing" and "Artificial Intelligence for Medical Imaging". Mohit Sharma, Data Scientist at Ericsson, Bangalore, India elaborated on the topic of "Data Science – CAPSTONE during the pre-conference workshop.

To enlighten our participants, five keynote talks were arranged for the second and third day. On the second day, three keynote talks on "Data Science and its applications", by Dileep A.D, Associate Professor, IIT Mandi, India, "What AI for Global Security?", by Eunika Mercier-Laurent, Chair IFIP TC12 (AI) and University of Reims Champagne-Ardenne France, and Dominique Verdejo, Chair IFIP WG12.13, France, and "Research Trends in Computer Vision" by Sunando Sengupta, Senior Researcher, Microsoft, UK, were delivered.

On the third day of the conference, two keynote talks on "From Features to Embeddings - Unified view of ML Pipelines for Text, Audio and Image Understanding", by Vinod Pathangay, Distinguished Member of Technical Staff - Senior Member, Wipro, India, and "Honeybee Democracy - Towards Collective Intelligence" by

Mieczyslaw L. Owoc, Professor Emeritus, Department of Business Intelligence in Management, Wroclaw University of Economics and Business, Wrocław, Poland, were delivered.

The conference received a total of 96 submissions from authors all over India and abroad, out of which 28 papers (an acceptance rate of 29%) were selected by the Program Committee after a careful and thorough review process. The papers were distributed across three paper presentation panels based on their domain. Each panel had a session coordinator and three session chairs (subject experts) to evaluate the research work. Based on the evaluation, one paper from each panel was declared as the best paper.

We would like to express our sincere thanks to IFIP and the Sri Sivasubramaniya Nadar College of Engineering for helping us in organizing this event. Special thanks go to Eunika Mercier-Laurent, University of Reims Champagne-Ardenne, France, for approving our conference from IFIP and Springer for accepting to publish the conference proceedings. We thank the authors for sharing their research and reviewers for their constructive suggestions.

March 2022

Lekshmi Kalinathan
Priyadharsini Ravisankar
Madheswari Kanmani
Manisha S.

Organization

Executive Committee

Chief Patron

Shiv Nadar — SSN Institutions, India

Patron

Kala Vijayakumar — SSN Institutions, India

General Chairs

V. E. Annamalai — Sri Sivasubramaniya Nadar College of Engineering, India

S. Radha — Sri Sivasubramaniya Nadar College of Engineering, India

Conference Chairs

T. T. Mirnalinee — Sri Sivasubramaniya Nadar College of Engineering, India

Chitra Babu — Sri Sivasubramaniya Nadar College of Engineering, India

Program Committee

Program Chairs

Lekshmi Kalinathan — Sri Sivasubramaniya Nadar College of Engineering, India

Priyadharsini Ravisankar — Sri Sivasubramaniya Nadar College of Engineering, India

Madheswari Kanmani — Sri Sivasubramaniya Nadar College of Engineering, India

Manisha S. — Sri Sivasubramaniya Nadar College of Engineering, India

Organizing Co-chairs

D. Thenmozhi — Sri Sivasubramaniya Nadar College of Engineering, India

J. Suresh — Sri Sivasubramaniya Nadar College of Engineering, India

V. Balasubramanian — Sri Sivasubramaniya Nadar College of Engineering, India

S. V. Jansi Rani	Sri Sivasubramaniya Nadar College of Engineering, India

Session Coordinators

N. Sujaudeen	Sri Sivasubramaniya Nadar College of Engineering, India
P. Mirunalini	Sri Sivasubramaniya Nadar College of Engineering, India
K. R. Sarath Chandran	Sri Sivasubramaniya Nadar College of Engineering, India

Session Chairs

Chitra Babu	Sri Sivasubramaniya Nadar College of Engineering, India
D. Venkatavara Prasad	Sri Sivasubramaniya Nadar College of Engineering, India
A. Chamundeswari	Sri Sivasubramaniya Nadar College of Engineering, India
R. Kanchana	Sri Sivasubramaniya Nadar College of Engineering, India
K. Vallidevi	Sri Sivasubramaniya Nadar College of Engineering, India
S. Saraswathi	Sri Sivasubramaniya Nadar College of Engineering, India
D. Thenmozhi	Sri Sivasubramaniya Nadar College of Engineering, India
J. Bhuvana	Sri Sivasubramaniya Nadar College of Engineering, India
B. Prabavathy	Sri Sivasubramaniya Nadar College of Engineering, India

Technical Program Committee

Mieczyslaw Lech Owoc	Wroclaw University, Poland
Eunika Mercier-Laurent	University of Reims Champagne-Ardenne, France
Ammar Mohammed	Cairo University, Egypt
Chua Chin Heng Matthew	NUS, Singapore
Wanling Gao	Institute of Computing Technology, China
Sanjay Misra	Covenant University, Nigeria
Shomona Gracia Jacob	Nizwa College of Technology, Oman
Latha Karthika	Brandupwise Marketing, New Zealand
Abhishek Kataria	Square, USA
Jyothi Chitra Thangaraj	CTS, USA
Vadivel Sangili	BITS Dubai, UAE
Venkatesh S.	Oracle, USA

Srikanth Bellamkonda	Oracle, USA
Vijaya Kumar B.	BITS Dubai, UAE
Kumaradevan Punithakumar	University of Alberta, Canada
Ashok Kumar Sarvepalli	Collabera, USA
Issac Niwas Swamidoss	Tawazun Technology and Innovation, UAE
Sundaram Suresh	IISc Bangalore, India
Chitra Kala	Anna University, India
Jayachandran	AI Labs@Course5i, India
Sivaselvan B.	IIITDM Kancheepuram, India
Umarani Jayaraman	IIITDM Chennai, India
Sriram Kailasam	IIT Mandi, India
R. Geetha Ramani	Anna University, India
Latha Parthian	Pondicherry University, India
Anbarasu	IIT Indore, India
Srinivasa Rao	IIT Kharagpur, India
Premkumar K.	IIITDM Kancheepuram, India
Ravindran	IIT Madras, India
Rahul Raman	IIITDM Kancheepuram, India
Munesh Singh	IIITDM Jabalpur, India
Peddoju Sateesh Kumar	IIT Roorkee, India
Raksha Sharma	IIT Roorkee, India
Veena Thenkanidiyoor	NIT Goa, India
Uma Maheswari	BITS Pilani, India
S. Sheerazuddin	NIT Calicut, India
K. Ruba Soundar	MSEC, Sivakasi, India
R. Sakthivel	Amrita School of Engineering, Coimbatore, India
Madhu Shandilya	MANIT Bhopal, India
Sheeba Rani J.	IISST Kerala, India
Vijay Bhaskar Semwal	MANIT Bhopal, India
V. N. Manjunath Aradhya	SJCE, Mysore, India
A. Priyadharshini	Coimbatore Institute of Technology, India
Cynthia J.	KCT Coimbatore, India
Rajalakshmi R.	VIT Chennai, India
K. Sathya	Coimbatore Institute of Technology, India
G. Manju	SRMIST, India
T. M. Navamani	VIT Vellore, India
B. S. Charulatha	REC Chennai, India
R. Jayabhaduri	SVCE, India
C. Jayakumar	SVCE, India

Contents

Comparative Analysis of Sensor-Based Human Activity Recognition Using Artificial Intelligence

Alagappan Swaminathan[(✉)] [iD]

Georgia Institute of Technology, Atlanta, GA 30332, USA
alagappan.swaminathan@gatech.edu

Abstract. Human Activity Recognition (HAR) has become one of the most prominent research topics in the field of ubiquitous computing and pattern recognition over the last decade. In this paper, a comparative analysis of 17 different algorithms is done using a 4-core 940mx machine and a 16-core G4dn.4xlarge Elastic Compute (EC2) instance on a public domain HAR dataset using accelerometer and gyroscope data from the inertial sensors in smartphones. The results are evaluated using the metrics accuracy, F1-score, precision, recall, training time, and testing time. The Machine Learning (ML) models implemented include Logistic Regression (LR), Support Vector Classifier (SVC), Random Forest (RF), Decision Trees (DT), Gradient Boosted Decision Trees (GBDT), linear and Radial Basis Function (RBF) kernel Support Vector Machines (SVM), K- Nearest Neighbors (KNN) and Naive Bayes (NB). The Deep Learning (DL) models implemented include Convolutional Neural Networks (CNN), Long Short Term Memory (LSTM), a combination of CNN-LSTM and Bidirectional LSTM. Neural Structure Learning was also implemented over a CNN-LSTM model along with Deep Belief Networks (DBN). It is identified that the Deep Learning models CNN, LSTM, CNN-LSTM & CNN-BLSTM consistently confuse between dynamic activities and that the machine learning models confuse between static activities. A Divide and Conquer approach was implemented on the dataset and CNN achieved an accuracy of 99.92% on the dynamic activities, whereas the CNN-LSTM model achieved an accuracy of 96.73% eliminating confusion between the static and dynamic activities. Maximum classification accuracy of 99.02% was achieved by DBN on the full dataset after Gaussian standardization. The proposed DBN model is much more efficient, lightweight, accurate, and faster in its classification than the existing models.

Keywords: Machine learning · Deep learning · UCI HAR · Inertial sensors · Divide and Conquer · Deep Belief Networks

ⓒ IFIP International Federation for Information Processing 2022
Published by Springer Nature Switzerland AG 2022
L. Kalinathan et al. (Eds.): ICCIDS 2022, IFIP AICT 654, pp. 1–17, 2022.
https://doi.org/10.1007/978-3-031-16364-7_1

1   Introduction

"Human Activity Recognition (HAR) has emerged as a major player in this era of cutting-edge technological advancement. A key role that HAR plays is its ability to remotely monitor people" [1, (K et al. 2021)]. Monitoring the activity of an individual is an important aspect in "health-related applications especially that of the elderly people" [2, (Kasteren et al. 2009)]. Smartphones along with several wearable devices have gyroscopes and accelerometers in them which can be used to classify physical activity. Video data is another approach to classifying physical activity.

Monitoring one's progress is a key element in any kind of education, whether in academic subjects or in physical activities. Recently activity trackers have become an omnipresent tool for measuring physical activities. It has been identified that existing activity trackers are inaccurate with their classification of even simpler activities like standing, sitting, walking upstairs, and downstairs. Dorn's 2019 study [3, (Dorn et al. 2019)] of the current commercial activity monitoring wearables like Fitbit indicated that the devices are inaccurate in their classification of the elliptical machine activity and biking and walking to an extent. A minimum accuracy of 93% was achieved for the activity of walking.

RubénSan-Segundo's 2018 study [4, (San-Segundo et al. 2018)] of human activity recognition, identifies that standing, sitting, walking upstairs, and walking downstairs have relatively lower accuracy as compared to biking and plain walking. The paper indicates that there exists a need to develop algorithms to classify the activity of the individuals with greater accuracy. The existing solutions aren't accurate enough in their classification and hence there arises a need to implement different models to attain better classification results to enable reliable and accurate personal activity tracking & learning. The author concludes that "Confusion is greater between static activities (sit and stand), and similar dynamic activities (walking, walking-upstairs, walking-downstairs, and biking) for smartphones" [4, (San-Segundo et al. 2018)]

2   Related Work

Negar Golestani [7, (Golestani and Moghaddam 2020)] has stated that "The recurrent neural networks can capture sequential and time dependencies between input data that results in a strong performance. The LSTM cells let the model capture even longer dependencies compared to vanilla cells. A deep architecture with an optimal number of layers enables the neural network to extract useful discriminative features from the set of input data and to improve the performance of the model". Huaijun Wang [8, (Wang et al 2020)] trained a combination of CNN, LSTM, CNN-BLSTM, CNN-GRU, CNN-LSTM on the UCI machine learning repository dataset named "Smartphone-Based Recognition of Human Activities and Postural Transitions Data Set" [11, (Dua, D. and Graff, C. 2019)] and achieved 95.87% with the standard CNN-LSTM model.

Uddin's 2021 study [9, (Uddin and Soylu 2021)] indicated that Deep Neural Structured Learning(NSL) achieved 99% accuracy for the classification of 20 activities that are part of the MHealth [12, 13] open-source dataset from the UC Irvine repository. Lukun Wang [10, (Wang, 2016)], concluded that a combination of sensors from the right and left leg using Deep Belief Network (DBN) yielded the highest average accuracy of 99.3% with differentiating the nineteen activities on their own dataset. An experiment was conducted by Akram Bayat [6, (Bayat et al. 2014)] on the classification of activities in a similar UCI dataset using the data from the triaxial accelerometer present within a smartphone. They achieved a maximum accuracy of 91.15% using a combination of MP, LogitBoost, SVM. Artificial Neural Networks were used for classification of similar signals from wearable devices with an accuracy of 89%. [45]

As evident from the literature survey presented in this section so far, and from the recent research papers in Table 1, It can be understood that a wide variety of algorithms have been implemented by researchers on this dataset over a long period of time. The algorithms all have varying values of accuracy from one author to the other and the results are not verifiable since the code is not open-sourced. In this paper, all the most successful algorithms in the HAR field (17 in number) have been experimented with in two different machines to confirm the results with extensive hyper-parameter tuning. The code will be open-sourced. The 17 algorithms experimented within this paper were chosen based on the accuracy of classification of other inertial sensor-based HAR data sets from the literature survey. There exists no function in the popular machine learning or deep learning libraries to implement DBN, yet the original model proposed by Dr. Hinton [35] achieves 99.02% accuracy indicating the capability of the model to excel in the classification of human activities.

The machine learning models chosen for this project were based on past research work cited in Table 1 in order to present the results of ML algorithms

Table 1. The results obtained by recent papers from 2019, 2020 & 2021 in comparison with the proposed model using the UCI HAR dataset [5]

S. No	Author	Model name	Accuracy (%)
1	[1, (K et al. 2021)]	Linear Kernel SVM	96.6
2	[14, (Sikder et al. 2019)]	CNN	95.25
3	[15, (Cruciani et al. 2020)]	CNN	91.98
4	[16, (Ullah et al. 2019)]	LSTM	93.19
5	[17, (Ramachandran & Pang, n.d., 2020)]	Transfer learning+CNN	94
6	[18, (Roobini & Naomi, 2019)]	RNN-LSTM	93.89
7	[19, (Khatiwada et al. 2021)]	CBO-RNN	90.10
8	[20, (Rabbi et al. 2021)]	SVM	96.33
9	[21, (Han et al. 2021)]	Stacked Discriminant Feature Learning (SDFL)	96.26
10	[22, (Oh et al. 2021)]	Semi-supervised active transfer learning	95.9
11	[23, (Huang et al. 2021)]	3-layer CNN	96.98
12	[24, (Xia et al. 2020)]	LSTM-CNN	95.78
13	[25, (Nambissan et al. 2021)]	3-layer LSTM	89.54
14	[26, (Tang et al. 2021)]	CNN using Lego Filters	96.90
15	[27, (Bashar et al. 2020)]	Dense Neural Network	95.79
16	[28, (Zhang & Ramachandran, n.d. 2020)]	Latent Dirichlet allocation	96.20
17	[29, (Nematallah & Rajan, 2020)]	1-D CNN +Stat features	97.12
18	This paper	Deep Belief Networks	**99.02**

for comparison with the more prominent and successful DL algorithms. The DL algorithms excluding DBN are a combination of CNN and LSTM, both of which excel when it comes to classifying time series data and activity recognition tasks.

3   Materials and Methodology

3.1   Data Collection

"The dataset was recorded by conducting an experiment on a group of 30 volunteer volunteers within an age bracket of 19–48 years. Each person performed six activities namely walking, walking upstairs, walking downstairs, sitting, standing, and laying while wearing a smartphone (Samsung Galaxy S II) on the waist. Using its embedded accelerometer and gyroscope, the Triaxial linear acceleration and angular velocity at a constant rate 50 Hz were recorded and the data was labeled manually using the video recordings of the experiments. The sensor signals (accelerometer and gyroscope) were pre-processed by applying noise filters and then sampled in fixed-width sliding windows of 2.56 s and 50% overlap (128 readings/window). The sensor acceleration signal, which has gravitational and body motion components, was separated using a Butterworth low-pass filter into body acceleration and gravity. The gravitational force was assumed to have only low-frequency components and therefore a filter with 0.3 Hz cutoff frequency was used. 561-feature vectors with time and frequency domain variables were extracted from the data." [5, (Anguita et al. 2013)]. The dataset was created and donated to the public domain by Davide Anguita [5, (Anguita et al. 2013)]. This is one of the most commonly used datasets to benchmark models in the field of Human Activity Recognition as evident from the list of recent research papers using the same dataset from Table 1. The large amount of recent research on this dataset makes it of paramount interest to HAR researchers around the world. All 561 features were used for the training and testing of the ML models. The length of each feature is 10299 rows of data. Some of notable features in the dataset are gravity, body acceleration magnitude, gyroscope magnitude, jerk magnitude along the different axes and along the time and frequency domain.

3.2   Preprocessing

The raw data was checked for null values, duplicates, and elements such as '()', '-',',' were removed from the dataset. "Features were normalized and bounded within $[-1,1]$" by Davide Anguita [5, (Anguita et al. 2013)] which were standardized between$[0,1]$ using Gaussian Standardization.

$$Standardization, Z = \frac{x - (\mu)}{\sigma} \tag{1}$$

where,

$$Mean, \mu = \frac{1}{N} \sum_{i=1}^{N} (x_i) \tag{2}$$

$$Standard\ deviation,\ \sigma = \sqrt{\frac{1}{N}\sum_{i=1}^{N}(x_i - \mu)^2} \qquad (3)$$

3.3 Methodology and Models

After collecting the data and pre-processing it, the data is segmented into train and test data with a 70:30 ratio. The data from 21 participants are allocated as train data and the data from 9 participants is allocated as test data. "From each window of data, a vector of features was obtained by calculating variables from the time and frequency domain" by Davide Anguita [5, (Anguita et al. 2013)]. The extracted features are then given to the ML classifiers and the models are evaluated and compared with the results of the other models. In the case of DL models, the data after segmentation is directly presented to the models without the step of feature extraction.

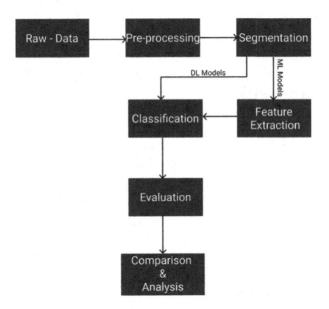

Fig. 1. Process flow chart adapted from [15] & [31]

Numpy [38] and Pandas [39] packages were used for pre-processing the data. Matplotlib [41] and Seaborn [40] packages were used for visualization. GBDT was implemented using Light GBM [42]. The other ML classifiers were implemented using Scikit-Learn [34].

The DL models (a,b,c,d) from Fig. 2 were chosen based on their ability to classify time-series data with great accuracy and the for their characteristics to perform very well with activity recognition tasks as evident from the related work in Table 1. The models e and f in Fig. 2 are unique in that only static or dynamic activities were given to these models and hence were trained for a

Model: "sequential"

Layer (type)	Output Shape	Param #
lstm (LSTM)	(None, 32)	5376
dropout (Dropout)	(None, 32)	0
dense (Dense)	(None, 6)	198

Total params: 5,574
Trainable params: 5,574
Non-trainable params: 0

(a) LSTM Model

Model: "sequential_3"

Layer (type)	Output Shape	Param #
conv1d_3 (Conv1D)	(None, 126, 64)	1792
conv1d_4 (Conv1D)	(None, 124, 64)	12352
dropout_5 (Dropout)	(None, 124, 64)	0
max_pooling1d_1 (MaxPooling1	(None, 62, 64)	0
flatten_1 (Flatten)	(None, 3968)	0
dense_4 (Dense)	(None, 100)	396900
dense_5 (Dense)	(None, 6)	606

Total params: 411,650
Trainable params: 411,650
Non-trainable params: 0

(b) CNN Model

Model: "sequential"

Layer (type)	Output Shape	Param #
conv1d (Conv1D)	(None, 126, 128)	3584
conv1d_1 (Conv1D)	(None, 124, 64)	24640
dropout (Dropout)	(None, 124, 64)	0
lstm (LSTM)	(None, 124, 128)	98816
dropout_1 (Dropout)	(None, 124, 128)	0
lstm_1 (LSTM)	(None, 124, 64)	49408
dropout_2 (Dropout)	(None, 124, 64)	0
max_pooling1d (MaxPooling1D)	(None, 62, 64)	0
flatten (Flatten)	(None, 3968)	0
dense (Dense)	(None, 100)	396900
dense_1 (Dense)	(None, 6)	606

Total params: 573,954
Trainable params: 573,954
Non-trainable params: 0

(c) CNN-LSTM

Model: "sequential"

Layer (type)	Output Shape	Param #
conv1d (Conv1D)	(None, 126, 32)	896
max_pooling1d (MaxPooling1D)	(None, 63, 32)	0
dropout (Dropout)	(None, 63, 32)	0
batch_normalization (BatchNo	(None, 63, 32)	128
conv1d_1 (Conv1D)	(None, 61, 64)	6208
max_pooling1d_1 (MaxPooling1	(None, 30, 64)	0
batch_normalization_1 (Batch	(None, 30, 64)	256
conv1d_2 (Conv1D)	(None, 28, 80)	15440
max_pooling1d_2 (MaxPooling1	(None, 14, 80)	0
dropout_1 (Dropout)	(None, 14, 80)	0
batch_normalization_2 (Batch	(None, 14, 80)	320
flatten (Flatten)	(None, 1120)	0
dense (Dense)	(None, 3)	3363

Total params: 26,611
Trainable params: 26,259
Non-trainable params: 352

(e) Dynamic CNN

Model: "sequential_1"

Layer (type)	Output Shape	Param #
conv1d_2 (Conv1D)	(None, 126, 128)	3584
conv1d_3 (Conv1D)	(None, 124, 64)	24640
dropout_3 (Dropout)	(None, 124, 64)	0
bidirectional (Bidirectional	(None, 124, 256)	197632
dropout_4 (Dropout)	(None, 124, 256)	0
bidirectional_1 (Bidirection	(None, 124, 128)	164352
dropout_5 (Dropout)	(None, 124, 128)	0
max_pooling1d_1 (MaxPooling1	(None, 62, 128)	0
flatten_1 (Flatten)	(None, 7936)	0
dense_2 (Dense)	(None, 100)	793700
dense_3 (Dense)	(None, 6)	606

Total params: 1,184,514
Trainable params: 1,184,514
Non-trainable params: 0

(d) CNN-BLSTM

Model: "sequential_1"

Layer (type)	Output Shape	Param #
conv1d_2 (Conv1D)	(None, 126, 50)	1400
max_pooling1d_2 (MaxPooling1	(None, 63, 50)	0
batch_normalization_2 (Batch	(None, 63, 50)	200
dropout_4 (Dropout)	(None, 63, 50)	0
conv1d_3 (Conv1D)	(None, 61, 100)	15100
max_pooling1d_3 (MaxPooling1	(None, 30, 100)	0
batch_normalization_3 (Batch	(None, 30, 100)	400
dropout_5 (Dropout)	(None, 30, 100)	0
lstm_2 (LSTM)	(None, 30, 32)	17024
dropout_6 (Dropout)	(None, 30, 32)	0
lstm_3 (LSTM)	(None, 30, 80)	36160
dropout_7 (Dropout)	(None, 30, 80)	0
flatten_1 (Flatten)	(None, 2400)	0
dense_1 (Dense)	(None, 3)	7203

Total params: 77,487
Trainable params: 77,187
Non-trainable params: 300

(f) Static CNN-LSTM

Fig. 2. Structure of the DL models LSTM(a), CNN(b), CNN-LSTM(c), CNN-BLSTM(d) the Divide and Conquer approach(e & f).

very large number of epochs. A max epoch value of 15000 and 500 was used for the training of the Dynamic CNN and Static CNN-LSTM models. The static CNN-LSTM was limited to 500 epochs due to its complex architecture. The unique architecture seen in Fig. 2 for highly successful Dynamic CNN and static CNN-LSTM was achieved after a great deal of experimentation with various combinations of deep learning models.

The architecture of the DL models can be found in the Fig. 2. The DL models were all implemented and evaluated using Keras [33] and Tensorflow [32]. NSL [37] is a wrapper from Tensorflow [32] constructed over Model(b) from Fig. 2. The unsupervised learning algorithm DBN invented by Hinton [35] was created using the DBN implementation made by Albertup [36].

3.4   Evaluation Metrics

The models were evaluated using the performance metrics accuracy, precision, f1- score & recall. The training and testing times were also recorded to analyze and compare with the other models that were used in this experiment.

"Accuracy: It is defined as the number of correct predictions over the total number of True Positive predictions by the model

$$Accuracy = \frac{TP + TN}{TP + FP + FN + TN} \tag{4}$$

where, TP = True positive class prediction, TN = True Negative class prediction, FP = False positive class prediction, FN = False negative class prediction.

Precision: It is the number of actual true predictions over total true predictions made by the model

$$Precision = \frac{TP}{TP + FP} \tag{5}$$

Recall: It is defined as Number of true predictions over actual number of true predictions made by the model

$$Recall = \frac{TP}{TP + FN} \tag{6}$$

F1- score: F1 Score is the weighted average of Precision and Recall [30].

$$F1Score = \frac{2 * (Precision * Recall)}{(Precision + Recall)}" \tag{7}$$

4   The Results

4.1   Hyperparameter Tuning

GridsearchCV and randomizedsearchCV from Scikit-Learn [34] were used for parameter tuning of the ML models. The parameters that yielded the maximum results are saved in Table 2. For the DL models, manual experimentation was done with the activation functions, model structure, layer counts, number of hidden units within each layer, batch size, max epochs, and the patience value in early stopping. Only the best model was saved and validation accuracy was

Table 2. Results of the hyperparameter tuning of the ML models that yielded the best results from the two instances.

Model name	Hyperparameter tuning	Best parameters	Accuracy (%)
Logistic Regression	'C': [0.01, 0.1, 1, 10, 20, 30], 'penalty': ['l2','l1']	'C': 1, 'penalty': 'l2'	95.89
Linear SVC	'C': [0.125, 0.5, 1, 2, 8, 16, 32, 64 ,40, 60, 52]	'C': 0.5	96.81
Random Forest	'n_estimators': (start = 10, stop = 201, step = 20), 'max_depth': (start = 3, stop = 15, step = 2)	'max_depth': 11, 'n_estimators': 130	91.96
RBF SVM	'C': [2, 8, 16],'gamma': [0.0078125, 0.125, 2]	'C': 16, 'gamma': 0.0078125	96.27
Linear SVM	'C': [2,8,16],'gamma': [0.0078125, 0.125, 2]	'C': 2, 'gamma': 0.0078125	96.44
Decision Tree	'max_depth': (start = 3, stop = 10, step = 2)	'max_depth': 9	87.55
GBDT	'max_depth': (start = 5, stop = 8, step = 1), 'n_estimators': (start = 130, stop = 170, step = 10)	'max_depth': 5, 'n_estimators': 160	94.1
K-NN	'n_neighbors' = 1 to 101	'n_neighbors' = 8	90.60
Naive Bayes	'var_smoothing': logscale (start = 0, stop = −9, num = 100)	'var_smoothing': 0.1	82.52

Table 3. Best hyperparameters of the DL models.

Model name	Best parameters	Accuracy (%)
LSTM	n_hidden_units = 32, Dropout layer rate = 0.5, Dense layer (units = 6, activation = 'sigmoid'), optimizer = 'rmsprop', batch_size = 16, patience = 100, epochs 500	93.38
CNN	2 layers of Conv1D (filters = 64, kernel_size = 3, activation = 'relu'), Dropout layer rate = 0.5, MaxPooling1D (pool_size = 2), Dense layer1 (units = 100, activation = 'relu'), Dense layer2 (units = 6, activation = 'softmax'), optimizer = 'adam', batch_size = 16, epochs = 500	94.06
CNN-LSTM	1st layer of Conv1D: filters = 128, second layer: filters = 64. Both Conv1D layers kernel_size = 3, activation='relu'. Dropout layer rate = 0.5 for all the dropout layers, LSTM layer 1 = 128 hidden units, LSTM layer 2 = 64 hidden units, MaxPooling1D (pool_size = 2), Dense layer1 (units = 100, activation = 'relu'), Dense layer2 (units = 6, activation = 'softmax'), optimizer = 'adam', batch_size = 16, patience = 100, epochs = 500	93.85
CNN-BLSTM	Same parameters and structure as CNN-LSTM but with Bidirectional LSTM	93.04
Dynamic CNN	1st layer of Conv1D: filters = 32, 2nd layer: filters = 64, 3rd layer: filters = 80, kernel_size = 3, activation = 'relu', kernel_initializer = 'he_normal' are the same for all the layers, MaxPooling1D(pool_size = 2), Dropout layer rate = 0.2, Dense layer (units = 3, activation = 'sigmoid'),optimizer = 'adam', batch_size = 10, patience = 500, epochs = 15000	99.92
Static CNN-LSTM	1st layer of Conv1D: filters = 50, 2nd layer: filters = 100, kernel_size = 3, activation = 'tanh', kernel_initializer = 'he_normal', MaxPooling1D (pool_size = 2), Dropout layer rate = 0.2 are the same both the Conv1D layers. LSTM layer 1: 32 hidden units, 2nd layer: hidden units = 80, activation = 'tanh', kernel_initializer = 'he_normal' are the same for both the LSTM layers, Dropout layer rate = 0.5 are the same both the LSTM layers. Dense layer(units = 3, activation = 'sigmoid'), optimizer='adam', batch_size = 10, patience = 100, epochs = 500	96.79
NSL + CNN	The same parameters of CNN wrapped in NSL with parameters multiplier = 0.2, adv_step_size = 0.05	91.72

monitored for change in every epoch. All the DL models were first experimented with the batch sizes [8,10,16,32,64 and 100] for 20 epochs and the batch size that returned the highest accuracy was used for further experimentation with the other parameters. The best parameters are saved in Table 3.

Table 4. Summary of the results obtained when the models were run in the 4-core, 16 gb RAM, 940MX GPU- local instance

Model name	Accuracy (%)	Training time (s)	Testing time (s)	Precision	Recall	F1-score
Logistic Regression	95.89	23.70	0.01	0.96	0.96	0.96
Linear SVC	96.81	123.68	0.009	0.97	0.97	0.97
Random Forest	91.96	848.18	0.07	0.92	0.92	0.92
RBF SVM	96.27	585.66	2.26	0.96	0.96	0.96
Linear SVM	96.44	48.96	0.56	0.97	0.96	0.96
Decision Tree	87.75	21.34	0.008	0.88	0.88	0.88
GBDT	94.1	2186.41	0.17	0.94	0.94	0.94
K-NN	90.60	203.20	1.16	0.91	0.91	0.91
Naive Bayes	82.52	84.09	0.07	0.86	0.83	0.81
LSTM	92.39	2044.74	1.32	0.89	0.88	0.89
CNN	94.06	510.84	51.97	0.92	0.92	0.92
CNN-LSTM	93.85	7116.62	53.70	0.92	0.91	0.91
CNN-BLSTM	93.04	13571.20	26.79	0.90	0.90	0.90
Dynamic CNN	99.92	2622.38	0.27	1.00	1.00	1.00
Static CNN-LSTM	96.73	1140.91	1.08	0.97	0.97	0.97
NSL + CNN	91.72	271.76	1.96	0.92	0.92	0.92
DBN	99.02	8643.82	0.61	0.99	0.99	0.99

Table 5. Summary of the results obtained when the models were run in the 16-core, 64 GB RAM, T4 GPU- G4dn.4xlarge EC2 instance

Model name	Accuracy (%)	Training time (s)	Testing time (s)	Precision	Recall	F1-score
Logistic Regression	95.86	9.08	0.01	0.96	0.96	0.96
Linear SVC	96.74	25.60	0.005	0.97	0.97	0.97
Random Forest	91.89	301.15	0.03	0.92	0.92	0.92
RBF SVM	96.27	119.89	1.31	0.96	0.96	0.96
Linear SVM	96.44	10.30	0.41	0.97	0.96	0.96
Decision Tree	87.55	9.93	0.006	0.88	0.87	0.87
GBDT	94.1	433.39	0.04	0.94	0.94	0.94
K-NN	90.60	55.92	0.55	0.91	0.91	0.91
Naive Bayes	82.52	18.56	0.04	0.86	0.83	0.81
LSTM	93.38	470.51	0.49	0.94	0.93	0.93
CNN	93.92	121.02	0.36	0.92	0.92	0.92
CNN-LSTM	92.26	644.65	1.07	0.91	0.91	0.91
CNN-BLSTM	92.77	1121.31	1.86	0.90	0.90	0.90
Dynamic CNN	99.56	2622.16	0.21	0.98	0.97	0.97
Static CNN-LSTM	96.79	2840.58	0.61	0.97	0.97	0.97
NSL + CNN	89.92	75.93	0.82	0.90	0.90	0.90
DBN	98.73	481.09	0.11	0.99	0.99	0.99

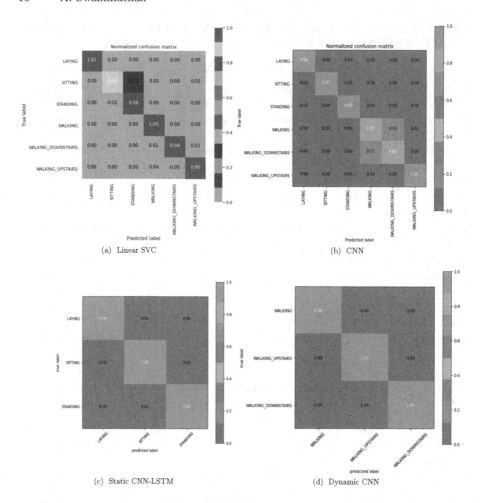

Fig. 3. Confusion Matrices of SVC(a), CNN(b) and the Divide and Conquer approach(c & d)

All the 17 models used for the comparative analysis were run on two instances to better understand the time constraints placed on the model by the computing power. The local instance used has a 4 core Central Processing Unit(CPU), 16GB of Random Access Memory(RAM), and a 940MX Graphical Processing Unit(GPU). The EC2 instance used is named G4dn.4xlarge and it has a 16 core Central Processing Unit(CPU), 64GB of Random Access Memory(RAM), and a Tesla T4 Graphical Processing Unit(GPU). The results of all the models from the local instance and the EC2 instance can be found in 4 and 5 respectively.

Among the ML classifiers, Linear SVC yielded the best result of 96.81%. It has been identified that the standard state-of-the-art DL models CNN, LSTM, and combinations of CNN and LSTM have difficulty distinguishing dynamic activities walking and walking upstairs. Also, that the ML classifiers are greatly confused between 2 static activities sitting and standing. This can be seen in the confusion matrix of Linear SVC in Fig(a) of Fig. 3 and CNN in Fig(b) of Fig. 3. As observed in these figures, it can be seen that the SVC is confused between sitting and standing and this confusion was seen among all the other ML models. The CNN model is confused between walking and walking upstairs which was also the same set of activities the other DL models excluding DBN were confused with. This identification of the activities that cause certain models' confusion, led to the Divide and Conquer approach to this problem.

The already segmented dataset was joined back and split based on the activity name. The activities sitting, standing, and laying were separated as static data and the other three activities were separated as dynamic. Both the static and dynamic dataset was then split into training and test data with an 80:20 ratio. The data was presented to deeper CNN and CNN-LSTM models for experimentation. The model architectures can be seen in Fig. 2. These models yielded much better results with reduced confusion between the activities as evident from Fig.(c) and Fig.(d) of Fig. 3. The Dynamic CNN yielded an average accuracy of 99.928% on the dynamic activity classification task whereas the proposed Static CNN-LSTM model yielded 96.79% accuracy on the classification of static activities.

4.2   Deep Belief Networks

Table 6. Results of the hyperparameter tuning of DBN

Model name	Hyperparameter tuning	Best parameters	Accuracy (%)
DBN	hidden units in the layer = ['32', '64', '128', '256', '512', '1024', '2048', '4096', '8192', '16384'], batch_size = ['32', '64', '128', '256', '512', '1024'], n_epochs_rbm & n_iter_backprop = ['100 & 100', '50 & 100', '50 & 200', '10 & 100']	hidden units in the layer = [2048], learning_rate_rbm = 0.05, learning_rate = 0.1, n_epochs_rbm = 10, n_iter_backprop = 100, batch_size = 32, activation_function = 'relu', dropout_p = 0.2	99.02

The DBN model was constructed with one layer of hidden units. The learning rate of the Restricted Boltzmann Machines (RBMs) and the Artificial Neural Networks(ANN) was kept constant at 0.05 and 0.1 respectively. The activation function was set to Relu. Manual experimentation was done with all the parameters and it was identified that changing the number of layers, the activation function and learning rates led to considerable overfitting and loss of valuable information. So, these parameters were kept constant. The batch size, hidden units within the layer, and epochs of pre-training with RBM, and Fine-tuning with ANN were experimented with in great detail. The tuning and best parameters are saved in Table 6. The plotting of Accuracy vs the batch size, hidden units & epochs can be found in subfigures(a, b & c) of Fig. 4. The confusion matrix with the classification accuracy for each activity is presented in the subfigure(d) of Fig. 4. The final model yielded an overall accuracy of 99.02% on the full dataset.

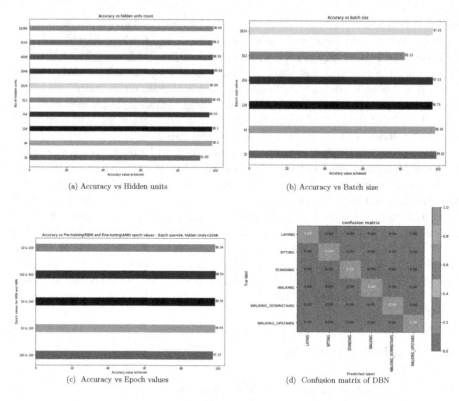

(a) Accuracy vs Hidden units

(b) Accuracy vs Batch size

(c) Accuracy vs Epoch values

(d) Confusion matrix of DBN

Fig. 4. Hyperparameter tuning of DBN (a,b,c) and the confusion matrix(d) of DBN

5   Conclusion

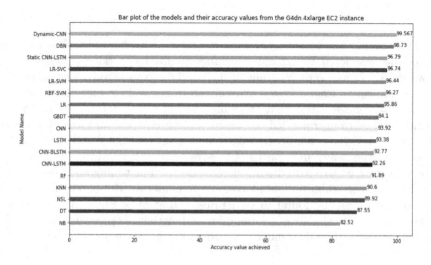

Fig. 5. Accuracy values of all the models run in the EC2 instance

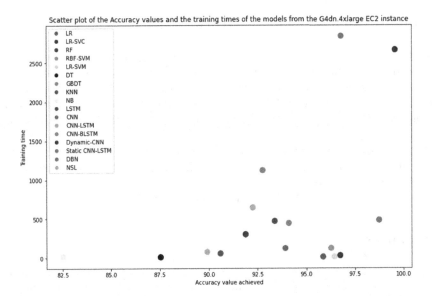

Fig. 6. Accuracy vs Training time of the models in the EC2 instance

It was identified that the omnipresent activity tracking wearable devices are inaccurate in their classification of simple activities like standing, sitting, walking upstairs, and downstairs. Inertial sensor data from smartphones were used for the classification task in this experimentation. The public domain dataset created by Davide Anguita [5] was used. 17 models were implemented on two machine instances of varying compute power. The confusion between static and dynamic activities with the ML and DL models was identified and the problem was solved using a Divide and Conquer approach to the problem. An accuracy of 99.92% was achieved with the proposed CNN model for the dynamic activities and 96.79% was achieved with the proposed CNN-LSTM model for the static activities. An "Unsupervised Probabilistic Deep learning algorithm" [35] named Deep Belief Networks was implemented on the full dataset and this model achieved an overall accuracy of 99.02% in the local instance. This final result is approximately 2% better than all the existing models as evident from the related work section. The comparison of the accuracy values of all the models run in the EC2 instance can be seen in the barplot presented in Fig. 5. The Accuracy vs Training time can be observed in the scatter plot presented in the Fig. 6. The DBN model implemented trains relatively quickly while achieving the maximum accuracy amongst all the models implemented.

6   Limitations and Future Work

The results of this experimentation can only be valid with this dataset. This dataset despite being frequently used as a standard dataset to confirm results and analysis does not have multiple complex activities. The results will also need to be confirmed with other public domain data sets using inertial sensors to confirm the findings and to add generalizability to the conclusions of this paper.

The experimentation with all the 17 models will be conducted on two other commonly used public domain data sets with complex activities in the future. The identified 2 data sets are PAMAP2 [43, (Reiss and Stricker 2012)]. from the UCI repository and the KU-HAR [44, (Sikder and Nahid 2021)].

References

1. Kasteren, T.V., Englebienne, G., Kröse, B.: An activity monitoring system for elderly care using generative and discriminative models. Pers. Ubiquit. Comput. **14**, 489–498 (2009). https://doi.org/10.1007/s00779-009-0277-9

2. Muralidharan, K., Ramesh, A., Rithvik, G., Prem, S., Reghunaath, A.A., Gopinath, M.P.: 1D Convolution approach to human activity recognition using sensor data and comparison with machine learning algorithms. Int. J. Cogn. Comput. Eng. **2**, 130–143 (2021). https://doi.org/10.1016/j.ijcce.2021.09.001

3. Dorn, D., Gorzelitz, J., Gangnon, R., Bell, D., Koltyn, K., Cadmus-Bertram, L.: Automatic identification of physical activity type and duration by wearable activity trackers: a validation study. JMIR Mhealth Uhealth **7**(5), e13547 (2019). https://doi.org/10.2196/13547

4. San-Segundo, R., Blunck, H., Moreno-Pimentel, J., Stisen, A., Gil-Martín, M.: Robust human activity recognition using smartwatches and smartphones. Eng. Appl. Artif. Intell. **72**, 190–202 (2018). https://doi.org/10.1016/j.engappai.2018.04.002

5. Anguita, D., Ghio, A., Oneto, L., Parra, X., Reyes-Ortiz, J.L.: A public domain dataset for human activity recognition using smartphones. In: 21th European Symposium on Artificial Neural Networks, Computational Intelligence and Machine Learning, ESANN 2013, Bruges, Belgium, 24–26 April 2013

6. Bayat, A., Pomplun, M., Tran, D.A.: A study on human activity recognition using accelerometer Dda from smartphones. Proc. Comput. Sci. **34**, 450–457 (2014). https://doi.org/10.1016/j.procs.2014.07.009

7. Golestani, N., Moghaddam, M.: Human activity recognition using magnetic induction-based motion signals and deep recurrent neural networks. Nat. Commun. **11**(1), 1551 (2020). https://doi.org/10.1038/s41467-020-15086-2

8. Wang, H.: Wearable sensor-based human activity recognition using hybrid deep learning techniques. Sec. Commun. Netw. **2020**, e2132138 (2020). https://doi.org/10.1155/2020/2132138

9. Uddin, M.Z., Soylu, A.: Human activity recognition using wearable sensors, discriminant analysis, and long short-term memory-based neural structured learning. Sci. Rep. **11**(1), 16455 (2021). https://doi.org/10.1038/s41598-021-95947-y

10. Wang, L.: Recognition of human activities using continuous autoencoders with wearable sensors. Sens. (Basel, Switzerland) **16**(2), 189 (2016). https://doi.org/10.3390/s16020189

11. Dua, D., Graff, C.: UCI Machine Learning Repository. Opgehaal van (2017). http://archive.ics.uci.edu/ml

12. Banos, O.: Design, implementation and validation of a novel open framework for agile development of mobile health applications. Biomed. Eng. Online **14**(Suppl 2), S6 (2015). https://doi.org/10.1186/1475-925X-14-S2-S6

13. Banos, O., et al.: mHealthDroid: a novel framework for agile development of mobile health applications. In: Pecchia, L., Chen, L.L., Nugent, C., Bravo, J. (eds.) IWAAL 2014. LNCS, vol. 8868, pp. 91–98. Springer, Cham (2014). https://doi.org/10.1007/978-3-319-13105-4_14

14. Sikder, N., Chowdhury, Md.S., Arif, A.S.M., Nahid, A.-A.: Human activity recognition using multichannel convolutional neural network. In: 2019 5th International Conference on Advances in Electrical Engineering (ICAEE), pp. 560–565 (2019). https://doi.org/10.1109/ICAEE48663.2019.8975649

15. Crucian, F., et al.: Feature learning for human activity recognition using convolutional neural networks: a case study for Inertial measurement unit and audio data. CCF Trans. Pervasive Comput. Inter. **2**(1), 18–32 (2020). https://doi.org/10.1007/s42486-020-00026-2

16. Ullah, M., Ullah, H., Khan, S.D., Cheikh, F.A.: Stacked Lstm network for human activity recognition using smartphone data. In: 2019 8th European Workshop on Visual Information Processing (EUVIP), pp. 175–180 (2019).https://doi.org/10.1109/EUVIP47703.2019.8946180

17. Ramachandran, K., Pang, J.: Transfer Learning Technique for Human Activity Recognition based on Smartphone Data. 18 (n.d.)

18. Roobini, S., Naomi, J.F.: Smartphone sensor based human activity recognition using deep learning Models. **8**(1), 9 (2019)

19. Khatiwada, P., Chatterjee, A., Subedi, M.: Automated human Activity Recognition by Colliding Bodies Optimization-based Optimal Feature Selection with Recurrent Neural Network. arXiv:2010.03324 [Cs, Eess] (2021)

20. Rabbi, J., Fuad, M.T.H., Awal, M.A.: Human Activity Analysis and Recognition from Smartphones using Machine Learning Techniques. arXiv:2103.16490 [Cs] (2021)

21. Han, P.Y., Ping, L.Y., Ling, G.F., Yin, O.S., How, K.W.: Stacked deep analytic model for human activity recognition on a UCI HAR database (10:1046). F1000Research (2021). https://doi.org/10.12688/f1000research.73174.1

22. Oh, S., Ashiquzzaman, A., Lee, D., Kim, Y., Kim, J.: Study on human activity recognition using semi-supervised active transfer learning. Sensors (Basel, Switzerland) **21**(8), 2760 (2021). https://doi.org/10.3390/s21082760

23. Huang, W., Zhang, L., Gao, W., Min, F., He, J.: Shallow convolutional neural networks for human activity recognition using wearable sensors. IEEE Trans. Instrum. Meas. **70**, 1–11 (2021). https://doi.org/10.1109/TIM.2021.3091990

24. Xia, K., Huang, J., Wang, H.: LSTM-CNN Architecture for Human Activity Recognition. IEEE Access **8**, 56855–56866 (2020). https://doi.org/10.1109/ACCESS.2020.2982225

25. Nambissan, G.S., Mahajan, P., Sharma, S., Gupta, N.: The variegated applications of deep learning techniques in human activity recognition. In: 2021 Thirteenth International Conference on Contemporary Computing (IC3-2021), pp. 223–233 (2021). https://doi.org/10.1145/3474124.3474156

26. Tang, Y., Teng, Q., Zhang, L., Min, F., He, J.: Layer-wise training convolutional neural networks with smaller filters for human activity recognition using wearable sensors. IEEE Sens. J. **21**(1), 581–592 (2021). https://doi.org/10.1109/JSEN.2020.3015521

27. Bashar, S.K., Al Fahim, A., Chon, K.H.: Smartphone based human activity recognition with feature selection and dense neural network. In: Annual International Conference of the IEEE Engineering in Medicine and Biology Society. IEEE Engineering in Medicine and Biology Society. Annual International Conference, vol. 2020, pp. 5888–5891 (2020). https://doi.org/10.1109/EMBC44109.2020.9176239

28. Zhang, Y., Ramachandran, K. M.: Offline Machine Learning for Human Activity Recognition with Smartphone. 6 (n.d.)

29. Nematallah, H., Rajan, S.: Comparative study of time series-based human activity recognition using convolutional neural networks. In: 2020 IEEE International Instrumentation and Measurement Technology Conference (I2MTC), pp. 1–6 (2020). https://doi.org/10.1109/I2MTC43012.2020.9128582

30. Ankita, R.S., Babbar, H., Coleman, S., Singh, A., Aljahdali, H.M.: An efficient and lightweight deep learning model for human activity recognition using smartphones. Sensors **21**(11), 3845 (2021). https://doi.org/10.3390/s21113845

31. Bulling, A., Blanke, U., Schiele, B.: A tutorial on human activity recognition using body-worn inertial sensors. ACM Comput. Surv. **46**(3), 33:1–33:33 (2014). https://doi.org/10.1145/2499621

32. Abadi, M., et al.: TensorFlow: Large-Scale Machine Learning on Heterogeneous Distributed Systems. arXiv:1603.04467 [Cs] (2016)

33. Chollet, F., et al.: Keras. GitHub (2015). https://github.com/fchollet/keras

34. Pedregosa, F., et al.: Scikit-learn: machine learning in python. J. Mach. Learn. Res. **12**(null), 2825–2830 (2011)

35. Hinton, G.E., Osindero, S., Teh, Y.-W.: A fast learning algorithm for deep belief nets. Neural Comput. **18**(7), 1527–1554 (2006). https://doi.org/10.1162/neco.2006.18.7.1527

36. albertbup. Deep-belief-network [Python] (2021). https://github.com/albertbup/deep-belief-network (Original work published 2015)

37. Gopalan, A., et al.: Neural structured learning: training neural networks with structured signals. In: Proceedings of the 14th ACM International Conference on Web Search and Data Mining, pp. 1150–1153 (2021). https://doi.org/10.1145/3437963.3441666

38. Harris, C.R., et al.: Array programming with NumPy. Nature **585**(7825), 357–362 (2020). https://doi.org/10.1038/s41586-020-2649-2

39. McKinney, W.: Data Structures for Statistical Computing in Python. 56–61 (2010). https://doi.org/10.25080/Majora-92bf1922-00a

40. Waskom, M.L.: seaborn: Statistical data visualization. J. Open Source Soft. **6**(60), 3021 (2021). https://doi.org/10.21105/joss.03021

41. Hunter, J.D.: Matplotlib: a 2D graphics environment. Comput. Sci. Eng. **9**(3), 90–95 (2007). https://doi.org/10.1109/MCSE.2007.55

42. Ke, G.: LightGBM: a highly efficient gradient boosting decision tree. In: Advances in Neural Information Processing Systems, vol. 30 (2017). https://papers.nips.cc/paper/2017/hash/6449f44a102fde848669bdd9eb6b76fa-Abstract.html

43. Reiss, A., Stricker, D.: Introducing a new benchmarked dataset for activity monitoring. In: 2012 16th International Symposium on Wearable Computers, pp. 108–109 (2012). https://doi.org/10.1109/ISWC.2012.13

44. Sikder, N., Nahid, A.-A.: KU-HAR: an open dataset for heterogeneous human activity recognition. Pattern Recogn. Lett. **146**, 46–54 (2021). https://doi.org/10.1016/j.patrec.2021.02.024

45. Sutharsan, V., et al.: Electroencephalogram signal processing with independent component analysis and cognitive stress classification using convolutional neural networks. In: Mahapatra, R.P., Peddoju, S.K., Roy, S., Parwekar, P., Goel, L. (eds.) Proceedings of International Conference on Recent Trends in Computing. LNNS, vol. 341, pp. 275–292. Springer, Singapore (2022). https://doi.org/10.1007/978-981-16-7118-0_24

A Survey on Cervical Cancer Detection and Classification Using Deep Learning

K. Hemalatha$^{(\boxtimes)}$ and V. Vetriselvi$^{(\boxtimes)}$

Department of Computer Science and Engineering (DCSE), College of Engineering Guindy,
Anna University, Chennai 600025, India
hemshema5@gmail.com, kalvivetri@gmail.com

Abstract. Cervical cancer is one of the main cause of cancer death, impacting 570,000 people globally. Cervical cancer is caused by the Human Papillomavirus (HPV), which causes abnormal cell growth in the cervical region. Periodic HPV testing in woman has helped to minimize the death rate in developed countries. However, due to a shortage of affordable medical facilities, developing countries are still striving to deliver low-cost solutions. The most commonly used screening test for early detection of abnormal cells and cancer is the Pap smear test. This paper explores existing deep learning model and recent research works done using publicly accessible Intel and Mobile-ODT Kaggle dataset for cervix detection and classification and ensures that these automated technologies helps pathologist in providing fast and cost-effective results.

Keywords: Cervical cancer · Cervix detection · Pap-smear · Image classification · Convolutional neural networks

1 Introduction

Cervical cancer is the second most prevalent carcinoma in women and one of the most common gynaecological malignancies, although the cure rate is about 100% if detected and treated early. Cervical cancer is a carcinoma that affects a woman's cervix. The cervical is the uterus's neck-shaped canal at the bottom. The pap test is a procedure that involves extracting a tissue from the cervix and examining it under a microscope. The pap test, on the other hand, is inaccurate, showing false negative rates ranges from 6% to 55% [1]. The Human papilloma virus test is a DNA test which identifies cervical cancer by relating it to a particular HPV type [1]. This test is usually not recommended because it has a significant false positive rate. Furthermore, such exams are quite expensive. Due to a shortage of affordable medical facilities, developing countries are still striving to deliver low-cost solutions. Cervical cancer accounts for 70% to 90% in several developing nations and are influenced by various factors like, including birth control pills, smoking, and exposure to diethylstilbestrol (DES) are the main causes of cervical cancer [4]. The endocervix and the ectocervix are two distinct areas of the cervix, with the endocervix closer to the female internal reproductive organ and the ectocervix closest to

© IFIP International Federation for Information Processing 2022
Published by Springer Nature Switzerland AG 2022
L. Kalinathan et al. (Eds.): ICCIDS 2022, IFIP AICT 654, pp. 18–29, 2022.
https://doi.org/10.1007/978-3-031-16364-7_2

the vagina. The cervical transformation zone is where cervical cancer generally occurs. The endocervix and ectocervix cells meet in the transformation zone of the cervix [3]. There are three main cervix types. Type1, Type2 and Type3 (a) Cervix type 1 the zone of transformation it is completely ectocervical and it is completely visible (b) Cervix type 2, which is the second type of cervix. The zone of transformation has a cervix that is endocervical and it is a component completely visible (c) Cervix type 3, the zone of transformation has a cervix that is endocervical it is a component not completely visible and is the most common type of cervix. Because of the type of treatment varies depending on the type of cervix, so it's crucial to know which one you have. Figure 1 shows the three forms of cervix types [3]. Deep learning models are used to identify the cervical types more appropriately.

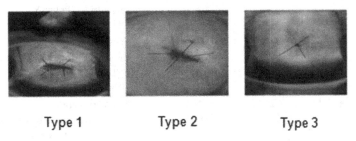

Type 1   Type 2   Type 3

Fig. 1. Images of the three different cervix types

Deep learning is supposed to assist pathologists in making more precise diagnoses by providing a precise analysis of cancerous lesions, as well as allowing for a faster clinical workflow. In this study, a survey is conducted on how prominent deep learning architectures are being employed in cervical cancer detection and classification. Also, to ensure that these automated technologies assist healthcare providers for fast and cost-effective results.

2 Related Work

Cancer is the world's most dangerous disease. Lung cancer, cervical cancer, breast cancer, ovarian cancer, sarcoma cancer, brain cancer, carcinoma cancer, and more types of cancer exist. Among several cancers, cervical cancer is the fourth most frequent in terms of death rate. If correct treatment/diagnosis is not received in a regular basis, it can be fatal and result in death. Many researchers used deep learning algorithms such as GoogleNet [1], CervixNet [2], CapsNet [3], Inception V3, ResNet50 and VGG19 [5], ResNet53 and ResNet 101 [6], AlexNet and SqueezeNet [7], MobileNet and VGG16 [9], RetinaNet [10], Densenet169 [13] and many more to predict cervical cancer.

A new fully automated deep learning pipeline for detecting and classifying cervical cancer in the cervical region was presented by Zaid alyafeai et al. [1]. With an accuracy rate of 0.68 on the intersection of union (IoU) measure, the pipeline starts with a detection method that locates the cervix section 1000 times faster than current date-driven methods. Cervical cancers may be classified using two compact convolutional neural network

Table 1. Comparison of existing algorithms

Authors	Year	Model	Description
J. Payette et al. [6]	2017	ResNet	To develop an image classifier for different types of cervixes
O. E. Aina et al. [7]	2019	AlexNet and SqueezeNet	To use a smaller CNN architecture to automate the classification of cervix types and deploy it on a mobile device
X. Q. Zhang et al. [4]	2019	CapsNet	To develop a deep learning model for classifying images of cervical lesions in order to improve diagnostic efficiency and accuracy
R. Gorantla et al. [2]	2019	CervixNet	Segmenting the Region of Interest (RoI) and then classify the RoI to improve cervigram contrast
Z. Alyafeai et al. [1]	2020	GoogleNet	To create a completely automated cervix detection and cervical image classification pipeline using cervigram images
P. Guo et al. [10]	2020	RetinaNet And Deep SVDD	In a smartphone-acquired cervical image dataset, an unique ensemble deep learning method is presented for identifying cervix images and non-cervix images
S. Dhawan et al. [5]	2021	InceptionV3, VGGand ResNet50	To create a cervical cancer prediction model that uses deep learning and transfer learning approaches to recognise and classify cervix image
Ankur Manna et al. [13]	2021	Inception V3, Xception and DenseNet169	Three Convolutional Neural (CNN) architectures were used to create an ensemble-based classification model

(CNN) versions. Cervical cancers are classified using self-extracted features learned by suggested CNN models with Area under curve (AUC) values of 0.82 and a 20-fold increase in the long it takes to recognise each cervix image, CNN-based classifiers outperformed previously created feature-based classifiers. Finally, cervigram images are used to train and assess the proposed deep learning pipeline. Sanjeev dhawan et al. [5] outlines an effort to build a cervical cancer prediction model that uses deep learning and transfer learning techniques to recognize and categories cervix images into one of three categories (Type 1/Type 2/Type 3). ConvNet, which will classify the cervix images, is created using the three models mentioned above: InceptionV3, ResNet50, and VGG19. The results of the experiment show that the Inception v3 model outperforms Vgg19 and ResNet50 on the cervical cancer dataset, with an accuracy of 96.1%.

Qing et al. [4] proposed a method were images of cervical lesions were diagnosed using a deep-learning algorithm. The research was subdivided into two halves. The lesions in the cervical images were segmented. Finally, the training set accuracy of the model was 99%, while the testing set accuracy was 80.1%, resulting in better results than earlier classification approaches and enabling for efficient prediction and classification of huge amount of data. Jack payettee et al. [6] experimented with a number of convolutional designs before settling on residual neural networks with batch normalisation and dropout. For each class, the loss was estimated using the multi-class method loss with a logarithmic scale. The dataset was made available by Kaggle which has 1481 training images, 512 test images. The author also used 4633 additional images for training and employed a variety of data sources due to the short size of the dataset strategies for augmentation.

Rohan et al. [2] suggested an innovative CervixNet methodology that involves image enhancement on cervigrams, segmentation was done to identify the Region of Interest (RoI) for selecting the appropriate treatment. A novel Hierarchical Convolutional Mixture of Experts (HCME) technique was presented for the classification problem. In the field of biomedical imaging, smaller datasets represent an inherent issue, HCME was capable of solving the problem of over fitting. The work was carried out on publicly accessible Intel Mobile-ODT Kaggle datasets, the suggested methodology outperformed all current methodologies, with an accuracy of 96.77%. The goal of Liming et al. [12] was to develop a deep learning-based visual evaluation algorithm which could automatically diagnose cervical cancer and precancer. Multiple cervical screening approaches and histopathologic confirmation of precancers were done on a population-based longitudinal cohort of 9406 women ages 18–94 years in Guanacaste, Costa Rica for 7 years (1993–2000). Tumors that was present even for up to 18 years were discovered. Using computerized cervical images for screening recorded with a fixed-focus camera, the deep learning-based technique was trained and validated. The image prediction score (0–1) could be categorized to balance sensitivity and specificity for precancer or cancer detection. On a two-sided basis, all statistical tests were performed. The findings support the idea of using modern digital cameras to do automated visual evaluations of cervical images.

Summary of the existing literature work is shown in Table 1. The author's name appears in column 1. The year is listed in column 2, while the author's model is listed in column 3. Column 4 displays the detailed description. The following section provides the materials and methods used for cervix detection and classification.

3 Materials and Methods

The research's purpose is to use deep learning technology to identify and classify objects in cervical images. Cervical cells are difficult to classify since the cytoplasm and nucleus are so distinct. In recent years, modern artificial intelligence techniques such as machine learning and deep learning have become increasingly popular in the medical health industry, and have a remarkable success with it. Further in this section we discuss the dataset used and about the object detection and classification techniques in previous research.

3.1 Data Collection

The dataset used for this research was provided by Kaggle as part of the competition [3]. Table 2 provides a breakdown of the dataset. Three types of cervix are reported in this data set are all classified normal (not malignant). It is difficult for healthcare providers to identify the transformation zones, Although the transformation zones are not always evident, some patient needs further testing while others do not. As a result, using an algorithm to make a choice will considerably increase the efficiency and effectiveness of pap smear tests.

Table 2. Intel mobile ODT Kaggle dataset [3]

Dataset	Type 1	Type 2	Type 3
Train	250	781	450
Additional	1189	1558	1886
Test	512 (unlabeled)		

3.2 Object Detection Techniques

CNNs and other deep learning-based techniques have dominated recent advances in object detection [1–11]. Faster R-CNN is a good example of a CNN-based object detection system that works well. (a) Two-stage object detectors, which includes a region proposal network followed by a classification and bounding box regression network, and (b) single-shot networks, which do everything in single network; examples of this method include YOLO [1], single shot detector, and RetinaNet [10]. Regardless, for feature extraction, each object recognition framework can use a number of base network designs, such as VGG [5], ResNet [5, 6, 11], GoogleNet [1], CervixNet [2], and CapsNet [4]. More training data is necessary as the number of parameters in architecture increases. Deep learning algorithms are frequently combined with transfer learning [5], data augmentation, and various optimization procedures to improve accuracy in specific applications.

3.3 Image Classification Techniques

In the field of computer vision and image processing, image classification has been a subject of great interest. The Convolutional neural network (ConvNet) is the frequently used deep learning network for classification of image [1–11]. The ConvNet has three layers, similar to a neural network: input layer, hidden layer, and output layer. It is a deeper neural network because it has hundreds of hidden layers. Raw image pixel values are given into the input layer, while neurons are given into the output layer according to the number of output classes. A cervigram image is taken as the input in the cervix classification problem, and the output is the possibility that the image belongs to Type I, II, or III. The final layer of the convolution layer is a fully connected layer with the SoftMax activation layer. Because the number of images in the dataset is insufficient to yield significant results, several models, including VGG [5], ResNet [5, 6, 11], GoogleNet [1], CervixNet [2], and CapsNet [4], rely on weights from pre-trained models to produce notable results. The deep learning models utilised for cervix detection and classification are described in detail in the next section.

4 Models Used

4.1 ResNet

The Residual Network architecture, developed by Microsoft Research in 2015. This architecture introduces the Residual Network concept to address the vanishing/exploding gradient issue.

In this network, it uses a technique known as "skip connections", which allows you to skip the first few layers of training and go straight to the final layer. The advantage of introducing skip connections is that regularisation will bypass any layer that reduces architecture performance. As a result, vanishing/exploding gradients are no longer an issue when training very deep neural networks. Residual blocks have three convolutional layers: a 1×1 layer with fewer filters, a 3×3 layer with the same number of filters, and a 1×1 layer with the same number of filters as the original input size to up sample and allow for the shortcut connection [6] (Fig. 2).

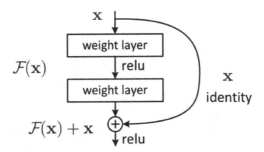

Fig. 2. Residual learning building block [6]

4.2 AlexNet and SqueezeNet

For the purpose of deploying the trained model, SqueezeNet, a compact network with reduced memory computation requirements and equivalent accuracy to AlexNet, was proposed [7]. AlexNet won the ILSVRC-2012 competition with a top-five testing error rate of 15.3%. The design is made up of five convolutional layers and three fully connected layers. AlexNet is a prominent object-detection network that could have a wide range of applications in the field of computer vision. In the future, AlexNet may be used for image processing more than CNNs. The accuracy of AlexNet and SqueezeNet are nearly same. SqueezeNet, on the other hand, is 500 times smaller and three times faster. Eight Fire modules follow the first convolution layer in the SqueezeNet before the final convolution layer is applied. Instead of using 3 by 3 filters, SqueezeNet uses 1 by 1, downsampling on convolution layers, a fire module, and a smaller number of input channels (3 by 3 filters). This results in smaller activation maps for convolution layers during late down sampling. The low memory and processing requirements make it ideal for embedded systems because it only needs roughly 5 MB of storage space. In addition for mobile app on iOS, it has the highest energy efficiency and throughput, while its processing rate on Android is even more faster.

4.3 Inception V3

In the process of image classification Inception v3 model is widely used. Inception model have been proven to be more computationally efficient than VGGNet, both in terms of the number of parameters generated and the cost [5]. The Inception v3 network model uses a convolution kernel splitting method to turn large volume integrals into compact convolutions. For example, a 3×3 convolution is split into 3×1 and 1×3 convolutions. The splitting strategy reduces the number of parameters, allowing the network training process to be accelerated while the spatial feature is retrieved more quickly.

4.4 GoogleNet

The ILSVRC 2014 competition was won by GoogleNet from Google. It had a 6.67% top-five error rate. The Inception Network was a key development in the study of Neural Networks, especially CNNs. GoogleNet is the term given to the first iteration of the Inception network. The GoogleNet image classification model has been enhanced to create the Cervix detection module [1]. It comprises of 24 convolutional layers followed by two fully connected layers, as seen in Fig. 3. GoogleNet is faster at training than VGG model.

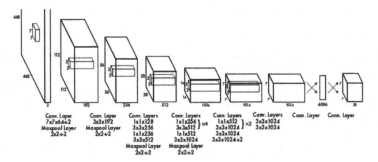

Fig. 3. Architecture of GoogleNet [1]

4.5 CapsNet

The capsule in the CapsNet neural network includes the entity properties and features, including position, deformation, velocity, reflectivity, colour, structure, and other feature space. This spatial information determines the consistency of orientation and size between features. Capsules are designed to record the probability of features and their variants rather than recording features of individual variants. As a result, the capsule's aim isn't just to detect features, but also to train the network to identify variants, because the same capsule can identify the same object class from multiple perspectives.

The CapsNet model [4] was used to classify cervical images. The cervical image training dataset was first fed into the capsNet classification model. After that, the image was put through the Primary Caps layer, which resulted in 16-D capsules. This layer generated 1600 vectors in total, with an output of 10×10 and 16×16. A 16 dimension was assigned to each vector. The DigiCaps layer was then applied. The output geometry of the DigiCaps layer was 32×32, indicating that there were three classes of 32-D capsule units. Then squashing non-linear function was used to transform this vector into the final output vector. Finally, 32×32 images was reconstructed using the completely connected layer FC (Fully Connected Layer).

4.6 CervixNet

CervixNet can handle a variety of noisy images. The CervixNet method [2] uses a flat field correction technique, imaging intensity modification methodology, Laplacian filter, and various morphological operations to automatically enhance the input image. The RoI is then retrieved from the improved image using Mask R-CNN segmentation. Cervix types are classified using a two-level hierarchical design based on the Hierarchical Convolutional Mixture of Expert (HCME) algorithm.

4.7 RetinaNet and Deep SVDD

Deep SVDD and RetinaNet, has been found to achieve state-of-the-art multiclass object detection performance on public datasets like the Intel & Mobile ODT kaggle dataset and COCO. Also implementing Deep SVDD, a method that involves training a neural network to extract and encode common features from training data. Deep SVDD [10] is a deep learning architecture-based one-class classification algorithm. The deep architecture's purpose is to extract features from input images and determine common aspects that may be used to represent the goal category. The architecture has three convolutional modules with $32 \times 5 \times 5$ filters, $64 \times 5 \times 5$ filters, and $128 \times 5 \times 5$ filters, as well as a 256-unit dense layer. RetinaNet [10] is said to perform better when dealing with class imbalance and focusing on difficult, misclassified samples. To assist in achieving higher level of training convergence.

5 Discussion

Existing research uses deep learning models such as ResNet, AlexNet, SqueezeNet, CapsNet, CervixNet, GoogleNet, RetinaNet, Inception V3 and others to predict cervical cancer. The applied models are compared on the advantages, disadvantages and various evaluation metrics as shown in Table 3. The performance of the detection and classification modules are compared to that of existing state-of-the-art models.

Ensemble of three models InceptionV3, DenseNet 169 and Xception provide complementary information, than the individual performance of each model and overfitting is more likely to occur in dense networks as it has high number of connections, which reduces computational efficiency and parameter efficiency thus the ensemble models acheives an accuracy of 96.36%

AlexNet and ResNet have around 60M parameters in common, however their top-5 accuracy differs upto 10%. However, training a Residual network necessitates a large number of computations (about ten times that of AlexNet), requiring more train time and energy. The vanishing gradient problem is a difficulty that ResNet has to deal with it. AlexNet use dropout layers to solve the over-fitting problem. In a comparison of AlexNet with ResNet on a cervix image dataset, AlexNet had a 62.6% accuracy. When comparing GoogleNet to AlexNet, GoogleNet has about 4 million parameters and identifies in 33 s per epoch, but AlexNet takes 54 s and is less accurate. The key distinction between of GoogleNet is that it combines inception models, whereas the Inception network looks deep in terms of the number of layers and expands. The VGG model was created primarily to lower the overall number of parameters with fewer trainable parameters, learning is faster and over-fitting is less likely to occur. When compared to all previous models, DenseNet has roughly 12.8 million parameters, eases the vanishing-gradient problem, and considerably reduces the amount of parameters with very much quicker computation time. Figure 4 shows the comparison of existing models with there accuracies.

Table 3. Result analysis of applied techniques

Model	Advantages	Disadvantages	Results
ResNet	In classification of images, the ResNet model outperformed humans and training a deeper network normally takes more time	Training a deeper network normally takes more time	Accuracy: 58%
AlexNet and SqueezeNet	It just requires a small amount of computing RAM and can be used on mobile devices	The AlexNet model took around an hour to train, whereas the squeezeNet took about 22 min	AlexNet Accuracy: 62.6% SqueezeNet Accuracy: 63.3%
CapsNet	Detects specific cervical type more accurately based on images	Capsule Network performs poorly on complex data and exhibits many of the same issues as CNNs initially	Training Accuracy: 99%, Test Accuracy: 80.1%
CervixNet	Robust model to diverse noisy images and imaging acquisition conditions	Not suitable for all types of image classification	Accuracy: 96.77%, Precision: 96.69%, Specificity: 98.36%, Sensitivity: 96.82%, F1 Score: 0.97
GoogleNet	The GoogleNet model has less parameters as well as more robust	Google Net is theoretically limitless	Accuracy: 82%
RetinaNet andDeep SVDD	On publicly available Datasets RetinaNet has been reported to reach state-of-the-art performance for multi-class Object detection	The duration of RetinaNet training varies between 10 and 35 h	Accuracy: 91.6%
InceptionV3, VGG and ResNet50	In comparison to its peers, Inception v3 has the lowest error rate	Inception V3 is made up of a series of complicated inception modules arranged in a hierarchy	Accuracy: 96.1%
InceptionV3 and Xception, DenseNet	Ensemble of three models provide complementary information, than the individual performance of each model	Overfitting is more likely to occur in dense networks	Accuracy: 96.36%

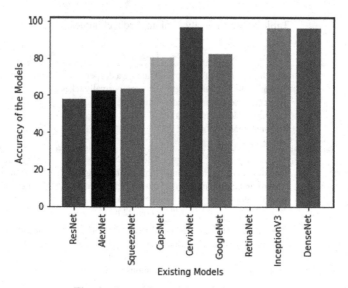

Fig. 4. Comparison of the existing models

6   Conclusion

The study presented in this paper shows how various deep learning methods were developed and used to solve the problem of cervix detection and classification. The experiment's algorithm is built on a convolutional neural network approaches, the model's effectiveness by fine-tuning the value of several parameters resulting in progressive improvement. After reviewing of all the applied models, it was discovered that CervixNet, DenseNet and Inception V3 provides 96.77%, 96.36% and 96.1% better prediction than all other models. The survey shows that a relatively modest design can achieve outcomes comparable to a non-human expert, automated technologies helps pathologist in providing fast and cost-effective results. We will continue to optimise and change the structure of the various network models addressed in this survey as future research to achieve better accuracy.

References

1. Alyafeai, Z. Ghouti, L.: A fully-automated deep learning pipeline for cervical cancer classification. Expert Syst. Appl. **141**, 112951 (2020)
2. Gorantla, R., Singh, R.K., Pandey, R., Jain, M.: Cervical cancer diagnosis using CervixNet a deep learning approach. In: Proceedings of the 2019 IEEE 19th International Conference on Bioinformatics and Bioengineering, BIBE 2019, pp. 397–404 (2019). https://doi.org/10.1109/BIBE.2019.00078
3. Samperna, R., Boonstra, L., Van Rijthoven, M., Scholten, V., Tran, L.: Cervical cancer screening (Kaggle competition)
4. Zhang, X.Q., Zhao, S.G.: Cervical image classification based on image segmentation preprocessing and a CapsNet network model. Int. J. Imaging Syst. Technol. **29**, 19–28 (2019)

5. Dhawan, S., Singh, K., Arora, M.: Cervix image classification for prognosis of cervical cancer using deep neural network with transfer learning. EAI Endorsed Trans. Pervasive Heal. Technol., 169183 (2018). https://doi.org/10.4108/eai.12-4-2021.169183

6. Payette, J., Rachleff, J. Van De Graaf, C.: Intel and MobileODT cervical cancer screening Kaggle competition: cervix type classification using deep learning and image classification. https://www.kaggle.com/c/

7. Aina, O.E., Adeshina, S.A., Aibinu, A.M.: Classification of cervix types using convolution neural network (CNN). In: 2019 15th International Conference on Electronics, Computer and Computation, ICECCO 2019, pp. 2019–2022 (2019). https://doi.org/10.1109/ICECCO 48375.2019.9043206

8. Guo, P., Xue, Z., Rodney Long, L., Antani, S.: Cross-dataset evaluation of deep learning networks for uterine cervix segmentation. Diagnostics **10**, 44 (2020)

9. Barros, H.: Deep learning for cervical cancer diagnosis: multimodal approach. https://www.mobileodt.com/products/evacolpo/

10. Guo, P., et al.: Ensemble deep learning for cervix image selection toward improving reliability in automated cervical precancer screening. Diagnostics **10**, 451 (2020)

11. Pfohl, S., Triebe, O., Marafino, B.: Guiding the management of cervical cancer with convolutional neural networks

12. Hu, L., et al.: An observational study of deep learning and automated evaluation of cervical images for cancer screening. J. Natl. Cancer Inst. **111**, 923–932 (2019)

13. Manna, A., Kundu, R., Kaplun, D., Sinitca, A., Sarkar, R.: A fuzzy rank-based ensemble of CNN models for classification of cervical cytology. Sci. Rep. **11**, 1–18 (2021)

Counting Number of People and Social Distance Detection Using Deep Learning

P. C. D. Kalaivaani[✉], M. Abitha Lakshmi[✉], V. Bhuvana Preetha[✉],
R. Darshana[✉], and M. K. Dharani[✉]

Department of CSE, Kongu Engineering College, Perundurai, India
{kalairupa,dharani.cse}@kongu.ac.in,
abithamatheswaran2012@gmail.com, bhuvanapreetha.v@gmail.com,
darshana26799@gmail.com

Abstract. With its enormous spread, the continuing COVID-19 corona virus pandemic has become a worldwide tragedy. Population vulnerability rises as a result of fewer antiviral medicines and a scarcity of virus fighters. By minimizing physical contact between people, the danger of viral propagation can be reduced. Previously, the distance between two individuals, as well as the number of people breaching the distance, could be computed with alarm. In the proposed methodology, including the existing features in the previous methodology and additionally, the total number of people present in a given frame, as well as the number of people who were violated,non-violated are tallied. The most crucial step in improving social separation detection is to use proper camera calibration. It will produce better results and allow you to compute actual measurable units instead of pixels It can also figure out how many people are in a certain location. As a result, the suggested system will be useful in identifying, counting, and alerting persons who are breaching the specified distance, as well as estimating the number of people in the frame to manage corona spread.

Keywords: COVID-19 · Social distancing detection · Object detection: you only look once (YOLO) · COCO · CCTV

1 Introduction

Since December 2019, the Corona Virus, which originated in China, has infected several countries around the world. Many healthcare providers and medical professionals are researching for proper vaccines to overcome the corona virus. Despite the fact that no development has been reported, Not very efficient. The community is exploring for new ways to stop the infection from spreading. Liquid from such an infectious person's nostrils or throat leaks when they sneeze, cough, or talk. It spreads quickly and has an impact on others.

© IFIP International Federation for Information Processing 2022
Published by Springer Nature Switzerland AG 2022
L. Kalinathan et al. (Eds.): ICCIDS 2022, IFIP AICT 654, pp. 30–43, 2022.
https://doi.org/10.1007/978-3-031-16364-7_3

Social distancing refers to strategies for preventing the transmission of a virus by limiting people's physical contact, such as avoiding crowds in public areas and keeping a safe distance between them. If done early on, social distance can help stop the virus from spreading and the pandemic illness from reaching its peak. For the last few years, computer vision, machine learning, and deep learning has shown promise in a variety of real-world scenarios. We employed an overhead perspective to create a framework for social distance monitoring in this study.

The above viewpoint provides a larger field of vision and eliminates occlusion difficulties when computing the distance between people, making it a valuable tool for social distance monitoring. Object detection has become common place in recent years, thanks to the usage of video surveillance, object tracking, and pedestrian identification. As a result, the goal of the project is to create a deep learning domain that can identify social distance from above. YOLOv3 object recognition is used by the system to recognise people in a video sequence.Using the detected bounding box information, the pre-trained detection algorithm locates persons. The bounding box is established, and the Euclidean distance is used to calculate the centroid's pairwise distances of persons. A tracking algorithm is also utilized to identify persons and quantify the amount of people in video sequences, allowing for the tracking of those who cross the social distance threshold or beyond it.

2   Existing System

In the existing system, the people are detected using YOLOv3 object detection algorithm with coco dataset. The distance between the people is calculated using Euclidean distance, who are all in the frame and also the minimum distance is given to identify whether the computed distance is exceeding given distance. if the distance is lesser than the given value, then person is added to the violated set and the bounding box colour for that person will get changed to red. If the person is maintaining the exact distance (i.e. the distance is exact as given value or greater than the given value) then the person's bounding box will be green to indicate that they are maintaining the social distance (Fig. 1).

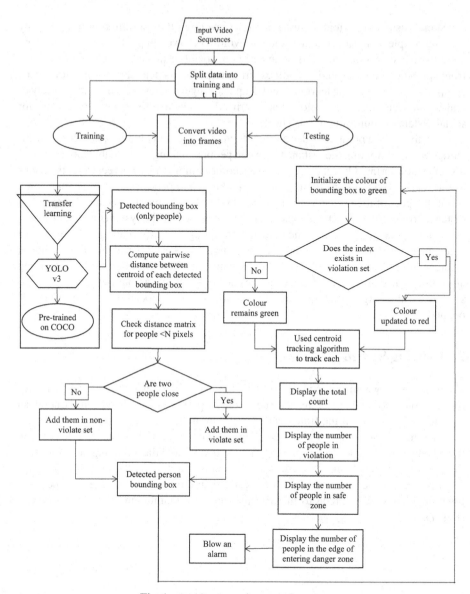

Fig. 1. Architecture of proposed system

3 Proposed System

As a result of the pandemic crisis that has taken over the world and made certain things worse, vaccines designed for contagious diseases have not shown to be effective, and thus social distance has arisen as one of the most efficient ways to stop the corona virus from spreading. Physical separation from one another is referred to as social distancing. Around the world, cases are increasing at an alarming rate, prompting societal seclusion.

We can compute and summaries the distances between individuals in the monitor who breaches the social distancing, utilizing CCTV cameras, and so keep track of human activities in public spaces. People who congregate in large numbers at religious sites might exacerbate the situation. Recently, all nations across the globe have been on lockdown, forcing residents to stay at home. However, as time passes, people will begin to attend more public and religious sites, thus in those conditions, the proposed system of monitoring social distance would be helpful all over the world. It is possible to track human beings and compute the distance between them in pixels, as well as set the standard-maintained distance to be followed, and get an overview of people who are all breaking the law and concerned authorities who can take action, using computer vision and deep learning, as well as the installed CCTV.

The proposed method uses the YOLO, COCO dataset to detect just persons in order to determine whether or not they are keeping a healthy distance. This keeps track of those who continue to break the social distance rules established by public health officials. An alert will sound if persons break the social distance inside the observation area. It also counts the number of persons who enter a restricted area and keeps track of the number. For each position, a distinct colour is allocated. A green colour box emerges when a person is in the safe zone. When he/she is about to enter a dangerous zone the colour of the box turns to orange to note that they are about to enter the danger zone.The box changes into red colour if the person enters into the danger zone.

4 YOLO Detector

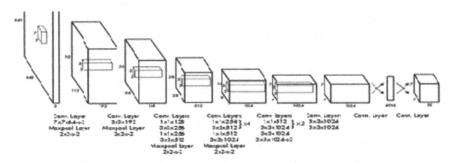

Fig. 2. Architecture of YOLO

You Only Look Once (YOLO), a single convolutional neural network that differs from earlier detectors, frames objects as a regression issue to spatially distributed bounding boxes and related class probabilities directly from full images in a single evaluation. The YOLOv1 structure, which has two completely linked layers and 24 convolutional layers (Fig. 2).

YOLOv3 features three distinct scales, each of which predicts three anchor boxes. For the four bounding box offsets, one objectless, and 80 class predictions in the COCO dataset, the tensors at each scale are N N [3 (4 + 1 + 80)]. To determine the priors,

YOLOv3 uses the k-means cluster (anchors). The width and height of the pre-selected 9 clusters are (1013), (1630), (3323), (3061), (6245), (59119). In order to detect objects, each group is given to a distinct feature map. Furthermore, YOLOv3 does feature extraction using a new network named Darknet-53. It still contains some shortcut connections, but it's a lot bigger today, with 53 convolutional layers compared to the YOLO 9000's 19 levels and the YOLO v1's 24 layers. This new network extracts features more precisely than Darknet-19, which was used in YOLOv2, or previous networks like ResNet-101 or ResNet-152.

5   COCO Dataset

COCO is a large-scale image dataset that may be used for segmentation, object detection, human key-point detection, stuff, and captioning. Its collection contains 1.5 million item instances and approximately 200K annotations, 330K images. All of the products are categorised into 80 groups. It contains Python, Matlab, and Lua APIs for loading, analysing, and displaying all picture annotations.

6   Centroid Tracking Algorithm

The bounding box is sent with (x, y) coordinates for every seen object in each and every frame in the centroid tracking technique. These bounding boxes can be generated by any object detector, such as Haar cascades, colour threshold + contour extraction, and so on. It is definitely necessary to compute bounding boxes for each frame in the video series. The centroid is computed and each bounding box is given a unique ID once the (x, y) coordinates of bounding boxes are allotted. For each continuous frame, the centroid of the object is calculated using the bounding box principle. Using a unique ID for each object detection, on the other hand, can negate the purpose of object tracking. To fix this, we'll try to link the new object's centroid to that of an existing object. To do so, we'll use the formula,

$$D = \sqrt{(x_2 - x_1)^2 + (y_2 - y_1)^2} \tag{1}$$

to calculate the Euclidean distance between the two observed objects. We suppose that the observer will travel between the frames at first, but that the frame centroids will be the shortest, i.e. smaller than all other object distances. As a result, by comparing the centroids to the shortest distances between succeeding photos, the object tracker may be easily constructed. We add a large number of items to our list in order to detect them. We add a big number of objects to our identification list. It essentially ensures If more objects are recognized and monitored than previously detected and monitored objects, the newly discovered objects must be registered. Creating bounding boxes for newly discovered objects can be used to register the new things that have been detected. Following the identification of people, we must measure the centroid of the bounding box and store it in the list, as well as keep track of the object's motions by using Euclidean's method to calculate the minimum distance.

To improve the effectiveness of a tracking system, it must be able to cope with the case where an object moves out of the frame or out of the field of view, which we do by de-registering the object. De-registering an object entails removing the object's unique ID and other details once it has passed out of sight or scope, which is determined by comparing the object to other objects in the frame. We assume an item has vanished or migrated out of the field when it does not fit another current object after a given number of frames, and the old object is de-registered (Fig. 3).

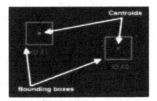

Fig. 3. Centroid tracking

7 Module Description

There are four modules in this module that involve and are completed by giving the pre-trained YOLO model. The modules are as follows:

- Module for detection
- Module for social distance
- Module for counting people
- Restriction module

7.1 Detection Module

Fig. 4. Detection of people

In the detection module, the frame or input video contains a pre-trained and pre-initialized yolo object identification module, as well as CNN output names and individual indexes to detect only the person in the video frame. The input frame's height and breadth are calculated using the frame measurements. Constructing the blob and then doing a frontal pass of YOLO object recognition using Open-CV. The class ID and confidence of the current object identification are retrieved after ensuring that the detected object is an individual and that the confidence rate (minimum) is met, and the filtering of identified objects begins. YOLO only gives the center (x, y) coordinates of the rectangular box after the width and height of the boxes have been scaled to the image's dimension. The top and left corners of the bounding box are obtained using the center of (x, y)- coordinates, and the values of bounding box coordinates, centroids, and confidences of a detected target are then adjusted. The centroid of a discovered individual is the centre of the bounding box, which is calculated by eliminating the centre of the bounding box (Fig. 4).

To limit the amount of overlapping bounding boxes, Non-Maxima Suppression is utilized. Finally, check that the frame has at least one detection and that the box coordinates in the result list have been retrieved and changed. The Social distance module received the trust of each person identified, as well as the bounding box and centroid of each person observed, for further processing and to determine whether or not the individuals are maintaining a safe distance. To limit the amount of overlapping bounding boxes, Non-Maxima Suppression is utilized. Finally, double-check that the frame has at least one detection and that the box coordinates in the result list have been retrieved and changed. The Social distance module received the trust of each person identified, as well as the bounding box and centroid of each person observed, for further processing and to determine whether or not the individuals are maintaining a safe distance.

7.2 Social Distance Module

The detection module is imported into the social distance module using the import command. For the whole input image, a single-stage network is built and employed in the structure to forecast the bounding box and class probability of identified items. Convolution layers are employed for feature extraction, while fully linked layers are used for class predictions in this design. The incoming frame is segmented into a region of S * S, usually referred to as grid cells, during human identification. The bounding box prediction and class probabilities are linked to the grid cells, as well as predicting whether the person's bounding box's center is in the grid cell or not:

$$Conf(p) = Pr(p) \times IOU(pred, actual) \qquad (2)$$

Pr(p) in Eq. (2) denotes whether or not the person is within the bounding box that has been detected. Pr(p) has a value of 1 for true and 0 for false. IOU stands for Intersection Over Union of the actual and anticipated bounding boxes (pred, actual). It is defined as follows:

$$IoU(pred, actual) = area\frac{BoxT \cap BoxP}{BoxT \cup BoxP} \qquad (3)$$

The anticipated bounding box is shown as BoxP, while the intersection region is shown as BoxT. Each recognized individual in the input frame is assigned to an appropriate region, which is forecasted and decided. To get the best bounding box, the confidence value is used to specify the bounding box following prediction. h, w, x, y are estimated for each anticipated bounding box of a detected person, with w, h determining width and height and x, y defining bounding box coordinates. As seen in Eq. 2, the model generates the expected bounding box values (4)

$$b_x = \sigma(t_x) + c_x$$

$$b_y = \sigma(t_y) + c_y$$

$$b_w = p_w e_w^t$$

$$b_h = p_w h_h^t \tag{4}$$

In Eq. (4), bx, by, bw, bh are expected coordinate bounding boxes, with x, y indicating the centre of the coordinates and w, h representing the width and height of the coordinates. The network output is defined by tw, th, tx, ty, whereas the grid cell's top-left coordinates are defined by cx, cy. Anchor height and weight are represented by ph and pw. Using a threshold value, the high confidence data are processed, while the low confidence values are rejected. Non-maximal suppression is used to acquire the finalized position values for such observed bounding box. For each bounding box that is detected, a loss function is calculated. Confidence, classification, and regression are the three functions that make up the provided loss function. If an object is discovered in each grid cell, the classification loss is determined as the squared error of the conditional class probabilities for each detected object.

$$\ell_{cls} = \sum_{j=0}^{s^2} l_{ij}^{obj} \sum_{c \in class} l_i^{obj} \left(p_i(c) - p *_i(c) \right)^2 \tag{5}$$

In Eq. (5), if the person is discovered in grid cell I 1obj ij = 1, else equals 0. p I denotes the conditional class probabilities for each class c in grid cell i. (c). The errors in the predicted bounding box sizes and positions of identified items are estimated using the localization loss. The bounding box is expanded to include the detected item, which is a human. It is defined as follows:

$$\ell_{loc} = \lambda_{coord} l_{ij}^{obj} \sum_{i=0}^{s^2} \sum_{i=0}^{n} [(x_i - x_i^*)^2 + (y_i - y_i^*)^2 + \left(\sqrt{w_i} - \sqrt{w_i^*}\right)^2 + \left(\sqrt{h_i} + \sqrt{h_i^*}\right)^2] \tag{6}$$

1 object is equal to 1 in the preceding equation if the jth bounding box in grid cell I is used for object detection, else it is equal to 0. The model predicts the square root of the bounding box height and breadth, rather than width and height. The scaling parameter coord is used in Eq. (6) to forecast bounding box coordinates. In the ith cell of the detected bounding box, the expected positions are represented by xi, yi, hi, and wi, meanwhile the observed positions of the bounding box in the ith cell are given by h I w I x I y i. The loss function of the anticipated bounding box with coordinates x, y is measured by Eq. (6). 1objij is used to denote the potential of the detected person in the jth bounding box. When (j = 0 to B) is used as a predictor for each grid cell I = 0 to S2),

the function in Eq. (6) calculates sum over each bounding box. Finally, the confidence loss, which is provided in Eq., is determined (7)

$$\ell_{conf} = \Sigma_{i=0}^{s^2}\Sigma_{j=0}^{B} \, l_{ij}^{obj} \, (C_i - C_i^*)^2 \tag{7}$$

1noobjij is defined as the complement of 1objij in Eq. (8). In cell I and noobj, the bounding box's confidence score C_i^* is being used to scale down the loss during background detection. In the great majority of cases, bounding boxes don't really include any objects that pose a class imbalance problem, hence the model is trained to identify background rather than identify objects. For the jth bounding box in grid cell I and 1objij, the loss is reduced by a factor noobj (default: 0.5), where the confidence score is defined as C_i^*, and is equal to 1 if the jth bounding box in cell I is responsible for object identification; otherwise, it is equal to 0.

$$\ell_{conf} = \lambda_{noobj} \, \Sigma_{i=0}^{s^2} \, \Sigma_{j=0}^{B} \, l_{ij}^{noobj} \, (C_i - C_i^*)^2 \tag{8}$$

The centroid of each identified person bounding box, displayed as green boxes, is being used for distance calculation once individuals are spotted in sequence video frames. Using the identified bounding box coordinates, the center of the bounding box is determined (x, y). The Euclidean distance is then used to compute the distance among each detected centroid. Compute bounding box centroids for each consecutive frame in the video stream, then calculate the distance between each pair of identified bounding box centroids. A list is built to retain the centroid information of each identified individual, as well as a threshold value to determine whether two people are fewer than the specified N pixels apart. The bounding box of the detected individual is set to green. If the identified person crosses the given social distance, the information about that person is added to the violation set and checked in the violation set. The bounding box's colour is changed to red if the person's current index is in the violation set. At addition, in each frame of the video stream, the centroid tracking approach is used to track the recognised humans. People that are detected as being too close to a predefined social distance threshold can likewise be tracked using this technology. The output of the model is the total count of people who break the social separation rule (Fig. 5).

Fig. 5. Detection of people with bounding box

7.3   People Counting Module

The people counting module consists of four sections and they are,

- Total count of people
- Number of people in the green zone
- Number of people in the orange zone
- Number of people in the red zone

The Total count of people Section counts the length of and the total number of people information will get 8tored and displayed in the output frame. Initially, all the objects will get detected using YOLOv5 object detection algorithm, then using COCO dataset only the people are detected from all the detected objects and bounding box will be computed only for the person object also the colour of bounding box assigned as green for all the detected person. In the next 8tep, the person's information who are violating the mentioned default minimum di8tance will get added to the violation set and then the bounding box colour. All updated to red to indicate they are in the danger zone. When two or more people are on the verge of violating the specific distance but haven't yet violated the distance, are considered to be in the orange zone and the count is calculated also the bounding box updated to orange. The remaining people in the frame are maintained the social distance is counted and they are considered to be in in the green zone and also the colour of the bounding box is updated to green (Fig. 6).

Fig. 6. Count of people with zone

7.4   Restriction Module

In the restriction module, a hall is set up where a group of people enter with or without maintaining social distance. The number of people entering the hall is also been calculated using people counting algorithm. When one or two people enter the hall with appropriate distance they are considered to be in the green zone. i.e., two or more people with provided minimum distance between them, when two or more people are on the verge of violating the specific distance but haven't yet violated the distance considered to be in the orange zone. When two or more people completely violating the specified distance then they are considered to be in the red zone, they will also be alerted with an Alarm. The count which is calculated in the people counting module results in the total number of people in that frame and it also blows an alarm when the number of people exceeds than the allocated number of people in order to maintain the specific distance in the hall. When an alarm is blown, it doesn't alert just the person who is violating it but the entire team whose present in the particular hall which will be beneficiary to alert the environment. These are the basic functions that the restriction module performs which enhances the outlook on the proposed project, which will be helpful in the current pandemic situation to avoid the spread with the first and foremost procedure, which is to maintain social distance (Fig. 7).

Fig. 7. People altering with distance

8 Results and Conclusion

In social distance monitoring, indoor data sets with video sequences taken from an above perspective are used. Data collection is always divided into two parts: 70% training and 30% testing. People in a scenario can freely move around the area, with just radial distance and camera location influencing their visual appearance.The example frames show that a person's visual appearance varies depending on the dataset, and that people's heights, postures, and scales fluctuate.The implementation also makes use of Open-CV. The results of the pre-trained testing are discussed in the first paragraph, while the results of the specified detection model after transfer learning and training on the overhead data set are discussed in the second paragraph. In most cases, the model is confirmed by comparing identical video sequences. This section contains a model performance evaluation as well as a comparison to several deep learning models. This section includes a performance evaluation of the model as well as a comparison to different types of deep learning models. The Euclidean formula was used to calculate the distance between the persons in the frame, and the distance was calculated between the current object's centroid and the new object's centroid. The formula is provided in Eq. 9 below (Fig. 8).

$$D = \sqrt{(x_2 - x_1)^2 + (y_2 - y_1)^2} \tag{9}$$

A variety of video sequences are used to analyse the test outcomes. The characters are free to wander around the scene in the video scenes. The size of the individual will change depending on where they are. Because the model only evaluates the human (person) class, a pre-trained model will only detect an object that has the appearance of a human. As illustrated by the green colour rectangles, the pre-trained model delivers results and also recognises varied sizes of person bounding boxes. People are designated as sustaining social distance thresholds along green rectangles in sample frames. The

Fig. 8. Sample output of the proposed framework

model is put to the test when numerous persons enter a scene. However, if the person's appearance changes, the model will produce false positives. This could be because when a pre-trained model is utilised, an individual's look from such an overhead view changes, that could be misleading for the present model. With over 500 sample frames, an overhead data set was employed to train the model. For training the given model, the epoch size 40 and each batch size 64 were chosen. After the model had been trained, a new layer was created and added to the pre-trained model. The results of the experiments show that transfer learning improves detection outcomes significantly. It may also be visualised accurately, with the model detecting people in various scenic locations (Fig. 9).

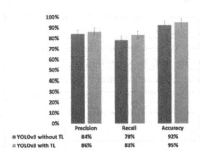

Fig. 9. Model (YOLOv3) precision, recall, and accuracy with and without transfer learning.

The utility of a framework for socially distancing tracking that used a deep learning method and an overhead perspective was assessed using a number of quantitative indicators in this study. Precision, recall, and accuracy were used to evaluate the detection model's effectiveness. Transfer learning significantly enhanced the outcomes for the overhead view by data set, according to the findings. Several deep learning models have a low false detection rate, ranging from 0.7–0.4% without any training, indicating about deep learning's performance (Fig. 10).

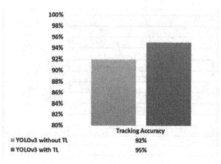

Fig. 10. Accuracy of tracking with the YOLOv3 detection model, which has been pre-trained and trained.

9   Conclusion and Future Work

The need for self-responsibility rises drastically when we examine the world following the COVID-19 outbreak. The scenario will mostly focus upon adopting and adhering to the WHO's set of precautions and restrictions, more specifically as the duty of someone who will entirely rely on themselves rather than the government. COVID-19 is disseminated through intimate contact with sick persons, therefore social distance is unquestionably the most important factor. To oversee wide regions, a distinct and successful method is required, and this survey article concentrates on that. Authorities will employ CCTV and drones to monitor human activity and regulate huge gatherings, averting law enforcement breaches. People will be signalled with a green light as long as they stay a safe distance. As the throng becomes larger, a red light will emerge on the CCTV, alerting the designated police to the location and allowing the issue to be addressed. Because supervising a large crowd is difficult, this survey can aid in the management of situations before they go out of control. As a consequence of this strategy, police on-the-ground operations will be reduced, allowing them to focus only on overseeing circumstances in locations where conditions are unfavourable, allowing them to utilise time effectively and conserve energy for equitable situations.

As a result, persons can be detected using the pre-trained YOLOv3 paradigm. The model has been trained on a separate data set and then attached to the prior model. Transfer learning has never been employed in a deep learning-based detecting approach for checking social distance from above, to our knowledge. The detection model provides bounding box details, including centroid coordinates.

The pairwise centroid distances between recognized bounding boxes are computed using the Euclidean distance. An estimation of physical separation to the pixel is performed to examine for social distance violations between persons, and a threshold is set. A violation threshold is used to evaluate if the distance result exceeds the set minimum social distance. In addition, to keep track of persons on the scene, a centroid tracking approach is used. The framework effectively detects people strolling too close together and violates social separation, according to experimental data; furthermore, the transfer learning methodology improves the detection model's entire efficiency and accuracy. Without transfer learning, a pre-trained model outperforms a detecting accuracy of 92%, but with transfer learning, the model generates a detecting accuracy of 95%. The tracking accuracy of this particular model is 95%.

In the future, the project could be improved for a variety of indoor and outdoor environments. The group of individuals who are breaking the social distance threshold will be tracked using a variety of detection and monitoring methods. We can build this model/system in real time with CCTV and required hardware pieces in future development.

References

1. Redmon, J., Divvala, S., Girshick, R., Farhadi, A.: You only look once: unified, real- time object detection. In: Proceedings of the IEEE Conference on Computer Vision and Pattern Recognition (2016)
2. Redmon, J., Farhadi, A.: YOLO9000: better, faster, stronger. In: Proceedings of the IEEE Conference on Computer Vision and Pattern Recognition. arXiv:1612.08242 [cs.CV] (2017)
3. Syed Ameer Abbas, S., Oliver Jayaprakash, P., Anitha, M., Vinitha Jaini, X.: Crowd detection and management using cascade classifier on ARMv8 and OpenCV-Python. In: 2017 International Conference on Innovations in Information, Embedded and Communication systems (ICIIECS). Mepco Schlenk Engineering College, Sivakasi (2017)
4. Joy, J.J., Bhat, M., Verma, N., Jani, M.: Traffic management through image processing and fuzzy logic. D. J. Sanghvi College of Engineering (2018)
5. Bhave, N., Dhagavkar, A., Dhande, K., Bana, M., Joshi, J.: Smart signal-adaptive traffic signal control using reinforcement learning and object detection, Department of IT, RAIT, Nerul, Maharastra (2019)
6. Syed Ameer Abbas, S.,. Anitha, M, Vinitha Jaini, X.: Realization of multiple human head detection and direction movement using raspberry Pi. In: Electronics and Communication Engineering, Mepco Schlenk (2017)
7. Article on Object detection with 10 lines of code by Moses Olafenwa on June 16 (2018). https://towardsdatascience.com
8. World Health Organisation: WHO Corona- viruses Disease Dashboard, August 2020. https://covid19.who.int/table
9. Rezaei, M., Azarmi, M.: DeepSOCIAL: social distancing monitoring and infection risk assessment in COVID-19 pandemic. Institute for Transport Studies, The University of Leeds (2020)
10. Keniya, R., Mehendale, N.: Real-time social distancing detector using Net19 deep learning network (2020)
11. Punn, N.S., Sonbhadra, S.K., Agarwal, S.: Monitoring COVID- 19 social distancing with person detection and tracking via fine-tuned YOLO v3 and deep sort techniques, 06 May 2020
12. Redmon, J., Farhari, A.: YOLOv3: an Incremental improvement (2018)
13. Singh Punn, N., Sonbhadra, S.K., Agarwal, S.: Monitoring COVID-19 social distancing with person detection and tracking via fine-tuned YOLO v3 and deepsort techniques (2020)
14. Figure 2 is taken from https://www.researchgate.net/figure/YOLO-architecture-YOLO-architecture-is-inspired-by-GooLeNet-model-for-image_fig2_329038564
15. Sathyamoorthy, A.J., Patel, U., Savle, Y.A., Paul, M., Manocha, D.: COVID-robot: monitoring social distancing constraints in crowded scenarios. https://arxiv.org/abs/2008.06585v2

Analysis of Age Sage Classification for Students' Social Engagement Using REPTree and Random Forest

Jigna B. Prajapati[✉]

Acharya Motibhai Patel Institute of Computer Studies, Ganpat University, Gujarat, India
jignap15@gmail.com, jigna.prajapati@ganpatuniversity.ac.in

Abstract. Study and analysis of train dataset along with various ML algorithms is used widely in different sectors. The accuracy parameters can be clarified to have prediction of different score levels. This study covers the extension work of Students' social engagement during covid-19 pandemic. The study was initiated with students' social connection during the pandemic. We had compared various machine learning algorithms with its performance about the engagement of students in various social network. After studied, analyzed & compared, we derived that the most of students' social engagement found in WhatsApp, YouTube & Instagram. The current study is foreseeing age wise social media connection. It correlates between student & their social engagement during the pandemic phase. In which age group, which social media is one of the most popular one. This study focuses on age wise classification using Machine Learning. In this paper, the decision-making classification is compared. The Reduced Error Pruning Tree (REPTree) and Random Forest algorithm is implemented on train dataset with diverse nodes. The attributes are focused as age & time spent on social media as per necessity of study. This paper includes the study and analysis of RAE & RMSE along with ML tree approach. The discoveries of this study can lead better classification in regards of students' age and duration which they have spent on social media for derived social platform.

Keywords: Social media platform · REPTtree · Random forest · Classification · Decision tree

1 Introduction

The virtual social engagement is drastically increasing day by day. The people are using n number of social platforms to make them engage in various community. This is also adopted by the student fraternity for social connection & entertainment to be in various social media platform.

© IFIP International Federation for Information Processing 2022
Published by Springer Nature Switzerland AG 2022
L. Kalinathan et al. (Eds.): ICCIDS 2022, IFIP AICT 654, pp. 44–54, 2022.
https://doi.org/10.1007/978-3-031-16364-7_4

In our previous work entitled "Performance Comparison of Machine Learning Algorithms for Prediction of Students' Social Engagement", we have studied Students engagement in different social media platform. Such platform as LinkedIn, Facebook, Instagram, Reedit, Snapchat, Talklife, Telegram, Twitter, WhatsApp, YouTube & etc. along with various attributes. The attributes covers the broad spectrum of student's routine activities during the pandemic. The data set collected with residence area, age, duration hours for online class, rating of online class experience, medium for online class, duration hours for self-study, duration hours for fitness, duration hours for sleep, duration hours for social media, social media platform, duration hours for TV, Number of meals per day, Change in weight, Health issue during pandemic, Stress busters, Time utilized for more than 1200 instances. After analyzed suitably we derived that the most of students were majorly engaged in WhatsApp, YouTube & Instagram [1].

Our dataset consists of student from all age group. Social engagement has been found to be associated with many factors for the student crowd as social connection, entertainment, relaxing phase, news-updates of others and many more. The dataset collected for this study includes residence region, age, online class hours, online education rating, online education medium, self-study hours, fitness hours, Sleep hours, social media platform used, time spent on specific social media, TV hours, meals plan, weight increase or decrease, health issues recorded, noticeable stress factor. The different age group many have their own reasons for joining & remaining on various social platforms. The WhatsApp, YouTube & Instagram have been found the most popular connection network during the last study [1]. Here the focus on age wise connection popularity for well said social media platform. The collected data is processed by feature abstraction, feature alignment very initially. After this noise removing, splitting & labeling of data is being done well during the study about the most popular social media platform [1]. To study accurately about the popular social medial in particular age wise group, the data are again pre-processed as per necessity of REPTree and Random Forest algorithm.

REPTree and Random Forest algorithm are effective to support decision making in real time environment [2]. These are supervised chaffier and popular to derive various classification results [3, 4].

2   Related Work

To analyze & process the data in well structural manner, studied the various researchers work in the same domain. The ML algorithm have been used find appropriate results from heterogeneous sectors [5].

Li Yang has discussed about cardiovascular disease prediction for the area of eastern China. They have used random forest mechanism. His results shown that random forest produced more sound for significant improvement for CVD prediction [6]. Joske Ubels has discussed about prediction of treatment benefit using random forest . They have used failed clinical drug trials [7]. Fabián Santos has discussed about the evaluation of forest change drivers using random forest for Northern Ecuadorian Amazon. This approach demonstrated the advantages for integrating remote data & remote sensing-derived products [8]. Martin Hanko has discussed about traumatic brain Injury using random forest to predict mortality in patients. They have constructed data for 6-month mortality and derived enhanced prediction results [9]. Toby G Pavey has discussed about wrist-worn accelerometer data in concern random forest classifier. He claims accurate group level prediction for controlled conditions using random forest classifier for wrist accelerometer [10]. Eric S Walsh has discussed about estuarine system aligning with the spatial distribution of sediment pollution using Random Forest [11]. Ishwaran H has discussed regression, classification, survival Statistics in medicine by SE (standard errors) & CE (confidence intervals) for variable with random forest [12]. Eric Ariel L Salas has discussed about the forest image classification of agricultural systems using random forest. They have used airborne hyperspectral datasets [13]. Samad Jahandideh has used a random forest classifier for improvement of chances of successful protein structure determination [14]. F Chris Jones has discussed about the Random forests as cumulative effects models for a case study of lakes and rivers in Muskoka, Canada [15]. Shiyang Li has discussed about the nitrate concentration and load estimation using data mining techniques. This study was focused on different type of watersheds. They have predicted nitrate levels and derived that REPTee has given better performance in concentration and load results [16]. Sankaralingam Mohan has discussed about summertime ground-level ozone concentration to forecast O3 concentration for the surface level using REPtree [17]. Mahfuzur Rahman has discussed about delineating multi-type flooding in Bangladesh using stacking hybrid machine learning algorithms [18]. Elizabeth Goya-Jorge has discussed about the chemical-induced estrogencity in silico and in vitro methods [19], Sunil Saha has discussed about forecast of probability of deforestation about Gumani River Basin, India using random forest, REPTree & binary logistic regression [20].

3 Proposed Work

In the previous study, the process of data acquisition, feature abstraction, feature alignment, noise removal, splitting and labeling data has been carried out [1] but Some data structuring is necessary to apply REPTree and Random Forest algorithm. As mentioned in Fig. 1, feature selection is done in the reference of age wise, social media usage with time spent. The train dataset is prepared to be applicable in REPTree and Random Forest. This train dataset is using age of student with Time spent on social media. It analyzes & compares the derived result using Weka tool.

Classification algorithm implements classifier to maps input data to a particular category. It is an instance of supervised learning. Here, the train dataset is used to identify classification observations for age wise (social media platform & duration they spent on social media). Classification can help us to have step-by-step observation. This study uses the 10-fold cross validation to implements various models.

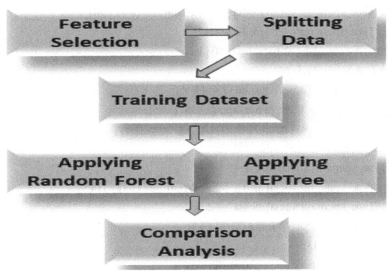

Fig. 1. Work flow diagram

3.1 RepTree

Reptree focus on each node which represents a decision based on input, and move to the next node & next until the predicted output. With the use of regression tree mechanism, it creates multiple node for specified tree in multiple iterations. Once the iteration done, it selects best one as representative. It is one of the fast decisions.

The REPTree is applied on dataset with 10 fold cross validation methods on different attributes wise Social media platform usage on Time spent and age. The REPTree classifies with predictive social media as mentioned in a. Predictive social media with size of 31 tree nodes.

3.1.1 Predictive Social Media Test Mode: 10-Fold Cross-Validation

Age $<$ 16.5
| Time spent on social media $<$ 1.75
| | Age $<$ 14.5
| | | Time spent on social media $<$ 0.25 : Youtube (8/4) [6/4]
| | | Time spent on social media $>=$ 0.25
| | | | Age $<$ 11.5 : Whatsapp (8/0) [3/2]
| | | | Age $>=$ 11.5
| | | | | Age $<$ 12.5 : Youtube (7/4) [4/1]
| | | | | Age $>=$ 12.5 : Whatsapp (39/18) [13/5]
| | Age $>=$ 14.5 : Whatsapp (42/27) [20/12]
| Time spent on social media $>=$ 1.75
| | Time spent on social media $<$ 5.5
| | | Time spent on social media $<$ 2.5 : Youtube (37/18) [14/9]
| | | Time spent on social media $>=$ 2.5
| | | | Age $<$ 13.5 : Youtube (11/0) [3/1]
| | | | Age $>=$ 13.5
| | | | | Time spent on social media $<$ 3.5
| | | | | | Age $<$ 14.5 : Youtube (3/2) [2/1]
| | | | | | Age $>=$ 14.5
| | | | | | | Age $<$ 15.5 : Instagram (2/0) [3/1]
| | | | | | | Age $>=$ 15.5 : Youtube (3/1) [0/0]
| | | | | Time spent on social media $>=$ 3.5 : Youtube (7/2) [4/2]
| | Time spent on social media $>=$ 5.5 : Instagram (4/2) [1/1]
Age $>=$ 16.5
| Age $<$ 29.5
| | Time spent on social media $<$ 0.45 : Youtube (11/7) [10/7]
| | Time spent on social media $>=$ 0.45
| | | Time spent on social media $<$ 1.25 : Whatsapp (153/105) [70/48]
| | | Time spent on social media $>=$ 1.25 : Instagram (403/232) [215/128]
| Age $>=$ 29.5 : Whatsapp (50/24) [26/13]

Size of the tree : 31

There are various age grouping whose engagement in different social media. The node 1 to 20 shown in Fig. 3 present the age slots. If age group less than 16 and greater than 14, majorly used YouTube & WhatsApp. The greater than 16 and less than 29 age

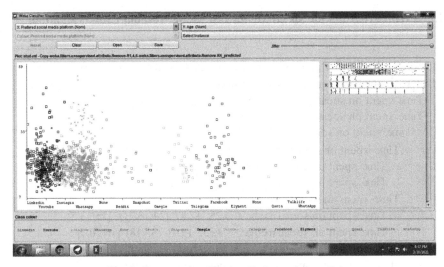

Fig. 2. Reptree visualize (predictive social media)

group is using Instagram also. Figure 4 display REPTree predictive age from weka 3.9 (Fig. 2).

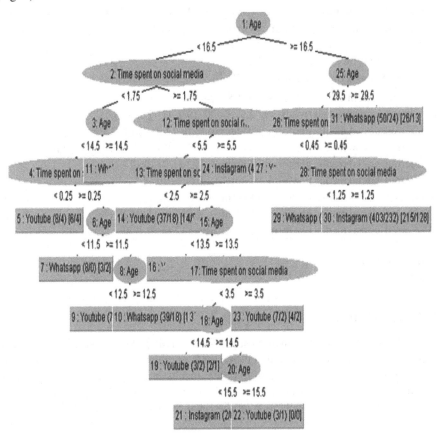

Fig. 3. Tree visualiser Reptree (age)

3.1.2 Predictive Rules for Social Media Test Mode: 10-Fold Cross-Validation

Prefered social media platform = Linkedin : 22.26 (41/8.51) [20/26.91]
Prefered social media platform = Youtube
| Time spent on social media < 0.15 : 16 (8/34.69) [4/10.81]
| Time spent on social media >= 0.15
| | Time spent on social media < 5.5
| | | Time spent on social media < 2.5
| | | | Time spent on social media < 1.5 : 18.59 (77/27.61) [33/25.99]
| | | | Time spent on social media >= 1.5 : 20.32 (55/55.4) [38/39.72]
| | | Time spent on social media >= 2.5
| | | | Time spent on social media < 3.25: 17.87 (24/13.71) [15/13.5]
| | | | Time spent on social media >= 3.25: 19.41 (33/18) [16/20.7]
| | Time spent on social media >= 5.5: 21.09 (6/6.33) [5/8]
Preferred social media platform = Instagram : 19.83 (246/9.19) [106/5.28]
Preferred social media platform = Whatsapp : 20.67 (211/46.49) [125/52.82]
Preferred social media platform = None : 18.41 (13/33.3) [4/74.81]
Preferred social media platform = Reddit : 18.8 (5/2.16) [0/0]
Preferred social media platform = Snapchat : 18.25 (7/15.14) [1/4]
Preferred social media platform = Omegle : 21 (0/0) [1/0.72]
Preferred social media platform = Twitter
| Time spent on social media < 2.5
| | Time spent on social media < 1.5 : 19.5 (4/2.75) [4/9.25]
| | Time spent on social media >= 1.5 : 24.33 (7/34.12) [2/8.59]
| Time spent on social media >= 2.5 : 20.82 (9/2.47) [2/3.09]
Preferred social media platform = Telegram : 19 (2/9) [1/0]
Preferred social media platform = Facebook : 23.48 (36/77.47) [16/29.36]
Preferred social media platform = Elyment : 22 (1/0) [0/0]
Preferred social media platform = None : 14 (1/0) [0/0]
Preferred social media platform = Quora : 20 (1/0) [0/0]
Preferred social media platform = Talklife : 20 (0/0) [1/0.02]
Preferred social media platform = WhatsApp : 12 (1/0) [0/0]

Size of the tree : 45

3.2 Random Forest

Random forest Tree is a mechanism of supervised Classifier. Here, nodes are divided in suitable subset randomly. Randomly chosen node are related with super node & same tree structure. Radom forest can help to derive results in classification and regression problems. The random trees classifier classifies each set of nodes with each tree in the forest and generates outputs as majorly voted. Suppose we discusses regression,

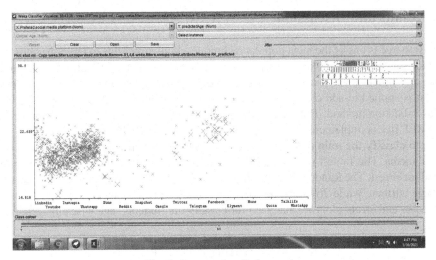

Fig. 4. Reptree (predictive age)

the average of the responses for all the trees in the forest is concentrated to focus more appropriate answer [17]. Figure 5 shows Random Forest algorithm with predictive social media factor. It focusses on the age on y-axis and social media on x-axis. The particular age group is using particular type of social media which is represented in broadly blue, green and red colors. Whatapp & youtube is being use mostly in age around 20.

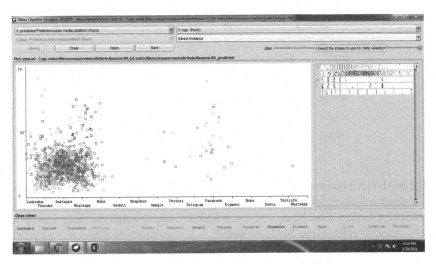

Fig. 5. Random Forest (predctive Social media for age)

4 Results and Discussion

Decision tree is one of the most popular techniques. Sometimes due to large structure of data size, the results are complex to analyze. REPTree can produce a simple tree structure. Such tree structure will work on accurate classification with Pruning Methods. Random Forest is applied to add on the age wise student engagement in social media using 10 folds validation method. The 66% of slit is used on dataset for further decision process. The REPTree is implemented with different attributes. The REPTree is structed with 31 node to classify the train dataset for age and 45 node tree structure for time spent on social media. The Table 1 shows the results for REPTree. Table 2 shows the results from Random Forest. The Random Forest suggested 165 node structure for same train data set for age attribute while 269 for time spent on social media attributes. The performance measures and errors plots are analyzed for classify the students' age wise engagement.

Table 1. Weka measurements: REPTree

Error	Social media	Age	Time spent
MAE	0.09	3.5	1.23
RMSE	0.22	5.46	1.7
RAE	94.58	99.98	93.64
RRSE	99.17	98.96	96.34

Table 2. Weka measurements: RandomForest

Error	Social media	Age	Time Spent
MAE	0.09	3.58	1.29
RMSE	0.22	5.54	1.79
RAE	94.1	99.98	93.05
RRSE	99.98	100	99.98

The MAE is popular for continuous variable data. The Lower value of Mean Absolute error is shown as better performance of predicted model. The RMSE is popular for high or low values for large errors. Again, the lower RMSE direct towards the more appropriate model results. The REPTree MAE & RMSE is low compare to Randon Forest in all factors as age wise social media platform time spent on social media.

5 Conclusion

The REPTree & Random Forest algorithms are popular for decision making in quick mode. These are is implemented on preprocessed dataset using weka 3.9. The REPTree is implemented with 31 node size to determine the age wise, social media wise and time wise usage as shown in tree visualizer Fig. 3. The different decision rules confirm that age greater than 16 and less than 29 majorly engaged in WhatsApp and YouTube. The decision rules defines also that age less than 16 are majorly engaged in YouTube. The same data is implemented with Random Forest and shown similar outcomes about age wise social engagement. Figure 5 is presentation of age and social media connection using Weka tool. The cluster created on greater than 5 and less than 29 which is concentrated around in between age number. Such age number supports the major engagement in WhatsApp & you tube as shown in Fig. 5. The study accomplishes the cluster on age nearby 16 is engaged in WhatsApp and you tube. The REPTree is more appropriate algorithm for implemented train dataset with minimum & structure node tree.

References

1. Prajapati, J.B., Patel, S.K.: Performance comparison of machine learning algorithms for prediction of students' social engagement. In: 2021 5th International Conference on Computing Methodologies and Communication (ICCMC), 8 April 2021, pp. 947–951. IEEE (22021)
2. Sheela, Y.J., Krishnaveni, S.H.: A comparative analysis of various classification trees. In: 2017 International Conference on Circuit, Power and Computing Technologies (ICCPCT), pp. 1–8 (2017). https://doi.org/10.1109/ICCPCT.2017.8074403
3. Shubho, S.A., Razib, M.R.H., Rudro, N.K., Saha, A.K., Khan, M.S.U., Ahmed, S.: Performance analysis of NB tree, REP tree and random tree classifiers for credit card fraud data. In: 2019 22nd International Conference on Computer and Information Technology (ICCIT), pp. 1–6 (2019). https://doi.org/10.1109/ICCIT48885.2019.9038578
4. Classification using REPTree. Int. J. Adv. Res. Comput. Sci. Manag. Stud. 2(10), 155–160 (2014)
5. Anguita, D., Ghio, A., Greco, N., Oneto, L., Ridella, S.: Model selection for support vector machines: advantages and disadvantages of the machine learning theory. In: The 2010 International Joint Conference on Neural Networks (IJCNN), pp. 1–8 (2010). https://doi.org/10. 1109/IJCNN.2010.5596450
6. Yang, L., Wu, H., Jin, X., Zheng, P., Hu, S., Xu, X., et al.: Study of cardiovascular disease prediction model based on random forest in eastern China. Sci. Rep. 10(1), 5245 (2020)
7. Ubels, J., Schaefers, T., Punt, C., Guchelaar, H.J., de Ridder, J.: RAINFOREST: a random forest approach to predict treatment benefit in data from (failed) clinical drug trials. Bioinformatics 36(Suppl_2), i601–i9 (2020)
8. Santos, F., Graw, V., Bonilla, S.: A geographically weighted random forest approach for evaluate forest change drivers in the Northern Ecuadorian Amazon. PLoS ONE 14(12), e0226224 (2019)
9. Hanko, M., Grendár, M., Snopko, P., Opšenák, R., Šutovský, J., Benčo, M., et al.: Random forest-based prediction of outcome and mortality in patients with traumatic brain injury undergoing primary decompressive craniectomy. World Neurosurg. 148, e450–e458 (2021)
10. Pavey, T.G., Gilson, N.D., Gomersall, S.R., Clark, B., Trost, S.G.: Field evaluation of a random forest activity classifier for wrist-worn accelerometer data. J. Sci. Med. Sport 20(1), 75–80 (2017)
11. Walsh, E.S., Kreakie, B.J., Cantwell, M.G., Nacci, D.: A Random Forest approach to predict the spatial distribution of sediment pollution in an estuarine system. PLoS ONE 12(7), e0179473 (2017)
12. Ishwaran, H., Lu, M.: Standard errors and confidence intervals for variable importance in random forest regression, classification, and survival. Stat. Med. 38(4), 558–582 (2019)

13. Salas, E.A.L., Subburayalu, S.K.: Modified shape index for object-based random forest image classification of agricultural systems using airborne hyperspectral datasets. PLoS ONE **14**(3), e0213356 (2019)

14. Jahandideh, S., Jaroszewski, L., Godzik, A.: Improving the chances of successful protein structure determination with a random forest classifier. Acta Crystallogr. D Biol. Crystallogr. **70**(Pt 3), 627–635 (2014)

15. Jones, F.C., Plewes, R., Murison, L., MacDougall, M.J., Sinclair, S., Davies, C., et al.: Random forests as cumulative effects models: a case study of lakes and rivers in Muskoka, Canada. J. Environ. Manag. **201**, 407–24 (2017)

16. Li, S., Bhattarai, R., Cooke, R.A., Verma, S., Huang, X., Markus, M., et al.: Relative performance of different data mining techniques for nitrate concentration and load estimation in different type of watersheds. Environ. Pollut. (Barking, Essex: 1987) **263**(Pt A), 114618 (2020)

17. Mohan, S., Saranya, P.: A novel bagging ensemble approach for predicting summertime ground-level ozone concentration. J. Air Waste Manag. Assoc. (1995) **69**(2), 220–33 (2019)

18. Rahman, M., Chen, N., Elbeltagi, A., Islam, M.M., Alam, M., Pourghasemi, H.R., et al.: Application of stacking hybrid machine learning algorithms in delineating multi-type flooding in Bangladesh. J. Environ. Manag. **295**, 113086 (2021)

19. Goya-Jorge, E., Amber, M., Gozalbes, R., Connolly, L., Barigye, S.J.: Assessing the chemical-induced estrogenicity using in silico and in vitro methods. Environ. Toxicol. Pharmacol. **87**, 103688 (2021)

20. Saha, S., Saha, M., Mukherjee, K., Arabameri, A., Ngo, P.T.T., Paul, G.C.: Predicting the deforestation probability using the binary logistic regression, random forest, ensemble rotational forest, REPTree: a case study at the Gumani River Basin, India. Sci. Total Environ. **730**, 139197 (2020)

Factual Data Protection Procedure on IoT-Based Customized Medicament Innovations

N. Srinivasan$^{(\boxtimes)}$ and S. Anantha Sivaprakasam

Department of CSE, Rajalakshmi Engineering College, Chennai 602105, India
{srinivasan.n,ananthasivaprakasam.s}@rajalakshmi.edu.in

Abstract. The basic values of urban communities in terms of smart worlds modified quickly by innovative industry. Mini sensor systems and smart access points are heavily involved to use the surrounding atmosphere data. In all industrial Internet of Things, it has been used to remotely capture and interpret real-time data. Since the Interoperable ecosystem collects and discloses data via unsecure government networks, an effective Encryption and Schlissel Agreement approach to avoid un authorized access is preferred. The Internet of Medical Stuff has grown into an expert technology infrastructure in the medical industry. The clinical symptoms of patients are obtained and analyzed. The clinical smart objects, incorporated in the human chest, must be studied functionally. In exchange, it will use smart mobile devices to provide the patient medical records. Although the data obtained by patients is so delicate that it is not a medical profession, the safety and protection of medical data becomes a problem for the IoM. Thus, a user authentication protocol based on anonymity is chosen to solve problems in IoM about the security of privacy. A reliable and transparent facial recognition user identification scheme is introduced in this paper in order to ensure safe contact in smart healthcare. This report also indicates that a competitor cannot unlawfully view or remove the intelligent handheld card as a lawful user. A comprehensive review on the basis of the model of random oracles and resource analysis is given to illustrate medical applications' security and resource performance. Moreover, the proposed schema foresees that it has high-safety characteristics to develop intelligent IoM health application systems as a result of the performance review. This application system uses ADV routing protocol. For the analyses of routing protocols with the NS3 emulator, experimental research was performed here. The findings obtained have been comparable to several other protocols in terms of the packet processing, end-to-end delay, throughput speeds and overhead route for the proposed SAB-UAS.

Keywords: National Health Service (NHS) of the UK · Wireless Sensor Network (W-WSN) · A holistic PHY and MAC layer protocols for LPWAN · WGAc then attempts to measure mx. P · World Health Organization (WHO)

© IFIP International Federation for Information Processing 2022
Published by Springer Nature Switzerland AG 2022
L. Kalinathan et al. (Eds.): ICCIDS 2022, IFIP AICT 654, pp. 55–70, 2022.
https://doi.org/10.1007/978-3-031-16364-7_5

1   Introduction

A Computer Stuff Internet consists of different sensing devices or technologies objects connecting to the corporate networks to share information. The physical items or equipment may be an electronic book, digital ticket or pocket, or may include detector, mobile screen, screen, robot or car. In IoT, connective items or artefacts should be made smartly without human intervention to take an ingenious decision (Al Turjman et al. 2019), The aim of IoT is therefore to incorporate a physical structure based on the machine to enhance the precision of social-environmental processes.

Gartner Inc. (Choi et al. 2016) forecasts the world's accessibility to about 8.4 billion IoT computers. IoT devices typically can be semi-structured or unstructured in nature.

Das et al. (2019), which may be an integral 5V broad data property of length, speed, variety, truthfulness and quality. In the database, e.g. on again and efficient storage media, the generated data volume is stored (Das 2014). In today's world, IoT efficiency is adopted by technical growth, which reaches a high standard of manufacturing and completes the job by fewer attempts. And therefore, our universe is more convergent to the IoT. In the various industries such as shipping, power/services, logistics, engineering, mines, metals, oil, gas and aviation (Das 2009), IoT integration can also be implemented.

It can be described as the next wave of innovation for maximising environmental capital according to market research and academic experts. IoT encourages sound decision and data processing to transform the manufacturing properties by using a sensor or virtual objects. The companies thus connect smart equipment or machines in order to estimate that by 2021 the IoT market would hit 123.89 billion dollars (Farash et al. 2017). The development of wireless technology has recently worked extensively to build numerous sensor-based technologies, for example environmental testing, automotive, electronics, military health, Internet connectivity (Gope et al. 2018), drone delivery, etc. (Hameed et al. 2018).

A medical electronic device has a wireless network of medical sensors with low memory capacity, bandwidth and processing power (Huang. et al. 2013), and lightweight tools. A heterogeneous network of wireless physical body areas typically consists of medical sensors such as ECG, blood pressure, oximeter, temperature, etc. They sensed and gather physiological data on patients to move on to smart medical devices, i.e. iPhones, Laptops, PDAs, implantable medical devices, etc., using a wireless communication channel (Li et al. 2018).

Therefore, the health care professional should read and consider a wider examination evaluation, as the analysis of data is required to be one of the main concerns for the introduction of wireless communications technology, namely remote access to gateways, telephone and medical sensors (Odelu et al. 2015). Health sensors are used to read the biochemical details in the patient's body. A medical specialist may use an authenticated wireless gateway to access the sensing data. The problem of user verification, which becomes a major research field for wireless communication, is thus tackled.

A standard IoM model for the hospital setting as seen in Madhusudhan and Mittal (2012), to evaluate safety and efficiency problems. This involves patients, healthcare personnel, medical sensors, a device archive, a portal and a server that provides tremendous software advantages, including large-scale medical surveillance, emergency medical surveillance and response. Due to the insecurity of data sharing across public networks, medical sensor privacy is so necessary in order to avoid data exploitation. Patient's protection and privacy in the healthcare application framework. A 2013 World Health Organization (WHO) study revealed "the global shortage of human health staff in the coming decades to hit 12.9 million" (Al Turjman et al. 2019), The decline was primarily due to a drop in the participation of college students who are entering the occupation, ageing of current employees and a rising risk of not-communicable disorders of citizens such as cancer, cardiac disease, stroke etc. "Individualized and related wellbeing" has recently offered the healthcare sector a ray of hope for revolution.

New studies suggest that the National Health Service (NHS) of the UK save 7 billion pounds a year by minimizing the number of innovative hospital visits and admissions of critically sick people from a distant area using innovative technologies (Choi et al. 2016). The promise of customized and linked wellness is to give patients, physicians and medical personnel various benefits. In addition to tracking and tracking their own vital signs, insulin pumps and blood pressure mangoes, for example, enable doctors and physicians to monitor them on a remote basis. This is effective for patients, too, as they are treated immediately. In addition, patient engagement helps elderly adults, as they can preserve their safety at homes with no need for long-term, often depressing hospital stays.

Current smart phones can act as on-body coordinators and central systems for customized health monitoring and are fitted with a range of sensors including heart rate optics, blood glucose and pressure measurements, concentration of oxygen, oxygen and other vital signs, air temperature, pressure and moisture measurements, and so forth. In addition, built-in smart phone apps can be used to monitor everyday health. However, performance, safety and protection, economic productivity in the common use of mobile support technologies and generic open-source platforms remain a concern.

Wearable devices such as cameras, actuators, coordinators and doorways form a wearable Wireless Sensor Network (W-WSN), in a classical scheme of personal health surveillance where a coordinator is a main centralized controller that programme the on-body contact nodes and collect the information from those nodes. Normally, such signal can be accessed to the remote physician through gateways or bases through a wireless or wired network which can be considered off-body communication often. The overall design and enabled technology for linked safety and health implementations have, however, grown with the emergence of corporate communication (Das et al. 2019). Figure 1 shows healthcare model using IoT infrastructure.

It is necessary for other similar current surveys to illustrate the major gaps and responses to this study. Islam et al. cover IoT-based innovations in medical care in the new studies. The focus will be on exploring cutting-edge communication protocols, IoT-based software and applications, with particular emphasis on security concerns in IoT health systems. They also list some rules and standards on the implementation of different IoT technology to the field of healthcare. Another Alam et al. survey shows portable human resources technologies and implementations.

The standards cover many legacy technology and standards. Choudhary et al. survey [7] also deals with the heritage of simple digital technology. In addition to this, a legacy short-ride connectivity protocol is usable in the surveys by Lin et al. and Al-Fuquha et al. and the convergence of fog/edge and IoT computing and their implementations. A holistic

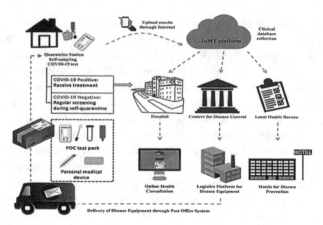

Fig. 1. Healthcare model using IoT infrastructure

PHY and MAC layer protocols for LPWAN solutions have been provided in Wang et al. survey's [10]. In addition, IoT's stability and privacy issues were also addressed by a variety of efforts. For example, Andrea et al. [11]'s survey work highlights security and data protection concerns related to IoT technology related to physical networks, communications, applications and authentication.

In the [12] survey, the IoT technology, security issues and counter-measures are also discussed. The Jianbing et coll. Survey highlights the problem of security in IoT fog computing. In addition, the Ida et al. survey presents IoT's security concerns and problems in relation with telemedicine and clouds. Botta et al. recognize the convergence of cloud infrastructure and IoT as well as the survey papers described previously. In addition, the Verma et al. survey introduces the insights on IoT data and various IoT information facilitators.

2   Literature Survey

A substantial attempt has been done to build reliable user security features, but there is no major result in achieving greater privacy and protection. As stated, the use of current cryptosystems cannot meet those security objectives. It is clear that an upgraded or expanded implementation of the authenticator to increase the security performance of all software applications is recommended. Very few articles have taken formal security and results monitoring architecture and evaluations into account in literature. From the other hand, some authentication programmes have been considered inadequate for the accomplishment of safety targets and their main characteristics. Therefore, a stable and effective user authentication framework does not provide distinctive standard of authentication.

There have been some improved implementations of authentication schemes for different applications but most schemes have been found to be unsuitable for security purposes. What matters is how to attain targets, such as two-factor authentication even though the smartcard has been misplaced or faulted and the password upgrade is protected. More complex topics have been tackled by Huang et al. (2013), Madhusudhan and others have recently discovered a problem of inflexibility for multiple crypto-system programming techniques. Double factor user authentication in the literature means that the user will invariably pick his/her password to draw the PS password equally. Because it is impractical, this presumption may trigger.

A confusion impact, A mistake effects. As an example, the preceding hypothesis states that an adversary Adv deleted the smartcard parameters. A risk of good Adv is defined exactly as in the attempt of an online attack. A two-factor technique, PS, is used to ensure the safest way to disseminate multiple attack vectors such as playback, concurrent session, off-line login imaging, etc., while a stable user authentication protocol is implemented. Specifically, Adv is supposed to penetrate the hazard of PS, which is not greater than when Adv and TO signifies a negligible-value to online impersonation assaults. User picked codes, on the other hand, are also far from uniformly spread. The suggested SAB-UAS system offers a flipped verifier that can deduce users' Smartcard depravity in due course. To provide a protective mechanism As a consequence, an online attack can be avoided to give evident intractability.

Several encryption schemes [11] have been implemented for data protection and protected contact. However, a login credentials verification protection problem requires retention of a password board for proof of the validity of the recipient. In order to store the password database, it needs an additional memory space. Several scientists have proposed an alternative signature or iris approach for fast overhead storage. As a norm, it gives you a storage advantage for the measurement of the intelligent card at many levels of protection.

A protected authentication mechanism on the basis of RSA and DH was implemented for WSNs by Watro et al. (2004), A hash-base dynamically authenticated device was proposed by Wong et al. (2006) to withstand many potentials, including man in the middle, repeat, falsification, and main character. Das et al. (2019) have however shown their devices vulnerable to a privileged security attacks and have even suggested an updated version in order to increase protection efficiency improvements.

Yoon and Kim (2013), recommended a method of biometric user verification to avoid flaws in security including low reparability, service denial and an imitation of sensors. Choi et al. (2016) found that Hwang and Kim had neglected to fix issues of security, including user inspection, user confidentiality, biometric identification, symmetric encryption disclosure, DoS assault, key withdrawals, and complete potential secrecy.

They also expanded the biometric user identification programme in order to increase security efficiencies and have also discovered that their solutions are simpler than most authentication mechanisms and key arrangements, Sadly (Choi et al. 2016), scheme's remains unclear about crucial impersonation assaults. Sadly, (Park et al. 2016), Since WSNs deal with different environmental structures; any enemy may deduce or collect sensor information physically from sensor memory. An opponent can attempt to destroy entire networks of medical sensors using extract knowledge from the sensor node. It is then evaluated as a possibility.

WSN and Clinical Communication Protocols flaws too. At first the key exchange protocol was implemented by Lamport. In the past, several protocols were suggested for authentication. The elliptic-curve cryptosystem was used by Chang et al. to develop a compact authentication system. To attain the property of future anonymity, they established an ECC authentication. The two-factor authenticator ECC for WSNs was constructed by Yeh et al. [25].

Even so, their scheme did not accomplish a satisfactory reciprocal authentication of the primary security objective. Shi et al. also considered the Yeh et al. scheme not safe. Subsequently, (Choi et al. 2016) showed that Shi et al. scheme will ensure key-sharing, robbing smart cards and sensor energy supply. The so-called sensory energy assault is critical to adding energy usage to limit a sensor node's lifespan. Choi et al. (2016) improved the Shi et al. scheme to resolve the problem of sensor energy depletion.

However, the customer anonymity and intractability of contact agencies were not maintained by their system. The RFID authentication protocol for IoT was proposed by Li et al. (2018). Their protocol allows for clear reciprocal authorization to safeguard the confidentiality of the reader, tag and application server in real time. Li et al. (2018) further expanded their authentication protocol to address the previous mechanism's security disadvantages, i.e. Healthcare based on IoT.

This upgraded version allows consumers more anonymity so that replay and data communication attacks cannot be avoided. Li et al. (2018) subsequently developed a tripartite User Authentication Protocol to provide evidence of the client confidentiality with the Chebyshev and Chaotic-Map. Hameed et al. (2018) proposed an integrative data integrity protocol on IoT-based WSNs via gateway control information i.e. the base station. Al Turjman et al. (2019) have developed surface architectures to facilitate interoperability, consistency and adaptability to mobile devices, IoT, and cloud services.

Al-Turjman et al. (2019) developed a cloud-centred IoT based cloud-based system for main agreements. Deebak et al. submitted a context-aware IoT hash-based RFID authentication. Li et al. (2018) have developed an ECC-based IoT environment authentication protocol to authenticate service access using the biometrics functionality. For potential IoT implementations Challa et al. implemented an ECC-based user authentication method. In relation to the non ECC authentication mechanism, however, their scheme requires more computational and coordination overhead. Wazid et al. have developed a stable lightweight IoT network cryptography. Their device uses biometrics, intelligent cards and passwords as a four approach to satisfy key agreements. Roy et al. subsequently proposed the new crowd sourcing IoT User.

Authentication Protocol Their scheme asserts that biometric models are anonymous to the recipient. Wazid et al. also created the latest verification method to assess the validity of medicinal, i.e. dosages, schemes for medical counterfeiting. Al-Turjman et al. proposed to assert the context-sensitive knowledge function a Seamless Shared Authentication Mechanism for IoT. The protective features and associated disadvantages of the literature have been well studied. Therefore, a secure encrypted face recognition method (SAB-UAS) for the IoT setting is presented. It summarises the methodology used in modern authentication systems, their downside, formal analysis and simulation.

Our research focuses on potential health-care systems and their technological demands and how new connectivity technologies enable those applications. The key goal is to examine and demonstrate the specifications of potential implementations and address whether or not these requirements can be fulfilled by new communication technologies. This survey further highlights the transparent obstacles and problems of science that will need to be tackled to satisfy these criteria. Park et al. (2016).

However, one aspect of the survey that is not addressed is privacy concerns in IoT, because we agree that this is largely a matter of legislation and regulations. Questions including collection, transmission and ownership of data of people would entail a new regulatory structure, which will also eliminate undue limitations on the IoT markets. At the same time, the framework should be enforced. In the general context, this category includes a scientific population, physicians, SMEs. The target demographic includes research population.

Section 4 elaborates the proposed system concept in details. Results and conclusion are depicted in Sect. 4. Section 5 portrays the conclusion.

3 Proposed System

The approach is modeled to provide full control of the contact channel here between communications institutions in real-time under the standard password authentication and key agreement protocol. Adv will allow the corrupting legitimacy of contact parties to deduce the hidden key for a long time to characterize the qualities of potential secrecy. ADV can also receive a previous session Keys for unauthorized exclusion investigation. Latest research has shown that extraction of protection parameters can be deducted from power attacks, programme lapses, and geometric modeling. The Decline of data breaches such as off-line password guessing and impersonating attacks may arise from confidential details. It is also clear that the session key may be intercepted for a malicious memory card attack on the smartcard. But the intruder will use a card reader to intercept stolen or misplaced storage key to read user's sensitive information (Watro et al. 2004). Figure 2 shows in data flow architecture.

This will allow the assailants to interrupt any protected trust model when adhering to extreme opponent principles. It uses robust encryption to shield against detrimental behaviors that breach some kind of client authenticator trivially. The care described above this is A malicious user can disrupt terminal access to a side-channel attack ongoing; In a short period of time, an attacker can leak sensitive information from a legal user. The study seeks to nullify overly conservative proposal that a smartcard would merely be an external storage card that users an integrated microprocessor, supporting protection devices, to conduct a cost-effective process.

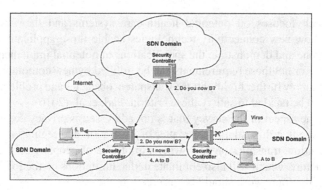

Fig. 2. Data flow architecture

The authentication system is completely insecure over public networks as a card; all authentication schemes based on memory cards were inherently insecure when used with unsafe terminals. The contingent recognition as non-tamper thus Safer than the severe presumption is represented is reactive. Regarding hazard C, this statement is claimed that it would not be very helpful to prove that it is actually true to its protection importance or not.

The password is checked on the other side to receive valuable knowledge from the relevant remote server that can lock the legitimate user account before executing the smartcard. If the above search is available so ADV should still use a malicious card reader to identify a user's password. The principal disagreement is that ADV is not established strictly in hazard C and D. As mentioned, the counter-protection can minimally presume that the duration of the lock approaches or does not surpass the threshold limit. It is expected to be possible with SAB-UAS ADV model.

Unencrypted described in Cartesian {DID {always to DAD} with quadratic times by {Substring, Igor} pair. It can accommodate possible functionalities such as offline password conjecturing and untraceable online password conjecture, etc. Please note that B is already clearly assumed as a hazard that does not take into account the user's protection function, while the Weep model suggested is stronger.

To integrate both previous and current assumptions in a realistic way to include a strong and stable signature scheme. As stated in (Yang et al. 2008), a constructive review reveals that the user authentication mechanisms on smart cards have a common collection of security features to ensure that strong authentication parameters are effective. Madhusudhan and Mittal (2012), Wong et al. (2006), showed that there are inconsistencies and inefficiencies in a previous sequence of safety assets, and so proposed nine separate protection priorities and ten wanted attributes.

As the protection objectives are based on semi tolerance, the device is set to be higher. However, significant security contradictions also remain problematic among the existing requirements. Smartcard Failure Attack is totally free. This is, unauthenticated users acquiring a valid User Card should be unable, even though the smart card was acquired or exposed to incur the hidden details, to easily alter their Smartcard password or to retrieve their Victim's password by online, offline, hybrid password guess or an impersonal key attack. The system of a computer.

Do not store a user-password verifier database or extract user passwords meaning. The scheme will endure multiple possible attacks including brute force password devaluation, playback, concurrent guessing, de-sync, burglary, key imitation, unauthorized key-share, crucial and known key-control. The device can handle many possible threats, including offline password guessing, replaying, concurrent guessing, de-sync, robbing authentication, key imitation, unauthorized key-share and crucial. During the device authentication process, the client and the server will create a common secret session key to secure data communication among real-time companies.

The scheme is not quickly delayed and synced, i.e. the server has to synchronise the time of its clock with the smartcard input devices and vice versa. The structure will attempt to attain the privilege of full confidentiality. It clearly points out that the criteria collection - C4 offers an attack scenario in which ADV has obtained access to the intelligent card while C5 has no access to the smartcard of the victim. C4 assumes a standard reissue for the smartcard user's access to it with the following requirements.

The oracle's random oracle model. The criteria - C5 is only based on basic assaults that the authentication mechanism linked with passwords will effectively be protected against stolen new attack channels, which are addressed in based authentication schemes.. The criteria is seen that the conventional authentication method removes redundancies and complexity to make it easy to cryptographic functions on the principle of coherence. In addition, real-time environmental systems will rely on the efficiency of the proposed authentication method. A broad comparison reveals that the proposed model of opposition is too complicated and the parameters are more rigorous.

Compared to current programmes practical and detailed, an ethical products data encryption framework with smartcard is introduced in this section. Three contact agencies in SAB-UAS, ME, mobile gateway connectivity WGAc and MS j, the medical specialist in particular. Before the Sab-UAs scheme starts, WGAc produces two main keys, including m x and my and transmits the long-hidden H key (SIDjie my) to MS.

The method of its introduction, WGAc then attempts to measure mx. p, known as a public portal key. P. This proposed scheme consists of three phases: user registration, device login, authorization and authentication declining prices. To substitute for the lack of the intelligent card or long-term key Divulgation, failure, or interfered intelligent cards should be withdrawn or repealed at cyclical stage annually. The systematic evidence in this segment is seen Random oracle template that illustrates the safety performance of the SAB-UAS system proposed. The importance of spiral model r2 and master hidden session keys mx and my off is to be defined in a collision free hash function.

This section uses the rationale of Burrows Abadi Needham to prove that the method suggested is entirely true and functional to prevent common threats to achieve healthcare systems' protection performance. This model is becoming a well-known formal cryptographic protocol used for cryptography research. The main notes and philosophical postulates of the BAN are identified. This attack uses WGAc to capture the data from the DC data centre Trying to get the legitimate user's access. The passwords of the current SAB-UAS scheme are safely passed to avoid privileged insider's attack. It is disguised to create a long-lasting private key using a one-way hash function. In addition, master keys like m x and my use BTi on Usr, for extracting for store Pi values.

Assume that the rightful person has misplaced his/her SMi smartcard and ADV attempts, using a power-analysis mechanism, to retrieve the legal details from U sr, such as ostemm, ni, vrin, h(). ADV will not however retrieve or infer hidden session keys, as the master keys such as mx and I'm unclear, to execute a parallel attack. The SAB-UAS structure then argues that the privileged insider assault would be resilient. User anonymity plays an important function in security applications frameworks.

Therefore, wireless networking and electronic technology must be improved. The planned SAB-UAS device protects User's classified knowledge, biometric prototype and manager—instead, my features—to secure the user identification UID. Furthermore, it is apparent that communications from the planned SAB-UAS framework retain biometric blueprint and digital certificates—the purpose of my key with symmetric encryption. Therefore, the derivative UID is impracticable in the proposed SABUAS scheme to achieve the user identity preservation function. ADV attempts to take care of some MS j sensor nodes that create communications with User in the proposed SAB-UAS system. However, as designed or constructed using Ni, ADV does not easily catch or forge message transmission MS3. In addition, MSj shares a USK session, which is not connected to KSU in any way, with WGAc. The suggested SAB-UAS framework then notes that ADV was unable to effectively operate this assault.

4   Results and Discussion

In general, the user identity storage character can only be six letters. As DES is generally referred to as unsafe, a 56-bit key size is not known to be stable. The efficiency comparison indicates, relative to other existing systems for contact bodies such as Usr, WGAc and MSj, that the delivery times of the proposed system are less costly. For the sake of practical use, the suggested SAB-UAS system is believed to be durable and reliable. Yet (Gope et al. 2018) are not entirely viable since they are vulnerable to an assault on synchronisation.

Even if no adversary tries to block the data packets in the wireless environment, the missing data packet between Usr, GAc and MSj cannot occur. It looks like it is a replay attacks problem. Assume that a last message clarification of the planned SAB-UAS system has been blocked or robbed because of the overdue time. The parameter pair—Adi ADINA—cannot be replaced by WGA. The data between Usr and WGAc can then be changed. This section illustrates the use of the NS3 simulation to realistic implementation of the proposed SAB-UAS in network parameters such as packet-related distribution, end-to-end connectivity retardation, data transmission throughput rate (first file bps) and overhead routing.

A well-reacted version labelled as NS-3.28 was built on the Ubuntu-14.04 LTS platform for the study of the above-mentioned parameters. This indicates the critical variables used in the Simulation of NS3 assuming that the area of network coverage is 80/80 qm for testing the medical sensor and system node at 25 m and 50 m distance. A mode of connectivity called Used to simulate network length by means of media access control <1800 s i.e. 30 min. Due to the design of the network i.e. routing the link state is preferred to be optimised ad hoc. It is used to include competitive exploration which requires careful connectivity in order to keep the regression model among communicators.

As shown in Fig. 3 comparison of throughput values, The efficiency of the communication routing in all surveillance systems is incredibly important to calculate. This research has been carried out with the use of packet size, node availability, and transfer range and coverage area. The efficient transmission packet distribution rate at the sink node is defined by the contact metric. It is clear that as the number of sensors increases, the PDR ratio of proposed SAB-UAS deteriorates significantly. In particular, a small deflexion in the proposed SAB-UAS and Wu et al. indicates a stronger packet distribution ratio than other authentication systems were obtained from the row applied.

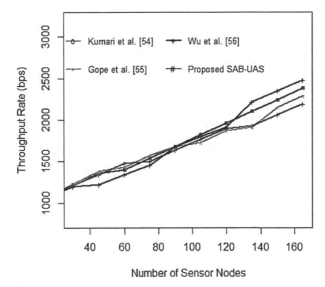

Fig. 3. Comparison of throughput values

In comparison, the signal congestion tended to occur as rows were inserted consistently. As a result, when the far distance transmitting message was recorded, the energy model described in the wireless environment began to drain more than expected. A threshold limit may be set on the receiver side to increase the delivery ratio to control the energy dissipation or avoid packet transfer while the connection is distant. During analysis the number of interconnections will increase if mobility ranges between ~4 ms and ~20 ms. The results are calculated. As a result, rare loss of packets and errors decline.

Interoperability efficiency of the connection. The time taken to reach the receiver from the source node is determined by the average data transmission packet. The findings of the examination demonstrate that the suggested delay increases in proportion to the number of contact nodes updated. As such, multiple transmitting messages are strongly mentioned that they are subject to further congest in the specified scenario. The rate of transmission may be calculated by the number of bits per unit of runtime. The overhead routing can be described as the total number of route packages divided by the total number of packets delivered effectively during the flexibility interval. The thorough study indicates that the current number of routing packets is used to efficiently deliver each incoming packets. In addition, to find this parameter.

The use of unused bandwidth to control network traffic during overhead routing. The consequence of the simulation shows that the OLSR protocol attempts to reduce overhead correspondence, as it holds a constructive routing table for managing daily "Hi" and "Topology search" communications. Figure 5 indicates that the new SAB-UAS results in lower overhead routing e.g. packet than other routing solutions such as packets routing. In OLSR, transmission routing is handled tacitly to increase the flexibility speed and network efficiency at 2 to 50 m/s. The today's telemedicine and eHealth platforms are targeted at user data collection and off-line information exchange among user and clinician, including wearable and mHealth solutions.

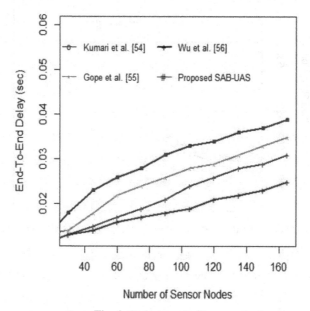

Fig. 4. Delay comparison

Figure 4 shows delay comparison, that means very soft QoS specifications for the efficiency of the contact networks and their output in real time. Advanced remote sensing technologies, such as the LoC systems mentioned above, do not set harsh criteria for the efficiency of the contact networks. For loop systems which involve such actuators, the situation is more complicated. In terms of physical walking assistance and slip avoidance, for instance, there is a need for technical help solutions for neuromuscular disorders. In reality, about 40 million people suffering from neurologic diseases and neurological disabilities in 2005 lead to over 90 million years of worldwide disability-adjusted life. Electrical stimulation is an innovative way to treat Parkinson's disease tremors in this respect. In addition, metronomically timing sound may lead to recovery from the off phases of Parkinson.

Any of the patients with stroke, ataxia, and traumatic brain injury will benefit from gait enhancement electric muscle stimulation. There are electrical stimulators that can activate the neuronal muscles of patients. Commercial foot drop systems, for example, can trigger heel structures while walking through electrical impulses and thus reduce the risk of patients deteriorating dramatically. The electrical pacemaker may be operated manually by a wireless connection or by a wireless press switch under the single in modern systems. The sixth Latencies do not surpass roughly according to human haptic feedback tests. 50 ms to have an intense response or suggestions sensation.

The established stimulators work however as discrete solutions without considering modifying ambient factors and other patient health criteria, either manually or on the single sensor. For example, the standalone treatment does not involve such scenarios, such as the existence of ice or stairs or the patient's elevated heart rate. Unreliable (wireless) contact between the single sensor and relaxation actuator can be counterproductive to the welfare of the patient, for example through network congestion. Connectivity strategies for wearable actuators therefore have to give such deterministic QoS standards, particularly though there are many external background outlets.

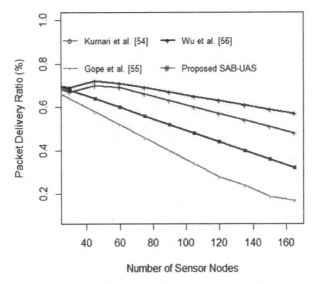

Fig. 5. Performance of the proposed system

An idea of an electric muscle relaxation device responsive to a contemporary context. In conclusion, the four above-mentioned case studies on more healthcare applications would force to the edge the current wireless communications and technology by integrating new types of sensors and their numbers, actuating capacities, the quantity of data and data speeds, ultra-low power utilizing, higher level of service and reliability. In addition, fresh, tougher communications service quality standards for WBAN solutions would possibly be created.

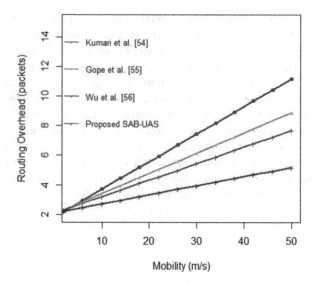

Fig. 6. Packet delivery ratio comparison

The key focus implementations of LPWA technology include but are not limited to smart cities, personal IoT application, smart grid, intelligent metering, organizational networks, industrial control, agriculture, etc. This involves a transition from legacy technologies to the current generation standards. Smart fitness and remote wellness control devices are one of the futures uses for LPWA technology (Yoon and Kim 2013). These systems are W-WSNs which transmit or receive information via long-range wireless links. In particular, for delay tolerant applications requiring low data rate, low energy usage and low cost, LPWA technologies are considered. One such instance is the remote observation of patients at home, which does not usually have tough time specifications. Figure 6 shows Packet delivery ratio comparison.

Moreover, the majority of Wireless standards are based on the topology of Star Network in which each node/sensor contacts the ground station directly. The consequence is that a node and base station have high asymmetric connections. The data flow, as in the majority of IoT applications, primarily consists of acknowledging or clear commands or actions from base station to access point and traffic from base station to node. This asymmetric design tends to render the base station all the complexity, contributing to simpler end devices that have a low cost and improved battery capacity. On the other extreme, this also has a negative effect on LPWA technology's scalability and quality of operation. Figure 5 depicts the performance of the proposed system in terms of Packet delivery ratio and Number of sensor nodes used in different protocols.

5   Conclusions

In this post, for the intelligent computerized medical framework using IoM a secure anonymous biometric user authentication method was suggested. The planned Bhai Structure demonstrates the structured authentication scheme, capital and strategies build to illustrate protection, storage and productivity. The previous evidence suggests that the framework proposed will shield a user's personal information from opponents and obtain the estate of a full confidentiality. Sometimes this study illustrates that the emerging Mara system greatly decreases the costs of storage, computing and connectivity for improving the productivity of all healthcare software networks in real-time. Moreover, the comprehensive casual and systematic safety review using the BAN rationale and a random oracle model shows that it offers stronger safety proof to defend multiple future IoM-based attacks. It is also seen that the proposed regime increases the resource utilisation of developing smart e-health networks, including energy, computing, and connectivity. The sensor nodes, including packet distribution, end-to-end latency and include at, were tested with an NS3 network simulator. It is shown that as the amount of data delivery increases proportionately by inclusion of sensors in a row, the proposed SAB-UAS system would become more congested. But in contrast to other digital certificates, even if the transmission of message increased proportionally, the suggested SAB-UAS could achieve a better packet distribution ratio, end to end, throughput rate and overhead routing for the specified scenario.

References

Al Turjman, F., Hasan, M.Z., Al-Rizzo, H.: Task scheduling in cloud-based survivability applications using swarm optimization in IoT. Trans. Emerg. Telecommun. Technol. **30**(8), e3539 (2019)

Choi, Y., Lee, Y., Won, D.: Security improvement on biometric based authentication scheme for wireless sensor networks using fuzzy extraction. Int. J. Distrib. Sens. Netw. **12**(1), 8572410 (2016)

Das, A.K., Wazid, M., Yannam, A.R., Rodrigues, J.J., Park, Y.: Provably secure ECC-based device access control and key agreement protocol for IoT environment. IEEE Access **7**, 55382–55397 (2019)

Das, A.K.: A secure and robust temporal credential-based three-factor user authentication scheme for wireless sensor networks. Peer Peer Netw. Appl. **9**(1), 223–244 (2014). https://doi.org/10.1007/s12083-014-0324-9

Das, M.L.: Two-factor user authentication in wireless sensor networks. IEEE Trans. Wirel. Commun. **8**(3), 1086–1090 (2009)

Farash, M.S., Chaudhry, S.A., Heydari, M., SajadSadough, S.M., Kumari, S., et al.: A lightweight anonymous authentication scheme for consumer roaming in ubiquitous networks with provable security. Int. J. Commun. Syst **30**(4), e3019 (2017)

Gope, P., Lee, J., Quek, T.Q.: Lightweight and practical anonymous authentication protocol for RFID systems using physically unclonable functions. IEEE Trans. Inf. Forensics Secur. **13**(11), 2831–2843 (2018)

Hameed, K., Khan, A., Ahmed, M., Reddy, A.G., Rathore, M.M.: Towards a formally verified zero watermarking scheme for data integrity in the Internet of Things based-wireless sensor networks. Future Gener. Comput. Syst. **82**, 274–289 (2018)

Huang, X., Chen, X., Li, J., Xiang, Y., Xu, L.: Further observations on smart-card-based password-authenticated key agreement in distributed systems. IEEE Trans. Parallel Distrib. Syst. **25**(7), 1767–1775 (2013)

Li, C.-T., Chen, C.-L., Lee, C.-C., Weng, C.-Y., Chen, C.-M.: A novel three-party password-based authenticated key exchange protocol with user anonymity based on chaotic maps. Soft. Comput. **22**(8), 2495–2506 (2017). https://doi.org/10.1007/s00500-017-2504-z

Madhusudhan, R., Mittal, R.C.: Dynamic ID-based remote user password authentication schemes using smart cards: a review. J. Netw. Comput. Appl. **35**(4), 1235–1248 (2012)

Odelu, V., Das, A.K., Goswami, A.: A secure biometrics-based multi-server authentication protocol using smart cards. IEEE Trans. Inf. Forensics Secur. **10**(9), 1953–1966 (2015)

Park, Y., Lee, S., Kim, C., Park, Y.: Secure biometric-based authentication scheme with smart card revocation/reissue for wireless sensor networks. Int. J. Distrib. Sens. Netw. **12**(7), 1550147716658607 (2016)

Watro, R., Kong, D., Cuti, S.F., Gardiner, C., Lynn, C., et al.: TinyPK: securing sensor networks with public key technology. In: Proceedings of the 2nd ACM Workshop on Security of Ad hoc and Sensor Networks, pp. 59–64 (2004)

Wong, K.H., Zheng, Y., Cao, J., Wang, S.: A dynamic user authentication scheme for wireless sensor networks. In: IEEE International Conference on Sensor Networks, Ubiquitous, and Trustworthy Computing (SUTC 2006), vol. 1, pp. 8-pp. IEEE (2006)

Yang, G., Wong, D.S., Wang, H., Deng, X.: Two-factor mutual authentication based on smart cards and passwords. J. Comput. Syst. Sci. **74**(7), 1160–1172 (2008)

Yoon, E.J., Kim, C.: Advanced biometric-based user authentication scheme for wireless sensor networks. Sens. Lett. **11**(9), 1836–1843 (2013)

Deep Learning Based Covid-19 Patients Detection

C. Paramasivam[(⊠)] and R. Priyadarsini[(⊠)]

Department of Electronics and Communication Engineering, Amrita School of Engineering,
Amrita Vishwa vidyapeetham, Coimbatore, India
`priyadarsini.r@spmvvengg.onmicrosoft.com`

Abstract. Radiographic variants in scanning of the chest has high deformity and-status from Polymerase Chain Reaction of Reverse Transcription evidence of the COVID-19 factor, which cannot be disputed by a UN agency, including a low rate of admission paid overtime in phases. We often summarize the relationship level tests that have broken down much back propagation neural network (BPN) to include Computed Tomography and COVID-19 tests, respiratory infection, or lack of morbidity. Through a mechanism known as chain rule, the (BPN) technique is employed to successfully train a neural network. Back propagation performs a backward pass through a network after each forward pass while modifying the model's parameters i.e., weights and biases. We often distinguish between the mean and created tests on the most critical 2nd and 3D learning models available. In addition to them with the most recent clinical information, corresponding degreed has gained zero terrorist organization.996 (95% CI: zero.989–1.00) Corona virus compared with Covid cases in each body Computed Tomography looks and determined 98% related affiliate qualifications associated with a minimum of 92%.

Keywords: Computed tomography · CNN · BPN · COVID · RT: PCR

1 Introduction

The SARS-CoV-2 animal infections were to magically spread to a big number of people in December 2019 after being contaminated by the rhinolophus bat. Huanan's food market in China, was targeted by the Covid malady (COVID-19) acquired by SARS-CoV-2. Covid was outraged by the process of malnutrition, heart problems, and mental disorders during critical patients, and in this way, many people died. The first and firstphase of treatment in the real world is crucial in reducing mortality.

[10] COVID-19 Detection Using Dynamic Fusion-based Federated Learning: To analyse medical diagnostic images, researchers developed a dynamic fusion-based supervised learning architecture of the system. Present a method for dynamically selecting participating clients and scheduling modeled fusion based on the participants' training time based on their local model performance. This method also produces a set of medical diagnostic picture datasets for COVID-19 recognition, that can be used for image analysis by the machine learning.

© IFIP International Federation for Information Processing 2022

Published by Springer Nature Switzerland AG 2022
L. Kalinathan et al. (Eds.): ICCIDS 2022, IFIP AICT 654, pp. 71–82, 2022.
https://doi.org/10.1007/978-3-031-16364-7_6

[11] Early detection of COVID 19 utilising multimodal imaging data and transfer learning may benefit in the development of medicine and disease control measures. This study illustrates how images from three of the most often used medical imaging types, X-Ray, CT scan and Ultrasound, can be used to detect COVID-19 using transfer learning from deep learning models. The idea is to provide a second set of eyes to overburdened medical professionals through smart deep learning picture classification algorithms.

Based on a preliminary evaluation of many prominent Convolution Neural Network (CNN) models, choose a decent CNN model. Then, Focus on the challenges (such as dataset size and quality) in using current publicly available COVID-19 datasets for developing useful deep learning models and how this negatively impacts the trainability of complex models, and propose an image preprocessing stage to create a reliable image dataset for compositing. The new approach tries to reduce visual noise so that deep learning models can focus on diseases with distinct features.

[12] Used an Deeper Learning-based System for Detecting COVID 19 Patients The outstanding performance of deep learning (DL) in numerous alternatives prompted us to use a DL-based approach for the COVID-19 detection system's computer-aided design (CAD) in this study. To identify and characterize whether the patients were normal or contaminated with COVID-19, they employed Res Net 50, a state-of-the-art segmentation technique based on Deep Learning. To assess the proposed system's robustness and efficacy, use two publicly available benchmark datasets (Covid Chest x ray-Dataset and Ch ex-Pert Dataset). The proposed system was trained on photos from 80% of the datasets and tested on images from 20% of the datasets. When cross-validation is employed, the performance is evaluated using a 10-fold cross-validation technique.

[13] They used a deep neural network to detect and categorize COVID 19 by using breast X-ray images: A machine learning strategy for detecting and categorizing COVID 19 from a chest x-ray with high accuracy has been suggested. The tensor stream related CNN technique is designed for identifying x-ray pictures. On the prepared dataset, the recommended model was trained and tested, The total accuracy for the 2-class categorization (COVID 19 vs. a healthy person)was discovered to be nintyfive%, with precision and recall rates of nintyfive%.

Photographs taking Computed Tomography model models on the chest showed a high degree of disability. The condition was highlighted in the RV-PCR ID of COVID-19, which, the UN agency, covers a limited range of initial limitations. In one study of 1,421 cases from China, exposure to COVID-19 in chest CT was determined to be 97% when the benefits of reverse transcription polymerase chain reaction were taken into account (RT-PCR). The opacities of the ground glass, varied green suspensions, merging, and opening of the indisputable assignment have a special model to take in computed tomography chest pictures with COVID-19 positive conditions.. Examination of twenty-one patients' communication certificates for the 2019 Covid found that fifteen (71%) had a combined CT scan of the chest. Twelve (fifty seven %) showed opacities in the ground glass, while seven (thirty three%) had flow disturbances. Six (29%) had contact with earth glass opacities, and four (19%) had irregular clearance.

There is no respiratory system; a different airway manages the acquisition of serous membranes and diseases. Fourteen patients (three out of 21) provided a standard Channel.

Data and CT scans are open from this current stage, showing the emergence of respiratory organs in a short period, from signal to forty-one days from the onset of the first phase (Fig. 1), during the patient's positive COVID-19 [1].

The emergence of the term COVID-19 in CT scans will be understood from different types of pneumonia. However, it effectively detects pre-existing weight loss of undeniable symptoms of the infections from the below table. By using AI and the continuation of critical learning, it is possible to distinguish between small types of respiratory diseases and create themes an essential experimental gadget.

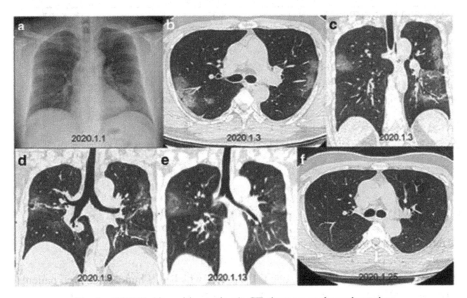

Fig. 1. COVID-19 positive patient's CT chest scans throughout time

On day 7, a scan showed opacities in the left and right brain's lower and upper lobes (after onset of symptoms). a, b, c Scans reveal multifocal bilateral ground glass opacities on the ninth day. d. On Day 15, the scans reveal a crossover and mutation of crushed glass opacities and consolidation. By the 19th day of the scan, antiviral treatment had cured the consolidations and ground glass opacities. f. Day 31: The scan confirms that the issue has been entirely resolved.

According to Shi et al., previous SARS study reported no lung cavitation, numerous effusions, or lymphadenopathy.

We often review other such tests, the notion of various back propagation neural network for preparing CT tests with COVID-19, respiratory infection, or lack of morbidity. COVID-19 has been related to a huge amount of cases in China, and this has emerged as a region that is making significant success in establishing major organisations to promote learning. This epidemic can provide exceptional Computational Tomography qualifications to promote the correction and security of direct COVID verification. The public database for CT scans is essential to reverse the AI model with the immediate effects of population growth throughout the epidemic. These essential learning strategies will help experts in realization and management.

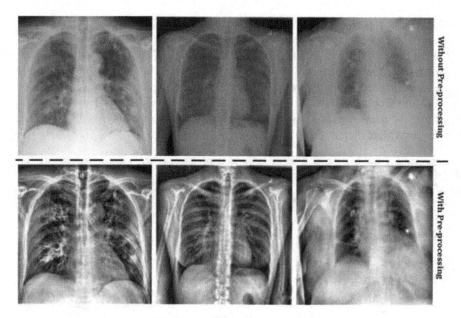

Fig. 2. X-ray pictures of patients

2 Methods

– Data set introduction

The test assembled 618 traverse zone CT tests, of which 219 hundred and ten patients were with COVID-19. Even if there were no picture markers on the CT film of the chest, COVID-19 patients were all screened or exposed to the RTPCR test section then disappeared. Furthermore, however, there is a difference of 2 days between CT data sets for any benefit from the comparison patient to verify the test classification.

The remaining 399 city trials were submitted due to a controlled evaluation collection. Among them, 224 of the 224 patients with influenza-A viral infection underwent CT tests with H1N1, H3N2, H5N1, H7N9, and one hundred and seventy-five CT trialsstrong individuals. Would have been of the 198 (90.4%) COVID-19 and (196), The remaining 96% [2] were influenza-A patients, mostly from the planned or correctional stages and 13.4% of cases were severe (severe) from the abuse phase. P) > Zero.05).

In addition, CT instances of influenza-A viral infection have been consolidated because, given China's current situation, it is critical to recall them from probable COVID-19 patients. 528 Computational Tomography trials (85.4%) came forward with an endorsement set of 189 COVID-19 patients with influenza A virus infection [3] and 145 samples stronger than them.

Due to the check set, the remaining ninety CT sets (14.6%) were used, with thirty COVID-19, thirty influenza-A virus infections, and thirty strong cases [3]. To boot, CT set tests were observed on those not related to the design phase.

Table 1.

	Incubation (days)	Transmission	Distribution	Consolidation	GGO nodule	Nodule	Bronchial wall thickening	Pleura Effusion
Human Corona (SARS-CoV2)	2–10	Droplet, airborne, contact	Peripheral, multifocal	+++	+	Rare	UC	Rare
Adenovirus	4–8	Respiratory, oral-fecal	Multifocal	+++	+++	Centrilobular	UC	C
RSV	3–7	Contact, aerosol	Airway	+	+	Centrilobular+++	C	C
Influenza	1–4	Droplet, airborne	multifocal	+	+	Centrilobular++	C	UC

– Process

Figure 2 below shows a sample of COVID-19 systematic reports during this evaluation. All in all, before that, CT footage was kept in separate areas with high aspirations. Second, a 3D Back Propagation Neural network model divides specific candidate image blocks. The middle image, mounted on two neighbors of each strong form, is assembled for subsequent methods. Finally, all photo patches are divided into three categories using the Picture Request model: Incompatibility-deflection, COVID-19, seasonal flu respiratory disease, and COVID-19 [2]. The image patch [2] is dominated by a strong comparison with the candidate type and the assurance score that all things think. Finally, a comprehensive evaluation report for the CT test is made subject to noise- orth theoretical limitations. The heart of neural network training is back propagation networks. It's a method of fine-tuning a neural network's weights based on the error rate of the preceding epoch (i.e., iteration). You can lower error rates and improve the generalisation of the model by fine-tuning the weights, making it more reliable.

– Location

Gathering ideas assumed that the image request model would understand the form and structure of the modified pollutants. Also, relatively good methods are used for forwarding weights to select capacities from the edge due to the relative field information of the aerial image's solutions. The purpose of the combination of contaminants found at

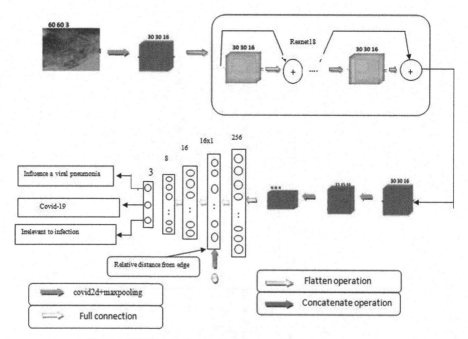

Fig. 3. COVID was detected using the following methodology

the serous membrane point can, without a doubt, be seen as COVID-19. The last useful methods of solving each solution. From the edge are as follows: 1) The best bottom practices win, ranging from the exold to the fixed center.

2) Remove the attached taxi-turvy. The position of the image of aspiration.3) The last right way from the edge is that the split is found in the reconciliation condition, separated from corner to corner by the second step.

– Network structure

Two back propagation neural network 3D request models were reviewed during this evaluation. There is a modest Resnet 23 based association. The inverted pattern is usually incorporated into the union structure by intercepting the field idea system in a fully-aligned layer to increase the final accuracy rate[1]. A simple Resnet-18 accessory structure was used for the complete extraction of the footage (Fig. 3) [1]. Pooling methods have been used relatively for dimensional decay of data to be confusing and improve hypotheses.

The convention layer's yield was flattened into 256 dimensional half vectors and modified shortly after 16 using the full-connection appendix in the dimensional section vector. For Locus Idea affiliation, the last unbelievable path of Edge-of-Honor 1 is immensely close, and sometime later, it is related to building a fully-aligned network. 3 full layers were then employed to obtain a final comprehensive result that was close to the confirmation score [1].

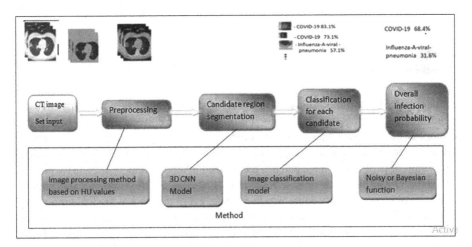

Fig. 4. Flowchart of a categorization process using a deep learning model

– Coaching method

One of the only proportional font disaster limits, cross-entropy, was employed in this examination. The adversity regard did not reduce or raise clearly once the age variety of preparing emphasis extended to quite one thousand, suggesting that the models joined well to associate degree overall ideal state while not plain overfitting. Figure 4 shows the design curves for the black eye regard as well as the correctness for two depiction models. On the differential readiness dataset and the principle reset, the association with the territory idea framework resulted in better execution.

3 Results

See Fig. 5.

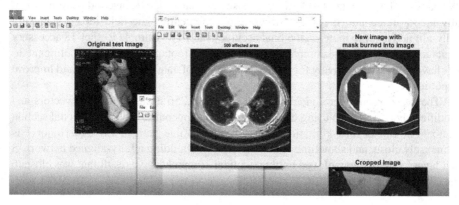

Fig. 5. selecting the frame of scanning picture to detect the pneumonia

– **Checking dataset**

Accurateness of the technique chooses, however, well, the figures are. Accuracy chooses the multiplication of a scale, or the amount of wishes are correct. The survey exhibits the number of positive results was procured. The F1 college uses a combination of accuracy and memory to work typical outcomes. One-piece image brokenness: associate degree entireness of 1710 image patches recognized in ninety models, together with 357 COVID19, 390 Influenza-A-disease respiratory diseases, and 963 illness safe (ground reality). To plan out that procedure was legitimate, every means was studied employing a disorder grid. 2 association structures were attempted: outside and outdoors the survey Table 1. Blueprint, victimization ResNet to get rid of options from CT footage and also the zone checking model, examined while not the region looking model, will positively understand COVID-19 cases from others, with preciseness a complete of eighty six.7%. Incredibly, it took a normal of beneath the 30 s on a CT set (with seventy layers) from information extensions to report (Figs. 6 and 7).

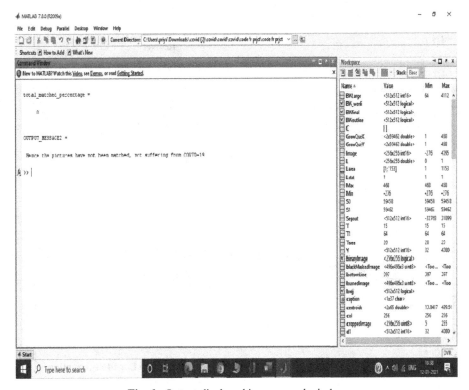

Fig. 6. Output displayed in command window

– Discussion

RT-PCR testing for a 2019-nCoV polymer may create associate degree proper investigation of COVID-19 from patients with Influenza-A malady respiratory disease. Regardless, destructive nucleic testing encompasses several burdens: time delay, low acknowledgment rate, and nonattendance of dextrose. Within the 1st place periods of COVID-19, a few patients could currently have the correct finding of pneumonic imaging. Regardless, there's nobody fluid, and during this means we tend to expertise alarming facet effects of RT-PCR testing from bodily cavity swabs. These patients don't seem to be poor down as suspected or insisted cases, don't seem to be separated or treated, and perhaps potential sickness sources. Impediments on the flip of events and end of China's insignificant gathering business have essentially influenced its economy and the effects of overall issue imports. With the final unfold of COVID19 at associate degree displeasing rate, a few of states ar in like manner declaring associate degree exceptionally delicate circumstance: taking evaluations, as an example, college terminations, travel impediments, and offers to folks operating isolated (notwithstanding key employees, e.g., clinical guardians, authorities, care workers). While not a doubt, even to the Fortune one thousand associations, it will undeniably not be the norm in business within

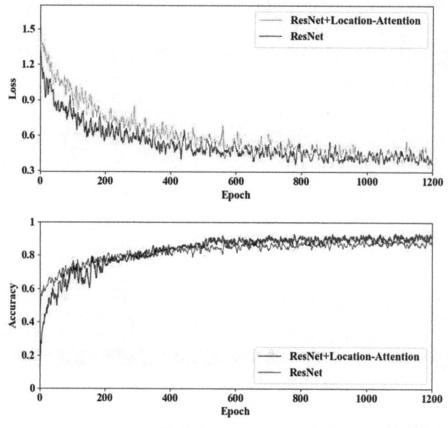

Fig. 7. Output waveforms

the related to barely any months. Market investigators have forewarned that the disorder may value $ one.1 trillion in lost overall compensation. Whereas the UN agency unambiguously recommends testing no matter variety suspects as would be cheap, this has not been followed by numerous countries that emit a control of being a result of a group action of resources/staff and nonattendance of RT-PCR testing. Here, CT imaging with AI will facilitate acknowledge COVID-19 by giving a speedier alternative different that lessen the unfold.

CT footage of COVID-19 cases shows a particular look, as analyzed in section one (Introduction). Regardless, it's not intentional or careful to understand COVID-19 from numerous ailments merely through regular eyes. In assessment, preliminary models dependent on all-around learning, as an example, those investigated here, are shown to administer unequivocal and powerful results by innovative digitization and film assessment.

– Limitations and any improvement

COVID-19 can be quickly and accurately identified from chest CT image collections using the start to finish evaluation technique utilised in this study, and its space rate is higher than those of RTPCR testing [1]. In any case, many distinctive assessments that take a look at AI models have beginning late found unbelievable motivation for revealing.

As an example, one assessment drove on existing models of all-around second and 3D obtaining, designing them with the newest clinical perception, and got terrorist group zero.996 (95% CI: zero.989–1.00) cases of Coronavirus versus Non-Covid with every pectoral CT channels. They determined ninety-eight (2%) affectability and 92% disposition. The initiative was multi-faceted, and it looked into every CT case on two levels: scheme A: Cases of handle and directed opacities in three - dimensional evaluation [1] (using business programming referred to as 3D pneumonic volume from RADLogics opposition.) and scheme B: as lately created second Analysis for every digit of the case to acknowledge and recognize tremendous degree unfold opacities as well as ground glass implants.

Similarly, they raised the Corona scale: a meter extent of respiratory organ opacities and an excellent strategy to screen a patient's headway once a while. This examination performed higher than the one investigated during this paper concerning space, affectability, and disposition, essentially since the structure used was unambiguously planned to understand the openness of COVID-19. As battlefront AI models will merely provide it, fast and actual knowledge examination for CT is prime for driving clinical specialists taking a goose at patients throughout this scourge.

– Different views

New prosperity connected procedures were applied against the unfold of the novel (COVID-19) in China. The latest development consolidates artificial thinking, suggestive programming, and robots for scattering in clinical centers, in line with clinical thought measurable wanting over firm GlobalData. Thanks to COVID-19 in China, the number of passings extended before the unfold of severe acute respiratory syndrome in 2003. During this attention-grabbing condition, technical school new organizations healing government workplaces, masters, and instructors to unfold the word. Unreal insight Developer Infirvision sent the Coronovirus AI resolution in China not terribly so much within the past, projected for pre-organization use to assist specialists with recognizing and screen specialists even additional effectively. As Johns Hopkins University moves during this field, GIS development has become a crucial contrivance to forestall the unfold of Covid data processing mistreatment. GIS development is significant as a result of it perceives zones wherever individuals say the contamination. Online media districts square measure adequate wellsprings of GIS data as advancement maps the zone of interest that individuals square measure talking concerning Covid. As needs are, preventive measures will be dead due to the higher track of the glow map zone and the normality of the sickness. Ten years back, it had been demanding to look at diseases; nowadays, with AI, AI, and GIS, data processing and sense extraction square measure ending up being additional priceless and even other superb in region contaminations. Essential concern: Preventive latency with the usage of AI is snappier nowadays.

References

1. Priyadarsini, R.: Pneumonia detection using mask RCNN (2021)
2. Butt, C., Gill, J., Chun, D., Babu, B.A.: Deep learning system to screen corona virus disease 2019 pneumonia. Applied Intelligence (2020)
3. Coronavirus disease 2019 (COVID-19) situation report – 51, WHO. https://www.who.int/docs/default-source/coronaviruse/situation-reports/20200311-sitrep-51-covid-19.pdf?sfvrsn=1ba62e57_10. Accessed 11 Mar 2020
4. Ai, T., et al.: Correlation of chest CT and RT-PCR testing in Coronavirus disease 2019 (COVID-19) in China: a report of 1014 cases. Radiology, 200642 (2020). https://doi.org/10.1148/radiol.2020200642. pubs.rsna.org (Atypon)
5. Chung, M., et al.: CT imaging features of 2019 novel coronavirus (2019-NCoV). Radiology, 200230 (2020). https://doi.org/10.1148/radiol.2020200230. DOI.org (Crossref)
6. Shi, H., et al.: Evolution of CT manifestations in a patient recovered (2020)
7. From 2019 novel coronavirus (2019-NCoV) pneumonia in Wuhan, China. Radiology, 200269. https://doi.org/10.1148/radiol.202020026. pubs.rsna.or (Atypon)
8. Koo, L., et al.: Radiographic and CT features of viral pneumonia. Chest Imaging (2018). https://doi.org/10.1148/rg.2018170048
9. Lessler, J., et al.: Incubation periods of acute respiratory viral infections: a systematic review. Lancet Infect Disk **9**(5), 291–300 (2019). https://doi.org/10.1016/S1473-3099(09)70069-6
10. Lu, Q.: Dynamic fusion-based federated learning for COVID-19 detection (2021)
11. Horry, M.J.: COVID-19 detection through transfer learning using multimodal imaging data (2020)
12. Nasser, N., Emad-ul-Haq, Q., Imran, M., Ali, A. and Al-Helali, A.: A deep learning-based system for detecting COVID-19 patients (2021)
13. Chakravorti, T., Addala, V.K., Verma, J.S.: Detection and classification of COVID 19 using convolutional neural networkfrom chest X-ray images (2021)
14. Chaudhary, S., Sadbhawna, S., Jakhetiya, V., Subudhi, B.N., Baid, U., Guntuku, S.C.: DetectingCovid-19 and community acquired pneumonia using chest CT scan images with deep learning (2021)
15. Karhan, Z., Akal, F.: Covid-19 classification using deep learning in chest X-ray images (2020)
16. Irmak, E.: A novel deep convolutional neural network model for COVID-19 disease detection (2020)

A Progressive Approach of Designing and Analysis of Solar and Wind Stations Integrated with the Grid Connected Systems

N. Balavenkata Muni[1][✉], S. Sasikumar[2][✉], K. Hussain[3][✉], and K. Meenendranath Reddy[4][✉]

[1] Department of Electrical & Electronics Engineering, Siddhartha Institute of Science and Technology, Puttur, India
nagugari@gmail.com
[2] Department of Electrical Engineering, Annamalai University, Annamalainagar, Tamil Nadu, India
ssasikumar77@yahoo.in
[3] Department of Electrical Engineering, SITCOE, Ichalkaranji, Kolhapur, Maharashtra, India
hussain16679@gmail.com
[4] SVR Engineering College, Nandyal, Andhra Pradesh, India
kvmsvist@gmail.com

Abstract. This article illustrates the advanced control strategy and analysis of hybrid resources integrated to grid system. The hybrid resources consist of solar station and wind station, which are integrated to the grid through AC bus system. The Perturb and observe algorithm (P&O) is used in Maximum Power Point Tracking (MPPT) to increase the power output of PV station and Wind station to maximum. The modeling and analysis of the system is improved using fuzzy based optimization is done in MATLAB/SIMULINK platform. The effectiveness of the system is improved using fuzzy algorithm during different load conditions. The improvements of grid voltage, grid power and grid currents can be observed during different load conditions from conventional PI controller to fuzzy logic controller. The advanced controller efficiently maintains the grid parameters stably in the hybrid systems.

Keywords: Solar station · Wind station · Fuel stack systems · Fuzzy controller and MPPT controller

1 Introduction

The renewable resources have significantly penetrated in the power system. The demand on the power system is increasing 20% every year therefore it is very important that the researchers shall be matured in the way of renewable energy resources for the growth of electrical requirements. In the last few decades the significance of these resources are increasing continuously along with the existing system owing to its effective and efficient

© IFIP International Federation for Information Processing 2022
Published by Springer Nature Switzerland AG 2022
L. Kalinathan et al. (Eds.): ICCIDS 2022, IFIP AICT 654, pp. 83–96, 2022.
https://doi.org/10.1007/978-3-031-16364-7_7

performance, when it is associated with the distributed generators. The choosing of DGs are important when it is integrating with existing AC grids.

Several research are in progress along with the energy systems like batteries, charging capacitors, super capacitors, magnetic energy storage systems, fuel cell electrolyzes etc. Recently, the doubly fed induction generators along with fuel cell electrolyze have produced and proved several advantages. Several control strategies have developed to control the output requirements. MPPT techniques with algorithms are incorporated to increase the power output and to maintain the voltage stability values with respect to the system specified values.

This paper has focused on the analysis of hybrid energy resources during the integration of AC grid systems. An advanced fuzzy controller has proposed the power management algorithm with different load condition.

In this article Sect. 2 represents the system description, Sect. 3 describes the design of controller for PV station, wind station and energy storage device, Sect. 4 describes bidirectional topologies for battery, Sect. 5 describes fuzzy process and Sect. 6 presents the analysis of the system with the proposed controller.

2   System Description

The proposed model consists of the design of PV, Wind stations and batteries are connected to 11 kW AC grid network, shown in Fig. 1. The three 2.4 kW small size solar stations are interconnected and connected to common coupling bus system. A doubly fed induction generator with constant speed is acted as wind station. The power output from PV panels and wind turbines are supervised by MPPT circuit with perturb & observe algorithm.

In this paper an optimization based controller is chosen to amplify the power output to the load so that stability of the system is maintained. Both linear and nonlinear loads are tested for steady state operations.

3   Design and Control of PV, Wind and Battery Systems

In this section, we illustrate the design and control of the PV station, wind station and battery storage device for integrating to common bus system at AC loads are connected.

3.1   Model and Operation of Wind Controller

A fixed speed DFIG is taken for wind station. There are two control modes based on wind speed and power output, i.e. pitch control mode and power limitation mode.

3.1.1   Pitch Control Mode

In this case control of wind speed comes into presence. If wind speed exceeds the rated value, it must bring to rated value to maintain the extracted power to the rated

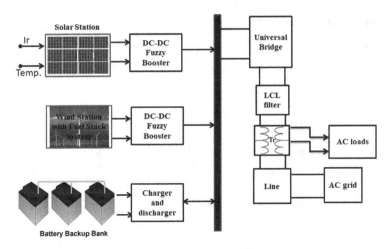

Fig. 1. System model

value. The power constant K_p and pitch angle α should be varies to maintain sufficient electromagnetic torque of wind turbine. The related expressions are shown below.

$$\lambda_{wind} = N_{turbine} * \frac{R_r}{v_{wind}} \tag{1}$$

$$K_p = p_{wind} * \frac{R_r}{0.5 * A_b * v_{wind}^3} \tag{2}$$

where $N_{turbine}$, λ_{wind}, p_{wind} and K_p are the rotational value, rotational value ratio, blow power and power gain respectively. The above all parameters are corresponding to rated value of wind speed. R_r and A_b are the radius of rotor and area blades respectively. By solving the above equation the pitch angle α can be obtained.

3.1.2 Power Limitation Mode

In this mode, the turbine speed is maintained to rated value. The power demand ad speed ratio is calculated as follows.

$$\lambda_{lim} = N_{t\,norm} * \frac{R_r}{v_{wind}} \tag{3}$$

$$K_{plim} = P_{m-rated} * \frac{R_r}{0.5 * A_b * v_{wind}^3} \tag{4}$$

$$\text{Pitch angle } K_p(\lambda_{lim}, \alpha) = K_{plim} \tag{5}$$

3.2 Model and Operation of PV Controller

3.2.1 PV Modeling

From Fig. 2, the generated current (I) by the PV cell is given by

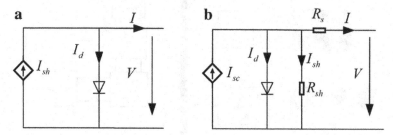

Fig. 2. Internal circuit of PV station

$$I = I_{sc} - I_d - I_{sh} \tag{8}$$

$$I = I_{sc} - \frac{V + IR_s}{R_{sh}} - I_o\left(e^{\frac{v+IR_s}{nV_T}} - 1\right) \tag{9}$$

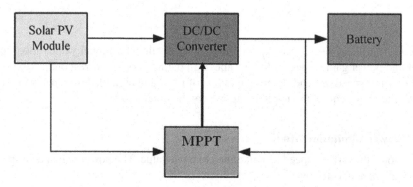

Fig. 3. PV model

The preferring maximum power from the PV panels with controller of procuring MPPT [8] is described by the known techniques. The rated voltage as Open circuit voltage and the rated current as short circuit currents are considered as backend parameters to measure the rated specifications of PV panels at different environmental conditions with MPPT techniques. There is another front end technology to measure the accurate panel voltage and currents with faster rate than the backend process. This advanced technology is known as Perturb and Observe algorithm (P and O) (Fig. 3).

3.2.2 P and O with MPPT Techniques

Particularly, this method is used to procure maximum power from the panels using MPPT procedures [9]. The output of PV panels are regularly measured and compared with previous values. The voltage increment or decrement is compared with the increase or decrease of powers procured from the panels if the increased voltage is proportional to increased power then the location of operating point is on left side. If the increased voltage is inversely proportional to the power then the operating point is located in right side. The perturbation is to be chosen such that the operating point shall be on left side. In this method, MPPT is used to measure the voltages of PV panels and battery. If PV voltage is reached to maintain which is more than battery voltage then the battery is charged to maximum so that the maximum power from PV panels are obtained. Otherwise the voltage is discharged from battery to maintain the PV power so that the maximum power is maintained at constant rate at any load conditions. In this paper, PWM cycle is used to trigger the charge controllers to extract the maximum power from PV panels. This MPPT technique with P and O algorithm is efficient easy technology to acquire the maximum power from PV and Wind systems.

3.3 Model and Operation of Bidirectional Controller for Battery

Lead acid batteries are the most used type of energy storage systems due to their reduced cost and high durability [10]. Due to their popularity, a large amount of research was performed to identify their properties and behavior for simulation purposes. Identifying the battery behavior is very important for the design of the converter. A large number of battery models were developed that include circuit based models such as [11]. It is important to identify a model for the battery that estimates the state of charge accurately to be able to assess the performance of the power converter operation. The battery model must be compatible with the charging and discharging profiles of typical lead acid batteries. The state of charge must be determined accurately for the purposes of converter control. A battery model is as found in [12] and can be summarized as follows.

The model diagram for the energy storage system is given in Fig. 4, which is connected for PV panels in the given grid connected power systems. Figure 5 and Fig. 6 are represented the respective characteristics during discharging and charging of energy storage system in the fuzzy based environment.

The integrating loop provides an estimation of the actual charge of the battery. The voltage is then calculated and supplied to the controlled voltage source that converts the signal into a power signal. Even Though this model has several simplifications, it still can provide an accurate response.

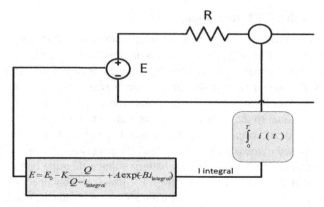

Fig. 4. Bidirectional battery controller

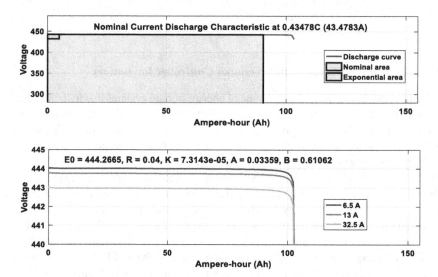

Fig. 5. Battery discharge characteristics during fuzzy environment

In order to verify the model, a test simulation will be performed and the charging and discharging characteristics will be examined. A short circuit test will demonstrate the discharge characteristics and an open circuit test will demonstrate the charging characteristics of the model. The model described is important for the simulation and gives an idea about the characteristics of the battery system.

In the Fig. 7, the performance of fuzzy logic controller in PV station is observed at different values of inputs and observed the different outputs. The duty cycle variation is observed in the Table 1 for different current and voltages with an irradiance of 1000 w/m2.

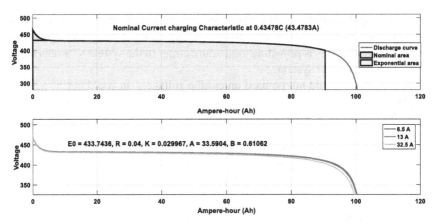

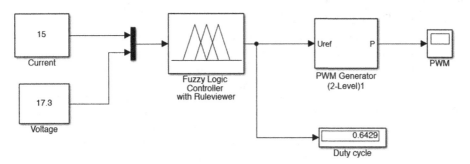

Fig. 6. Battery charging characteristics during fuzzy environment

Fig. 7. Simulink block for Fuzzy logic controller

4 Bi-directional Topologies

This section describes the various DC-DC converter topologies. They are (a) seclude type converter (ii) non-seclude type converter. The seclude type converter comprises of two Bi-directional type bridge converters which are separated by isolated transformers. The skeleton model converter type is given in [9].

Table 1. Duty cycle for input current and voltage

Current(A)	Voltage(V)	Duty Cycle(D)
3	15	0.9169
5	18	0.7293
10	25	0.6402
15	17	0.6429

The proposed converter type provides high efficacy with the coordination of transformer which acts as a protection cover for the grid system and energy storage system. The only exception in this is providing large transformer in the system to produce large efficiency to the converter system.

Such converters should not be used unless the voltage gain could not be achieved through non-isolated converters. The other type is the non-isolated converter which also has many topologies. The simplest design is the synchronous buck converter which is still used with electric vehicles as mentioned in [12].

In this design, the ideal voltage-boost-gain is calculated as follows:

$$\frac{V_o}{V_i} = \frac{1}{D}$$

The simulation block is shown in Fig. 8, is implemented in SIMULINK. The performance of fuzzy based PV controller is tested with different irradiance such as 1000 w/m^2 and 800 w/m^2. The variation of voltage, current and power is obtained in normal PI-MPPT and Fuzzy-MPPT environments, as shown in Table 2.

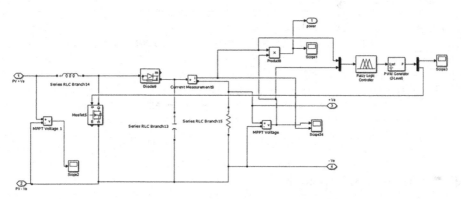

Fig. 8. Simulink block for fuzzy MPPT controller

Table 2. PV outputs

Irradiance (w/m^2)	PI MPPT			Fuzzy MPPT		
	Voltage (V)	Current (A)	Power (W)	Voltage (V)	Current (A)	Power (W)
800.0	528.3	3.9	2055.0	585.7	4.2	2465.9
1000.0	660.4	4.2	2783.0	658.3	4.5	2975.5

5 Fuzzy Logic Controller Operational Process

The process of fuzzy rules is incorporated into what is called a Fuzzy Interfacing System. It is classified into three cases, viz., Fuzzification, fuzzy process and Defuzzification which exercise the system inputs for the significant system outputs.

The linguistic variables denoted as NL, NM, NS, Z, PS, PM and PL stand for negative large, negative medium, negative small, zero, positive small, positive medium and positive large respectively. For both of the inputs, the equation of the triangular member function (MF) employed to determine the grade of membership values ($\mu A_i(x)$) is given by

$$\mu A_i(\text{TED}) = \frac{1}{b}(b - 2|TED - a|) \tag{10}$$

where 'a' represents the coordinate of the point at the grade of memberships (MFs) and 'b' is the width of MFs. $\mu A_i(\text{TED})$ is the grade value of membership. Figure 9 and Fig. 10 represent the fuzzy surface view and fuzzy membership triangles for the fuzzy rules respectively.

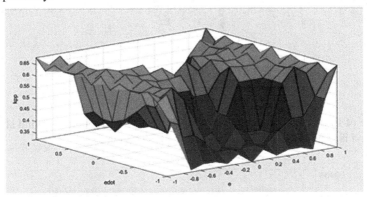

Fig. 9. Fuzzy surface view for fuzzy rules

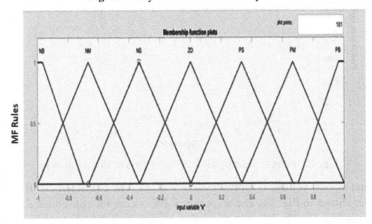

Fig. 10. Fuzzy membership plot

6   Simulation Analysis and Results

The simulation results of the test system, shown in Fig. 11 are presented with fuzzy logic approach. Table 1 detail the duty cycle for the input voltage and current of the energy storage device. Table 2 describes the variation of outputs PV station for different irradiance in the presence of conventional MPPT and fuzzy MPPT controllers.

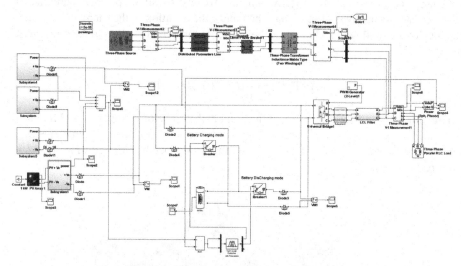

Fig. 11. Simulink model of overall system

In case of PI controller used in hybrid system, the power output of wind station is unstable and is oscillated 4.95 kw to 5 kw, as represented in Fig. 12.

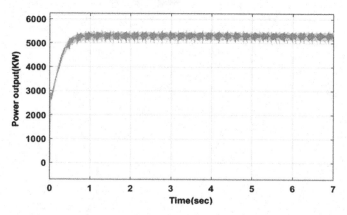

Fig. 12. Power output of wind station in case of PI controller

The wind station voltage is represented in Fig. 13. The characteristic of PWM for duty cycle is shown in Fig. 14. The response of PV station for conventional controller is presented in Fig. 15. The PV voltages and currents are slightly distorted initially and stabilized after 0.15 s, represented in Fig. 16.

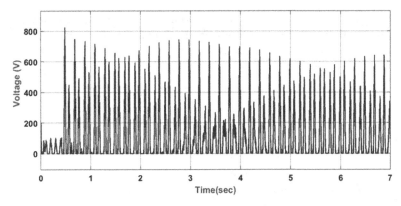

Fig. 13. Wind station voltage in case of PI controller

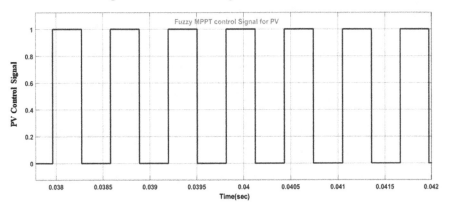

Fig. 14. PWM vs Duty cycle in case of fuzzy based PV controller

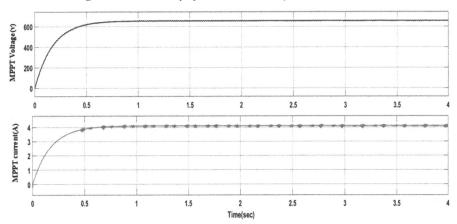

Fig. 15. PV MPPT voltage and current

When fuzzy MPPT controller with P&O algorithm is used in PV station, the distortion in load voltage and current becomes less at any load condition. The voltage and current are stabilized to 400 V and 19 A respectively, represented in Fig. 17.

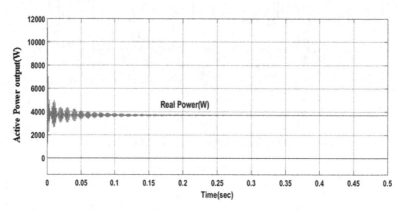

Fig. 16. Load Power (P) with fuzzy logic controller in PV station

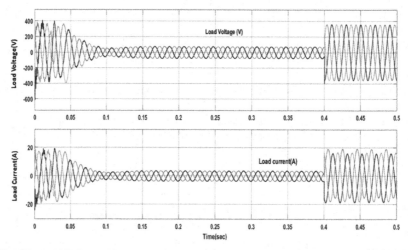

Fig. 17. Load Voltage (V) and Load Current (A) with fuzzy logic controller in PV station

Table 3 is the detailed comparison of PV voltage, current and Power during conventional MPPT and proposed MPPT approach. Table 4 gives the stable load parameters at any point of load.

Table 3. The variation of PV voltage, current and PV power in PI MPPT and Fuzzy MPPT

Time (sec)	PI MPPT			Fuzzy MPPT		
	Voltage (V)	Current (A)	Power (W)	Voltage (V)	Current (A)	Power (W)
0.25	427.7	3.0	1281.3	503.2	3.2	1586.8
0.5	528.5	3.7	1956.3	621.8	3.9	2422.7
0.75	552.4	3.9	2135.3	649.9	4.1	2644.4
1	558.0	3.9	2179.1	656.5	4.1	2698.6
2	560.0	3.9	2194.5	658.8	4.1	2717.7
3	559.5	3.9	2190.9	658.3	4.1	2713.2
4	559.6	3.9	2191.2	658.3	4.1	2713.6
5	559.9	3.9	2193.8	658.7	4.1	2716.7

Table 4. The effect of irradiance on Load values

Irradiance (w/m²)	Load Parameters					
	PI MPPT			Fuzzy MPPT		
	Voltage (V)	Current (A)	Power (W)	Voltage (V)	Current (A)	Power (W)
800.0	18.5	379	7011.5	18.5	385	7122.5
1000.0	18.52	388.4	7193.2	19.5	395	7702.5

7 Summary

The proposed fuzzy approach for the analysis of hybrid sources connected to the grid network. The proposed progressive approach is incorporated in PV station and inverter control and the outputs are compared with conventional mppt controller. The output voltage and active power of the loads are enhanced 10 to 15% in the presence of proposed approach, which are tabulated. The load distortion effect is considerably reduced and stable output parameters are obtained during the different loads.

The comparative results obtained with the progressive approach has proved that the fuzzy approach of improves the voltage stability with no distortion at any load. Further, the proposed controller has enhanced the transient stability such that the stable grid parameters are maintained.

References

1. Sun, Q., Xing, D., Yang, Q., Zhang, H., Patel, J.: A new design of fuzzy logic control for SMES and battery hybrid storage system. Energy Procedia **105**, 4575 (2017)
2. Himour, K., Ghedamsi, K., Berkouk, E.M.: Supervision and control of grid connected PV-Storage systems with the five level diode clamped inverter. Energy Convers. Manage. **77**, 98 (2014)
3. Muni, N.B., Sasikumar, S., Mandadi, N.: FRT performance enhancement of PV/Wind/Grid Connected hybrid Network by TMDBFCL using HBB-BC algorithm. Int. J. Adv. Sci. Technol. **28**(16), 308–319 (2019)
4. Bouharchouche, A., Berkouk, E., Ghennam, T.: Proceedings of the 8th EVER Conference on Ecological Vehicles and Renewable Energies (2013)
5. Sera, D., Mathe, L., Kerekes, T., Spataru, S.V., Teodorescu, R.: On the perturb-and-observe and incremental conductance MPPT methods for PV systems. IEEE J. Photovoltaics **3**, 1070–1078 (2013)
6. Rajesh, K., Kulkarni, A., Ananthapadmanabha, T.: Modeling and simulation of solar PV and DFIG based wind hybrid system. Procedia Technol. **21**, 667–675 (2015)
7. Muni, N.B., et al.: Improved fault blocking capability strategy using reverse blocking modular multilevel converters. J. Adv. Res. Dyn. Control Syst. **10**(06- Special Issue), 492–500 (2018)
8. Razavi, F., Abyaneh, H.A., Al-Dabbagh, M., Mohammadi, R., Torkaman, H.: A new comprehensive genetic algorithm method for optimal overcurrent relays coordination. Electric Power Syst. Res. **78**(4), 713–720 (2017)
9. Muni, N.B., et al.: FRT performance enhancement of PV/Wind/Grid connected hybrid Network by TMDBFCL using HBB-BC algorithm. Int. J. Adv. Sci. Technol. **28**(16), 308–319 (2019)
10. Sadi, M.A.H., Ali, M.H.: Lyapunov function controlled parallel resonance fault current limiter for transient stability enhancement of power system. In: North American Power Symposium (NAPS), pp. 1–6 (2018)
11. Pathare, M., Shetty, V., Datta, D., Valunjkar, R., Sawant, A., Pai, S.: Designing and implementation of maximum power point tracking (MPPT) solar charge controller. In: International Conference on Nascent Technologies in Engineering (ICNTE), pp. 1–5 (2017). https://doi.org/10.1109/ICNTE.2017.7947928
12. Atri, P.K., Modi, P.S., Gujar, N.S.: Comparison of different MPPT control strategies for solar charge controller. In: International Conference on Power Electronics & IoT Applications in Renewable Energy and its Control (PARC), pp. 65–69 (2020). https://doi.org/10.1109/PARC49193.2020.236559

A Survey on Techniques and Methods of Recommender System

Aanal Raval[1](✉) ⓘ and Komal Borisagar[2] ⓘ

[1] Computer Engineering Department, GSET, GTU, Ahmedabad, India
aanalraval@gtu.edu.in
[2] E.C. Engineering Department, GSET, GTU, Ahmedabad, India
asso_komal_borisagar@gtu.edu.in

Abstract. As prevalence is growing for social media, the value of its content is becoming paramounting. This data can reveal about a person's personal and professional life. The behaviors done on social media either frequent or periodic can comprehend fondness and attentiveness of users on certain matters. But accompanying it, diverse data from multiple sources with high volumes sets its foot in. Here becomes operational the usage of recommender systems. These are capable of providing customized assistance to users based on respective quondam behavior and preferences. Machine Learning and deep learning methods have proven to be a boon in these tasks of predictions with notable accuracy. This paper discusses existing techniques with its fors and againsts, concerns and issues, extrapolating the results and solutions, which in turn can help in better interpretation of current developments to pave a path for pioneering researches.

Keywords: Machine learning · Recommender systems · Deep learning · Reinforcement learning · Soft computing techniques

1 Introduction

It's an arduous task for business firms to recommend products and services according to their respective needs, withstanding the competition. Customized services satisfying individual demands, bearing vast and diverse information, can provide an enriched user experience. Recommender systems gratifies these needs with preeminent results. Different ML techniques can be used to forge flashing presciences of correlation between users and items, introducing tricky data representations, as well as deducing compendious facts regarding demographical, contextual, and geographical data along with those in patterns. This paper sets sight on to unearth different existing AI/ML techniques in recommender system, with reigning state of art methods and pinpointing current issues with the system with its futuristic evolution. The central concept of these system is to obtain a list or assemblage of commodities according to utility (in terms of user ratings) of items that user hasn't consumed, by defining the utility of particular item to user. The goal is to maximize this function. Predictions of these systems depend on different ML algorithms selected.

© IFIP International Federation for Information Processing 2022
Published by Springer Nature Switzerland AG 2022
L. Kalinathan et al. (Eds.): ICCIDS 2022, IFIP AICT 654, pp. 97–114, 2022.
https://doi.org/10.1007/978-3-031-16364-7_8

2 Traditional Methods

Recommender System resolves the prediction issue by filtering and assigning priorities to information according to their relevancy. Using different available methods, this paper broaches a survey on recommendation techniques.

2.1 Content-Based

Content- based filtering [1] uses similarities in features to make decisions, comparing user interests to product features. The products recommended are the ones which have most of the features overlapping. Here, two methods (or can be in combination) can be used. Firstly, two users will have same taste in the future, given they had similar preferences in past. There are two types: 1. Model-based systems deploy ML models to understand and predict ratings. Ratings can be given as if customers are regularly provided by any stock of functions from which they will filter items actually perceived foremost accordingly. Secondly, the set of regulations contains reports of the products preserving the advanced information provided by individuals instead of considering customers' data abilities together. Users' intuition about features that they identify frequently corresponds to products. First of all product features and user interest are allocated with some binary value or just an arbitrary number. Thereafter a formula is needed for the identification of products and user interests' similarities. This task can be implemented through dot product:

$\sum_{i=1}^{d} p_i u_i$ (where pi is the feature value of product and ui in column i is the value user interest).

Due to low volume of data, this model is extensible easily. This model, unlike other models, doesn't require data matching with another users', it nonpareil niche results particular to the current user.

However, this model requires enough domain knowledge from the people attributing product features. So, its accuracy is totally hooked in to that knowledge being precise. Cold start problem is a great issue.

2.2 Collaborative Filter Based

Collaborative filtering applications apply two types of techniques to show fresh products to the users [2].

The elemental perspective is to show new recommendations to the user separately and individually, simultaneously ratings will predict the buyer's interest in specific items. For any operative user *user$_a$,*

$$\text{Score}_a = 1/k_a \sum V_{aj} \in P_a |\hat{v}_{aj} - v_{aj}|$$

where $\hat{v}_{aj}$ of V_{aj} is the prediction of value for any item regarding user, attested value is denoted by v_{aj}, and k_a denotes the amount of predicted votes. Then score's average is taken. Evaluation of a model with how well it is consistent with given criterion is named average absolute deviation scoring. (Lower scores are more preferable).

Assigning currently operative users with an ordered list of recommended items is the second perspective which is regarded as a ranked list of things. An approach called ranked scoring is applied for the evaluation of the model and checking its consistency. The expected utility of ranked list is given by the sum over all items of utility merchandise. Probability of viewing any item by the user is given by,

$$R_a \equiv EU(\text{list}) = \sum_{(k=0)}^{n} max(v_{aj} - d_j, 0)P(\text{item}_j \text{ will be watched})$$

where k_a is number of predicted voted for each user $user_a$, the index j is sorting criteria for the votes, consistent with declining values of predicted votes v_{aj}. The initial item will be certainly watched. The probability of items being viewed from list declines down exponentially.

Collaborative filtering hinges on two presumptions:

- Assumption that user preferences remain constant for a specified time period.
- The chances that two users will have similar preferences in the future, given they had same taste in past.

There are two types:

1. Model-based systems use ML models to understand in order to estimate ratings, which could be either a real-valued number or a binary value. Hence, when we are given with a user and an item, the system estimates the rating that user will allocate to product. Diverse techniques such as neural networks, logistic regression, Bayesian networks or SVMs can be employed to perform prediction tasks. Matrix factorization methods are the recognized state of art technique in context of model-based recommender systems that deal with what matrix factorization outputs in due course harmonize given assortment of latent features.
2. *Memory-based systems:* This technique uses user ratings to calculate similarity amongst users or products for generating predictions. As examples of this approach, item-based (or user-based) top-N recommendations and CF (neighborhood-based) can be considered. In phrases of consumer desire, it generally expressed via way of means of categories. Explicit Rating, is a fee given via way of means of a consumer to an object on a sliding scale, Implicit Rating, indicates customers desire indirectly, together with web views and clicks of a page, buy records, willingness to a music track, and so on [3].
a) User Based CF: Here we have a rating matrix, n × m, where u_i, $(i = 1, \ldots n)$ is the user and p_j, $(j = 1, \ldots m)$ is the item. Now the requirement is to expect the score r^{ij} which depicts if goal person i did now no longer watch/charge an object j. The core idea is the calculation of the similarities among goal person i and all different users, then pick out the pinnacle X comparable customers, and then considering similarities as weights, taking the weighted average of ratings from X users. Some of the similarity measures, generally used are Pearson, Jaccard, Spearman rank, Mean Squared and Proximity impact popularity.

$$r_{ij} = R_i + [\sum_k similarities(u_i, u_k)(r_{kj} - R_k)/\text{number of ratings}]$$

b) Item Based: [3] It can be said that two products are identical in case they receive close ratings from a same user. Then, this information can be used to predict a goal user associated with any product by taking ratings' weighted average for mass of X similar products from current customer. One sure gain of Item-based-CF method is unlike the preferences of human beings, change of firmness of grading on a specific product overtime will not be remarkable.

Matrix Factorization starts with SVD, creating three matrices, n * r user feature, m * r product feature, and a diagonal matrix denoted by $\sum$ with of form r * r accommodating the singular value of original matrix. Now to reduce matrix $\sum$ to k dimensions, this k should be selected such that, within the original matrix R, A is apt of grasping the most of variance, for A to be approximation of R, A $\approx$ R, with minimization of error between A and R. When matrix R is dense, there are chances of factorization of U and V analytically. But in case it's sparse, missing values and its factors are to be found. So instead of using SVD, U and V can be directly found assuring the result matrix is the nearest approximation of R (R') instead of a sparse matrix, when U and V are multiplied back together. Non-Negative Matrix Factorization is generally used for this kind of numerical approximation. Now q_i is the vector of object i, p_u is the vector of the person u and the expected score for consumer u on object i is the dot Product of both mentioned vectors. This calculated value is represented in the matrix R' (with row u and column i) [3].

$$\text{Predicted ratings: } r'_{ui} = p_u^T q_i$$

To find optimal qi and p_u, the cost of errors is minimized using a loss function. r_{ui} is the genuine ratings taken from authentic user-item matrix. Now, An optimization procedure finds the optimal matrix P made via way of means of vector p_u and matrix Q composed via way of means of vector q_i so as to reduce the sum rectangular blunders between r'_{ui} (predicted ratings) and r_{ui} (true ratings). For overfitting issue of consumer and object vectors, L2 regularization is applied [3] . To feature bias term three steps are followed which include initially finding b_u by taking an average rating of item i minus μ, then averaging ratings of all items μ and finally finding b_i by averaging ratings by user u minus μ.

$$\text{Min}_{q,p} \sum \left(r_{ui} - p_u^T q_i \right)^2 + \lambda \left(||p_u||^2 + ||q_i||^2 \right)$$

Alternative Least Square is used for optimization of non-negative matrix factorization. Due to loss function being convex, global minimum can't be found, but can reach to an approximation by finding local minimum. For confining constants of user factor matrix, taking derivatives of loss function, adjusting item factor matrix, and its initialization to 0, and then, setting constants of item factor matrix along with adjusting user factor matrix, Alternative Least Square is used. This process is repeated until convergence [3].

2.3 Demographic-Based Recommender System

This technique categorizes the users taking attributes into account and makes predictions to recommend considering demographic classes [5]. Appropriate market analysis within a peculiar region escorted with a needed inspection is a must to accumulate information for creating categories. These strategies form collaborative like "people-to-people" correlations, however, utilize varying kinds of data. For this technique, the history of person scores like what during collaborative and content is not needed.

2.4 Knowledge-Based Recommender System

Knowledge-based advice works on useful understanding: they have got understanding approximately how a specific object meets a specific person's want, and may consequently purpose approximately the connection between a want and a piece of likely advice [5].

2.5 Utility-Based Recommender System

This type of recommender device based on utility does recommendations primarily by calculating every items' usage for the person. Now significant trouble for this sort of device is a way to create an application for person users. In application primarily based totally device, each enterprise can have a one-of-a-kind approach for arriving at a person particular application feature and making use of it to the items below consideration. The principal benefit of the usage of a application primarily based totally recommender device is that it could thing non-product attributes, inclusive of product availability and supplier reliability, into the application calculation. This makes it convenient to test the actual-time stock of the item and proclaim the same to the person [5].

2.6 Hybrid Recommender System [5]

It mingles the power of more than two Recommender machine and along with it eradicates any weak point which exist whilst simplest one recommender machine is used. Variety of methods wherein the structures may be fused, such as:

1. Weighted-Hybrid (Recommendation): The rating of an endorsed object is calculated from the outcomes of all that need advice strategies present inside the system.
2. Switching-Hybrid (Recommendation): switches among various recommendations techniques when one doesn't work.
3. Mixed Hybrid Recommender: It is used where a large number of recommendations are to be made. Here pointers from multiple approach are supplied together, so the person can pick from a huge variety of recommended items.

3 Review Work

Traditional systems were not competent to unstructured data and to preserve robustness [38]. In recent years, many techniques of computer vision, reinforcement learning, and many more are rapidly deployed for recommendation systems that yield better results than traditional ones.

3.1   Deep Learning Based Systems

First application of DL was seen in [39] where restricted Boltzmann machines (RBM) were used along with collaborative filtering [38]. Deep learning models that have been used are with two-layers cramped by RBM (Boltzmann machine) to search for the ordinal ratings [22] and an extension of mentioned work by the parameterization options [23]. Fact that recommender system includes all data like text, image, videos etc. motivated use of deep learning in recommendation systems [24]. MLP (multi-layer perceptron) [27] models both linear and nonlinear relationships with machine factorization for feature engineering [25]. To model the linear relation, in co-occurrence with matrix factorization, Neural collaborative filtering can prototype user-product non-linear relationships [26]. AutoRec combines an autoencoder and matrix factorization focusing learning of users' or products' non-linear latent representations [28]. AutoSVD++ is a hybrid for generation of item feature representations from item-content method that hybrids a matrix factorization and contractive autoencoder [29]. Both Long as well as short term user preferences are combined, along with user preference drift is taken to consideration with a fine tuning using hierarchical-attention-network [30, [31]. Used user-item embedding with perturbations with notion of Bayesian personalized ranking as an adversarial regularizer. User-item preferences are learnt from interactions, knowledge graphs [32], tags and images [33]. Apart from these, CNN based models like DeepConn, ConvMF, RNN, LSTM, GNN and GAN have also been applied [7]. Using computer vision, in-vogue models are Incetionv3, VGG-16, Xception, resnet-50 and VGG-19 [8]. Another common example can be seen of YouTube, Yahoo, Twitter and eBay prefer deep neural networks (DNNs), while Spotify choose convolutional neural networks (CNNs) [34].

3.2   Transfer Learning Based Systems

TL is a system mastering technique in which a version evolved for a project is reused considering it as the start line for a version on a 2d project [6]. Depicts some general examples of TL. To unsheathe knowledge from a number of data sources to help the task of learning for the target domain. Here are its classifications: (1) Transductive transfer learning. The initiating and goal actions are the same, however, their domain names are different. Area adjustment is its common application. The contradiction of supply area and the goal area may be as a result of either varying marginal distribution in feature spaces or characteristics feature spaces varieties themselves [7]. (2) Inductive transfer learning: Here both goal and initiating task differs from each other. If data labeling is present in the goal area, then its multi task learning, and if there is no labeled data, it is known as self-taught learning [7]. (3) Instance transfer: For target duties, the Reutilization of source domain knowledge is usually an ideal scenario. In maximum cases, the supply area records cannot be reused straightforwardly as it is. Rather, there are positive occurrences from the supply area that may be reused alongside goal records to enhance results. A good example of this is a modification to Adaboost [8]. (4) Unsupervised transfer learning. Goal duties are tasks of unsupervised learning, else all settings are the same as in inductive transfer learning. For source or for the target domain, data isn't labelled [7]. (5) Parameter Transfer: models for associated duties share prior distribution of hyperparameters or some parameters. Differing multitask learning, where both the origin and goal responsibilities used to learn simultaneously, for transfer learning, weightage to the loss of the target domain is added to improve overall performance is preferred [8]. (6) Feature-representation transfer: This method focuses on minimizing domain divergence and reduction of error rates by spotting good feature representations which can be in turn used from the source to target domains. Depending upon the accessibility of labeled data, feature-representation-based transfers may apply supervised or unsupervised methods [8]. (7) Relational Knowledge Transfer: This attempts to handle the data which is dependent and not identically distributed. In different words, statistics, in which every statistics factor has a relation with different statistics points; for instance, social community statistics makes use of relational-knowledge-transfer techniques.

Cross Domain Recommender System using transfer learning is capable of handling data sparsity problem, providing a rich, personalized and diverse recommendations.

a. In CDRS (with side information) [7], it is supposed that a part of secondary information such as social information, user-generated information, or product attributes entities is present. CMF (Collective matrix factorization) works on eventualities having a user-object score matrix and an object-characteristic (attribute) matrix for the identical organization of objects are available, thereby factorizing those matrices through parameters of object sharing because the objects are identical. Tag-informed collaborative filtering (TagiCoFi) [9, 10, 11] is an approach where a score of a user-object matrix and user-tag matrix for similar arrangements of customers are used. First user similarities are restored from shared tags are utilized for helping the ranking matrix's factorization. TagCDCF (Tag cross-domain CF) is the extended version of TagiCoFi method where, each having facts from those matrixes, two separate area eventualities are there. To tweak the recommenders' overall functioning inside the goal area, TagCDCF concurrently integrates intra-area and inter-area correlations to matrix factorization

b. CDRS (containing non-overlapping entities) manages non-overlapping entities transfer at group level for duo domains. Items and Users group are clustered done and through group-level rating patterns achieve knowledge sharing [7].

c. CDRS with fully or partially overlapping entities [12, 13, 14, 7] suppose that entities among two domains-source and goal overlap and are connected with constraints. This factorizes two matrices of user and item domain with the help of sharing a part of parameters after factorizing. Collective Transfer Factorization (TCF) has evolved for the application of implicit facts present within the source domain to predict explicit comments i.e. H. of rankings in the goal domain. CDTF (Cross-domain triadic factorization) is used to person-item domain tensor to merge implicit and explicit feedback of each. Due to full overlapping of users, user factor matrix is similar by connecting all existing domains. CBMF (Cluster-based matrix factorization) attempts to push the CDTF onto partially-overlapping entities. Because the correspondence between entities is partially available, a number of approaches are formulated combining users and/or items in two areas. For latent space matching, Unknown user/item likings are used. All these are shallow learning based. Both item and user features are mapped in target domain combining their features, knowledge transfer is done. GAN with an extra objective function is applied to differentiate embedding features of user-item differently [15, 16, 17]. Common framework of Cross Domain Recommender System along with Generative Adversarial Network is proposed to concern on all the three scenarios above [7].

Apart from these, two famous strategies for DL with transfer learning are [8]: 1. Pre-trained Off-the-shelf Models for extraction of Features: Crux is to use pre-trained model where weights of the model's layers are not updated during training with new data for the new task and weighted layers are used to to squeeze out features. 2. Fine-Tuning Off-the-shelf (Pre-trained Models): Here top layers of network are cut off and replaced with supervised objective of target task, and the network is fine-tuned with backprop using labels for target domain till validation loss begins to increase. Bottom n layers are frozen when those are not updated during backprop and target task labels are scarce as well as overfitting needs to be avoided or fine-tuned when those are updated during backprop and target labels are in abundance [8].

Here are mentioned some types of Deep transfer learning [8].

1. Domain Adaption: Domain adaption is commonly related to layout where the marginal probabilities among the initial and goal domains varies. To transfer the learning, there is a built-in drift (or shift) in both goal and source domains data distributions that needs refinements.
2. Domain Confusion: Different sets of features are captured by variety of DL network. To ameliorate the transferability across domains, this information can be used to recognize domain-invariant features.
3. Multitask learning: In this model, a variety of tasks are learned at same time without any dissimilitude between the source and targets. Unlike in Transfer learning where learner in the beginning doesn't have idea about target task, here, learner acquires information about multiple tasks instantaneously.
4. One-shot learning: Here inference is made for targeted output dependent on just one or little amount of training examples.
5. Zero-shot learning: Clever judgments are made at time of training phase to discover and use additional information to know unseen data.

[40] Used DL along with transfer learning and dimension reduction to extract features from images, thereby applying these features for user-based collaborative filtering for movie recommendations [41]. Proposed value-based neighborhood model using cosine similarity and Pearson coefficient along with KNN to predict customer loyalty addressing sparsity. [42] introduced a method with zero-shot learning and autoencoders to learn language naturally to learn item-item representations. [43] summarizes a feedback loop for understanding user ratings using their discoveries and blind spots, thereby using bias and fairness for next iteration. [44] uses cross convolution filters consisting of variable embedding, cross convolution filters and rating prediction module to effectively capture nonlinear user-item interactions [45]. Proposes a deep learning model with deep autoencoders to unveil nonlinear, nontrivial, and hidden user-item relations for multicriteria purposes [46]. Employs autoencoders to extract trust relation features and tag information from user-user and user-tag matrices. In turn, these extracted features are deployed for the calculation of similarity values among users for the prediction of unseen items

[47]. Proposed a method of embeddings for item and user representations for learning non-linear latent factors. The method used varying the values of epochs resulting increase in weight decay along with decrease in learning rate for improving efficiency [48]. Compared CDAE (Collaborative Denoising Auto Encoders), DAE-CF (Deep Auto Encoders for Collaborative Filtering) and DAE-CI (Deep Auto Encoders for Collaborative Filtering using Content Information), showing DAE-CF with best results [49]. Represented a model combining Hidden Markov Model with ANN. [50] used deep learning with collaborative filtering. They extracted aspects by word embedding and POS tagging, thereafter generation of ratings through LDA and lexicon based methods.

3.3 Active Learning Based Systems

The task of AL is to select the data precisely so that ML algorithms can perform better. The labelling process is always time consuming especially for online systems, active learning can successfully be applied here. Some of the active learning methods are query-by-committee, uncertainty sampling, anticipate model change, variance reduction, expected reduction in error, and density-weighted methods [7]. Its techniques include rating impact analysis and bootstrapping, decision trees and matrix factorization.

[51] Used Gibbs sampling with Bayesian Nonnegative Matrix Factorization for movie recommendations [52]. Proposed an approach for recommendation where initially users were clustered to different groups, then a representative from each was chosen. Using one-hot representation of selected users, the likeliness of users to rate the item and its variance was calculated. Using these ratings, further estimation was done [53]. Proposed a framework called POLAR++ which used Bayesian neural networks to grab the uncertainties in user preferences for the selection of articles for feedback. For task of article recommendation, attention based CNN was used for similarity between user preference and recommended articles.

3.4 Reinforcement Learning Based Systems

Reinforcement Learning (RL) is a form of ML technique that allows an agent to study in an participative surroundings via way of means of trial and blunders the usage of remarks from its personal movements and experiences [18]. This makes use of remunerates and penalties as alerts for negative and positive behavior. In desire to build up an optimal solution, the agent has to explore new states along with maximization its universal reward simultaneously. This is known as Exploration vs Exploitation trade-off. To stabilize both, the great standard approach may also contain quick time period sacrifices. Therefore, the agent is ought to acquire sufficient data to make the best possible verdict for future.

- Markov Decision Process [18]: An MDP is made up of a set of possible actions A(s) in each state, a set of finite environment states S and a real valued reward function R(s) with a transition model denoted by $P(s', s|a)$. In most of cases, real world environments don't have any advance knowledge of environment dynamics.
- Q-learning [18] It is a model-free approach which is commonly applied for creating a self-playing PacMan agent. It better believes in concept of overhauling Q values

indicating value of executing action a in state s.

$$Q(s_t, a_t) \leftarrow (1 - a).Q(s_t, a_t) + a(r_t + \gamma . \max_a Q(s, a))$$

$Q(s_t, a_t)$ is the old value, a is learning rate, r_t is reward, γ is discount factor and $\max_a Q(s, a)$ is the estimated value of optimal feature. The term $(r_t + \gamma . \max_a Q(s, a))$ represent learned value.

- Deep Q-Networks (DQNs) [18] uses NN to predict Q-values. But its disadvantage is it only can tackle with low-dimensional and discrete action spaces.
- SARSA [18] is an on-policy technique which grasps the value based on its existing action a extracted from its existing policy.
- DDPG (Deep Deterministic Policy Gradient) [18] is a off-policy, actor-critic and model-free algorithm that efficiently solves this problem by comprehending policies in high dimensional with continuous action spaces. This model has proven to be state of art on music recommendations. Here the action of the DDPG learner contains a particular song selected from a huge pool. Following this action, by representing every tune with the help of a set of continuous features expanding the action space from discrete to continuous, it became possible and easy to scale up the number of candidate songs it can accommodate, while maintaining its accuracy and diversity [19].
- A multi-armed bandit is a complex slot device in which in place of 1, many levers are present for a gambler to select, each lever having a distinctive reward. Unknowingly to gambler, probability distribution associated to each lever is varying [20]. Solutions to this problem includes action value function, regret, upper confidence bound, decayed epsilon greedy, softmax exploration, epsilon greedy approach, greedy approach and regret comparison [20].

[54] Proposed a multi-agent reinforcement learning framework, Temporary Interest Aware Recommendation (TIARec) which can grasp and differentiate between normal preferences and seasonal ones [55]. Partitioned a graph to three sub-graphs, item graph, user graph and user-item mapping graph. User and item graphs are pre-trained for user-item interaction graph with the help of the proposed model RsAM-GNN (Attentive Multi-relational Graph Neural Network) learning relations through attention layer. To sample top-k similar neighbors from the graph, RNS (Reinforced Neighbor Sampler) was employed to discover the threshold of optimal filtering. To recommend, Graph Neural Network model based on aggregation is initiated with user-item embeddings [56]. Used DQN based technique to recommend item list [57]. Proposed deep Q learning with double exploration-based framework- DEN-DQL to get offline data trained for news recommendations. [58] uses RL based dynamic algorithm for user feedback. To maximize click-through rate (CTR) and cumulative estimated reward (CMR), Q learning was used. And finally, to provide top-k online context-aware recommendations, frameworks named Cross Feature Bandits (CFB) and Sequential Feature Bandits (SFB) were proposed [59]. Introduces deep reinforcement learning based deep hierarchical category-based recommender system (DHCRS) with two deep Q- networks, one for selection of item-category relationship and another for selection of item to recommend thus addressing problem space with large action [60]. Introduces reinforcement learning based recommender systems in mobile edge computing to improve throughput [61]. Utilizes Markov decision processes (MDP) with RNN for cold start and warm start scenarios.

3.5 Soft Computation Techniques: Fuzzy System and Evolutionary Algorithms

Fuzzy Logic: In content based systems, fuzzy logic is applied for profiling the customers and matching of relevant items. In memory based, it is utilized for profile uncertainties in customer tastes. Besides, fuzzy sets can express precariousness in item features, lacking item descriptions, subjective user feedback on those items and for item portrayals [7]. These systems are capable of measuring user interest consistency and it's imprecision from given vast data [7].

Evolutionary Algorithms: The use of evolutionary algorithms can be seen in problem of multi objective optimization, considering multiple performance like accuracy, diversity, etc., to combine outputs of multiple recommender system. These systems are also applied to create user-item profiles to find core users and for handling ratings by combining a multi objective optimization problem to a single objective problem [7]. Ratings also include Pareto optimal recommendations or tasking multiple criteria as constraints. These algorithms also play a role in recommender's security.

[62] Made use of soft rough sets along with context-aware recommender systems for the task of video recommendations [63]. Proposed rough C-means type methods for recommendations of e-commerce sites as well as video streaming services [64]. Using game-theoretic N-soft sets, Attempted to solve the issue of missing ratings by three-way classification.

3.6 NLP Based

NLP can be used in recommender systems for topic modelling, in ratings, item description from text, identifying user generated tags, for user reviews, summary generation and extracting many more relevant features. Relevant techniques in NLP for above tasks include vector models (tf-idf, BM25), LDA, embeddings (Glove, Word2vec, FastText), semantic text similarity, summarization, encoders (BERT) [8] and many more [21]. Recommendations with features like item experience and sentiment of users to provide pertinent items according to user query have been evolved [7]. The neural embedding algorithm in order to deduce product similarity correlations, and multi-level item organization for application on personalized ranking, has been associated with a CF framework [7].

[65] Used CNN, VGG-16 architecture along with NLP techniques for product recommendations [66]. Makes use of Neural Network Language Models for semantic analysis of documents [67]. Developed a recommender system to recommend papers and articles that are related to particular topic of interest [68]. Employed Named Entity Recognition based extractor along with rule based and grammar based extractors to obtain the logical relations. Based on these data, similarity scores were calculated and then considered for recommendation.

4 Summary

See Table 1.

Table 1. Summary

Paper	Dataset	Methodology	Task
[39]	Yelp, Amazon, Beer	DeepCoNN	User-item recommendation
[40]	MovieLens	Transfer learning with collaborative filtering	Movie recommendation
[41]	Amazon movies review	Value based neighborhood model	Customer loyalty
[42]	(sample created)	Zero shot learning with auto-encoders	Learn item-item representations
[43]	MovieLens	Feedback loop setting	Study behavior of iterative recommender system
[45]	Yahoo movies, TripAdvisor	Auto encoder based model	Learn non-trivial and non-linear user item relations
[47]	MovieLens, FilmTrust, Book-Crossing	Embeddings	Address gap in collaborative filtering
[48]	(created from MOOC)	DAE-CF	E-learning recommendations
[49]	MovieLens	Hidden Markov model with ANN	Pattern recognition for recommendations
[50]	Amazon, Yelp	MCNN model with Weighted aspect-based opinion mining	Product recommendations
[51]	MovieLens	Bayesian Nonnegative Matrix Factorization	movie recommendation
[52]	movielens-Imdb	Active learning with clustering	To address cold start
[53]	Dataset from AMiner, Patent Full-Text Databases of the US Patent, Related-Article Recommendation Dataset (RARD)	Bayesian neural network with attention based CNN	Article recommendation
[56]	–	DQN based technique	Product recommendation
[57]	–	DEN-DQL	News recommendation
[58]	News article data	RL based dynamic algorithm	News recommendation
[59]	–	deep reinforcement learning based deep hierarchical category-based recommender system (DHCRS)	Action- space issue
[60]	–	reinforcement learning based	Mobile edge computing
[61]	–	Markov decision processes (MDP) with RNN	Address cold start problem

(*continued*)

Table 1. (*continued*)

Paper	Dataset	Methodology	Task
[62]	LDOS-CoMoDa	Context-aware recommender system	Context relevancy in video
[63]	NEEDS-SCAN/PANEL, d MovieLens	Rough C-means based methods	A recommendation in e-commerce
[64]	LDOS-CoMoDa, InCarMusic	N-soft sets based method	Context aware recommendations
[65]	Amazon Apparel database	NLP with CNN based methods	Product recommendation
[66]	Patent documents (sample)	Neural network language models	Document/article recommendations
[67]	–	Weighted average and text similarity	Document/article recommendations
[70]	MNIST, Fashion-MNIST	Bandit algorithms	Learning in non-stationary environments
[71]	MovieLens-100K	Auto-encoders based methods	Movie recommendations
[72]	Microsoft Academic Graph	Metric for exposure bias with text embeddings	Research papers recommendations

5 Observations and Research Gap

The primary issue faced by the recommender system is the requirement of bulky data. A good system needs data from a catalog or from other form for which then it must capture and analyze user behavioral events. Thus it can become a chicken and egg problem [35]. Another issues include changing trends and user preferences or sometimes include a large number of variables. The next problem of Cold Start crop up as a novel customer or product sets foot in the market. Another problems includes synonymy (n item is represented with more than two distinct labels or filings with similar intents), Shilling attacks (if a baleful user or competitor sets foot in the mechanism, inflicting fictitious grading (ratings) for particular items to astray information of popularity of product), overspecialization (Meagre content analysis leading to CB recommendation not to suggest novel items.), grey sheep (in run-of-mill C-F systems where users' viewpoint don't meet with any of type), sparsity, scalability, the problem of latency (it occurs when fresh products are included in dataset over and over again), unavailability of benchmark dataset and context awareness (users' short and long term history with geographic locations) [36].

Issue of Data Sparsity exists when collaborative filtering-based methods are employed. This problem to some extent can be solved if preferences relations are used instead of absolute ratings [69]. Time awareness regarding popularity span [69], or seasonal products are also important. Besides, online data are a vast pool including structured as well as unstructured data. Personalization of search engine for relevancy is also an important aspect. Sometimes decisions along with recommendations come to roles where the system is like Resource Allocation Problem [69].

6   Future Directions and Conclusion

Traditional recommender system supposes that user preferences are static over time, thus their history documentation are equally weighted. But as these changes over time-based on individual experiences personal preferences, or popularity-driven influences, this behavior of big data is known as concept drift. To resolve this issue, instanced decay and time window was introduced [7], which determines the data weights occurrences alongside the time-period and compares to that old data which weighs not so much as. Considering time as one more dimension of data, a method which operated on dynamic matrix factorization was introduced, apart from penalizing the old data. Still, these methods cannot regulate the increase in bias resulting due to the directions of change. Another topic includes long-tail, where items are unpopular, are rarely noticed by the users. With the development of these systems, the development of privacy conserving and secure systems are in demand. While some systems are too complex to understand or lack clarity in explanation. This challenging limitation can be solved using integrating visualization to recommender systems.

Current advancements in recommender system aims at assisting decision making support system that includes vast information involving items metadata, user profiles, user reviews, images and social networks. This paper reviews different available techniques in recommender system, issues faced in it. Apart from these, it also identifies the gap in current existing systems and paves path for future directions.

References

1. What is content-based Filtering? Towards Data Science. https://www.educative.io/edpresso/what-is-content-based-filtering
2. Collaborative Filtering, Science Direct. https://www.sciencedirect.com/topics/computer-science/collaborative-filtering
3. Introduction to collaborative Filtering, Towards Data Science. https://towardsdatascience.com/intro-to-recommender-system-collaborative-filtering-64a238194a26
4. A Simple Introduction to Collaborative Filtering, built-in beta. https://builtin.com/data-science/collaborative-filtering-recommender-system
5. Classifying Different Types of Recommender Systems, bluepi. https://www.bluepiit.com/blog/classifying-recommender-systems/#:~:text=There%20are%20majorly%20six%20types,system%20and%20Hybrid%20recommender%20system
6. A Gentle Introduction to Transfer Learning for Deep Learning, Machine Learning Mastery. https://machinelearningmastery.com/transfer-learning-for-deep-learning/#:~:text=Transfer%20learning%20is%20a%20machine,model%20on%20a%20second%20task.&text=Common%20examples%20of%20transfer%20learning,your%20own%20predictive%20modeling%20problems
7. Zhang, Q., Lu, J., Jin, Y.: Artificial intelligence in recommender systems. Complex Intell. Syst. **7**, 439–457 (2021)
8. A comprehensive hands on Guide to Transfer Learning with Real World Applications in Deep Learning, towards data science. https://towardsdatascience.com/a-comprehensive-hands-on-guide-to-transfer-learning-with-real-world-applications-in-deep-learning-212bf3b2f27a
9. Zhang, Q., Hao, P., Lu, J., Zhang, G.: Cross-domain recommendation with semantic correlation in tagging systems. IEEE, July 2019

10. Zhen, Y., Li, W.J., Yeung, D.Y.: TagiCoFi: tag informed collaborative filtering. In: RecSys 2009—Proceedings of the 3rd ACM Conference on Recommender Systems, pp. 69–76 (2009)
11. Hao, P., Zhang, G., Martinez, L., Lu, J.: Regularizing knowledge transfer in recommendation with tag-inferred correlation. IEEE Trans. Cybern. **49**, 83–96 (2017)
12. Pan, W., Yang, Q.: Transfer learning in heterogeneous collaborative filtering domains. Artif. Intell. **197**, 39–55 (2013)
13. Hu, L., Cao, J., Xu, G., Cao, L., Gu, Z., Zhu, C.: Personalized recommendation via cross-domain triadic factorization. In: Proceedings of the 22nd International Conference on World Wide Web (2013)
14. Mirbakhsh, N., Ling, C.X.: Improving top-n recommendation for cold-start users via cross-domain information. ACM Trans. Knowl. Discov. Data **9**, 1–19 (2015)
15. Hu, G., Zhang, Y., Yang, Q.: Conet: collaborative cross networks for cross-domain recommendation. In: Proceedings of the 27th ACM International Conference on Information and Knowledge Management (2018)
16. Wang, C., Niepert, M., Li, H.: RecSys-DAN: discriminative adversarial networks for cross-domain recommender systems. IEEE Trans. Neural Netw. Learn. Syst. **31**, 2731–2740 (2019)
17. Yuan, F., Yao, L., Benatallah, B.: DARec: deep domain adaptation for cross-domain recommendation via transferring rating patterns. In: Proceedings of the Twenty-Eighth International Joint Conference on Artificial Intelligence (2019)
18. Reinforcement Learning 101, towards data science. https://towardsdatascience.com/reinforcement-learning-101-e24b50e1d292
19. Recommender System with Reinforcement Learning, towards data science. https://towardsdatascience.com/recommendation-system-with-reinforcement-learning-3362cb4422c8
20. Reinforcement Learning Guide: Solving the Multi-Armed Bandit Problem from Scratch in Python, Analytics Vidhya. https://www.analyticsvidhya.com/blog/2018/09/reinforcement-multi-armed-bandit-scratch-python/
21. Raval, A., Lohia, A.: A survey on techniques, methods and applications of text analysis. IJCRT2105349, May 2021
22. Salakhutdinov, R., Mnih, A., Hinton, G.: Restricted Boltzmann machines for collaborative filtering. In: Proceedings of the 24th International Conference on Machine Learning (2007)
23. Truyen, T.T., Phung, D.Q., Venkatesh, S.: Ordinal Boltzmann machines for collaborative filtering. In: Proceedings of the 25th Conference on Uncertainty in Artificial Intelligence, pp. 548–556 (2009)
24. Zhang, S., Yao, L.: Deep learning based recommender system: a survey and new perspectives. ACM J. Comput. Cult. Herit. Artic. **1**(35), 1–35 (2017)
25. Cheng, H.T., et al.: Wide and deep learning for recommender systems. arXiv Prepr. (2016)
26. He, X., Liao, L., Zhang, H., Nie, L., Hu, X., Chua, T.S.: Neural collaborative filtering. In: Proceedings of the 26th International Conference on World Wide Web, pp. 173–182 (2017)
27. DeepFM: a factorization-machine based neural network for CTR prediction. In: International Joint Conference on Artificial Intelligence (2017)
28. Sedhain, S., Menon, A.K., Sanner, S., Xie, L.: AutoRec: autoencoders meet collaborative filtering. In: Proceedings of the 24th International Conference on World Wide Web (2015)
29. Zhang, S., Yao, L., Xu, X.: AutoSVD++: an efficient hybrid collaborative filtering model via contractive auto-encoders. In: Proceedings of the 40th International ACM SIGIR Conference on Research and Development in Information Retrieval (2017)
30. Ying, H., et al.: Sequential recommender system based on hierarchical attention network. In: International Joint Conference on Artificial Intelligence (2018)
31. He, X., He, Z., Du, X., Chua, T.S.: Adversarial personalized ranking for recommendation. In: The 41st International ACM SIGIR Conference on Research & Development in Information Retrieval (2018)

32. Yang, D., Guo, Z., Wang, Z., Jiang, J., Xiao, Y., Wang, W.: A knowledge-enhanced deep recommendation framework incorporating GAN-based models. In: 2018 IEEE International Conference on Data Mining, pp. 1368–1373 (2018)
33. Tang, J., Du, X., He, X., Yuan, F., Tian, Q., Chua, T.-S.: Adversarial training towards robust multimedia recommender system. IEEE Trans. Knowl. Data Eng. **32**(5), 855–867 (2019)
34. Deep Learning Based Recommender System, Sciforce. https://medium.com/sciforce/deep-learning-based-recommender-systems-b61a5ddd5456
35. 5 Problems with Recommender System, readwrite. https://readwrite.com/2009/01/28/5_prob lems_of_recommender_systems/?__cf_chl_jschl_tk__=pmd_UZyz8bgl.30UoIXhsuXwJ jqLz2DxSjHu4LMvOrVSuf0-1629638145-0-gqNtZGzNAlCjcnBszQh9
36. Khusro, S., Ali, Z., Ullah, I.: Recommender systems: issues, challenges, and research opportunities. In: Kim, K., Joukov, N. (eds.) Information Science and Applications (ICISA) 2016. LNEE, vol. 376, pp. 1179–1189. Springer, Singapore (2016). https://doi.org/10.1007/978-981-10-0557-2_112
37. Milano, S., Taddeo, M., Floridi, L.: Recommender systems and their ethical challenges. AI Soc. **35**, 957–967 (2020). https://doi.org/10.1007/s00146-020-00950-y
38. Liu, B., et al.: A survey of recommendation systems based on deep learning. J. Phys. Conf. Ser. ISPECE (2020)
39. Lei, Z., Vahid, N., Philip, S.: Joint deep modeling of users and items using reviews for recommendation. In: Proceedings of the Tenth ACM International Conference on Web Search and Data Mining (2017)
40. Hassen, A.B., Ticha, S.B.: Transfer learning to extract features for personalized user modeling. In: Proceedings of the 16th International Conference on Web Information Systems and Technologies (WEBIST 2020)
41. Srivastava, A., Bala, P.K., Kumar, B.: Transfer learning for resolving sparsity problem in recommender systems: human values approach. JISTEM J. Inf. Syst. Technol. Manag. **14**(3), 323–337 (2017)
42. Wu, T., et al.: Zero-shot heterogeneous transfer learning from recommender systems to cold-start search retrieval. arXiv:2008.02930v2 [cs.LG], 19 August 2020
43. Khenissi, S., Boujelbene, M., Nasraoui, O.: Theoretical modeling of the iterative properties of user discovery in a collaborative filtering recommender system (2020)
44. Lee, S., Kim, D.: Deep learning based recommender system using cross convolutional filters. Inf. Sci. **592**, 112–122 (2022)
45. Shambour, Q.: A deep learning based algorithm for multi-criteria recommender systems. Knowl. Based Syst. **211**, 106545 (2021)
46. Ahmadiana, S., Ahmadianb, M., Jalili, M.: A deep learning based trust-and tag-aware recommender system., Neurocomputing **488**, 557–571 (2021)
47. Kiran, R., Kumar, P., Bhasker, B.: DNNRec: a novel deep learning based hybrid recommender system. Expert Syst. Appl. **144**, 113054 (2020)
48. Gomede, E., de Barros, R.M., de Souza Mendes, L.: Deep auto encoders to adaptive E-learning recommender system. Comput. Educ. Artif. Intell. **2**, 100009 (2021)
49. Djellali, C., Mehdiadda: A new hybrid deep learning model based-recommender system using artificial neural network and hidden Markov model. Procedia Comput. Sci. **175** , 214–220 (2020)
50. Da'u, A., Salim, N., Rabiu, I., Osman, A.: Weighted aspect-based opinion mining using deep learning for recommender system. Expert Syst. Appl. **140**, 112871 (2020)
51. Ayci, G., Köksal, A., Mutlu, M.M., Suyunu, B., Cemgil, A.T.: Active learning with bayesian nonnegative matrix factorization for recommender systems. In: 2019 27th Signal Processing and Communications Applications Conference (SIU). IEEE (2019)
52. Zhou, J., Chiky, R.: Improving the attribute-based active learning by clustering the new items. In: 2019 IEEE World Congress on Services (SERVICES). IEEE (2019)

53. Du, Z., Tang, J., Ding, Y.: POLAR++: active one-shot personalized article recommendation. IEEE Trans. Knowl. Data Eng. **33**(6), 2709–2722 (2021)
54. Du, Z., Yang, N., Yu, Z., Yu, P.: Learning from atypical behavior: temporary interest aware recommendation based on reinforcement learning. IEEE Trans. Knowl. Data Eng. (2022)
55. Li, X., et al.: Pre-training recommender systems via reinforced attentive multi-relational graph neural network. In: IEEE International Conference on Big Data (Big Data) (2021)
56. Chen, H.: A DQN-based recommender system for item-list recommendation. In: IEEE International Conference on Big Data (Big Data) (2021)
57. Song, Z., Zhang, D., Shi, X., Li, W., Ma, C., Wu, L.: DEN-DQL: quick convergent deep Q-learning with double exploration networks for news recommendation. In: International Joint Conference on Neural Networks (IJCNN). IEEE (2021)
58. Kabra, A., Agarwal, A.: Personalized and dynamic top-k recommendation system using context aware deep reinforcement learning. In: IEEE 45th Annual Computers, Software, and Applications Conference (COMPSAC) (2021)
59. Fu, M., et al.: Deep reinforcement learning framework for category-based item recommendation. IEEE Trans. Cybern. (2021)
60. Rabieinejad, E., Mohammadi, S., Yadegari, M.: Provision of a recommender model for blockchain-based IoT with deep reinforcement learning. In: IEEE, 5th International Conference on Internet of Things and Applications (IoT) (2021)
61. Huang, L., Fu, M., Lia, F., Qu, H., Liu, Y., Chen, W.: A deep reinforcement learning based long-term recommender system. Knowl. Based Syst. **213**, 106706 (2021)
62. Abbas, S.M., Alam, K.A.: Exploiting relevant context with soft-rough sets in context-aware video recommender systems. In: IEEE International Conference on Fuzzy Systems (FUZZ-IEEE) (2019)
63. Ubukata, S., Takahashi, S., Notsu, A., Honda, K.: Basic consideration of collaborative filtering based on rough C-means clustering. In: IEEE, Joint 11th International Conference on Soft Computing and Intelligent Systems and 21st International Symposium on Advanced Intelligent Systems (SCIS-ISIS) (2020)
64. Abbas, S.M., Alam, K.A., Ko, K.-M.: A three-way classification with game-theoretic N-soft sets for handling missing ratings in context-aware recommender systems. In: IEEE International Conference on Fuzzy Systems (FUZZ-IEEE) (2020)
65. Sharma, A.K., Bajpai, B., Adhvaryu, R., Pankajkumar, S.D., Gordhanbhai, P.P., Kumar, A.: an efficient approach of product recommendation system using NLP technique. Mater. Today Proc. (2021)
66. Trappey, A., Trappey, C.V., Hsieh, A.: An intelligent patent recommender adopting machine learning approach for natural language processing: a case study for smart machinery technology mining. Technol. Forecast. Soc. Change **164**, 120511 (2021)
67. Sterling, J.A., Montemore, M.M.: Combining citation network information and text similarity for research article recommender systems. EEE Access, **10**, 16–23 (2021)
68. Zhang, T., Liu, M., Ma, C., Tu, Z., Wang, Z.: A text mining based method for policy recommendation. In: IEEE International Conference on Services Computing (SCC) (2021)
69. Kuanr, M., Mohapatra, P.: Recent challenges in recommender systems: a survey. In: Panigrahi, C.R., Pati, B., Mohapatra, P., Buyya, R., Li, KC. (eds.) Progress in Advanced Computing and Intelligent Engineering. AISC, vol. 1199, pp. 353–365. Springer, Singapore (2021). https://doi.org/10.1007/978-981-15-6353-9_32
70. Di Benedetto, G., Bellini, V., Zappella, G.: A linear bandit for seasonal environments. In: Proceedings of the 37th International Conference on Machine Learning, Vienna, Austria (2020). Amazon Science

71. Zhao, J., Wang, L., Xiang, D., Johanson, B.: Collaborative deep denoising autoencoder framework for recommendations. In: SIGIR 2019, Paris, France, July 2019. Amazon Science
72. Gupta, S., Wang, H., Lipton, Z., Wang, Y.: Correcting exposure bias for link recommendation. In: Proceedings of the 38th International Conference on Machine Learning, PMLR 139 (2021). Amazon Science

Live Social Spacing Tracker Based on Domain Detection

Divyank Agarwal, Monark Mehta, and Nirmala Paramanandham[✉]

Vellore Institute of Technology, Chennai, India
nirmalavp.ece@gmail.com

Abstract. Corona virus Disease-2019 (COVID-19) is caused by infection with the severe acute respiratory syndrome (SARS) coronavirus. In this COVID 19 pandemic, the virus spread bringing the whole world to downfall. This disease can spread through the slightest touch, breathing the same air or using same basic things like clothes, hairs, combs, etc. These viruses can live for hours even without a host body. To prevent the spread, the world was put on lockdown, and people were constrained to their homes, but human life cannot go on without interaction. We need a better way of preventing the spread of this type of disease. Hence a detector is proposed for measuring the social distance between the people. This social distancing detector can track the people who are not following social distancing norms, then that person can be tracked down or the person is marked in red and triggers the warning. In this paper we will be using Computer Neural Network (CNN) to process our images and videos because CNN is a type of artificial neural network that is used in recognition of large pixel videos or images. The proposed technique can be the perfect way to help the person while people carry out their daily tasks. From the evaluation parameters it is proven that the proposed technique yields better results when compared to the state- of- the art techniques.

Keywords: COVID-19 · Person detection · Social distancing

1 Introduction

Coronavirus and its variants like delta, omicron and IHU cause many deaths in all over the world. This virus does not show symptoms based on age. Anyone can be affected by this virus at any age. In this current pandemic, coronavirus cases are increasing day by day. Each country and state has been affected by this virus which has caused the demolition of many families. This virus can be spread over an infected person's mouth or nose to the recipient's mouth or nose. This could happen when the infected person sneezes, cough, speak, sing or breathe [1].

World Health Organization (WHO) have considered the implementation of public health and social measures (PHSM) which can result in decreasing the number of deaths and spreading of coronavirus. PHSM is not a rule of law; it has been implemented based on the cause of spreading coronavirus. It includes social distancing, avoiding crowded places, and cleaning hands regularly using any antibacterial soaps or sanitizers [2].

© IFIP International Federation for Information Processing 2022
Published by Springer Nature Switzerland AG 2022
L. Kalinathan et al. (Eds.): ICCIDS 2022, IFIP AICT 654, pp. 115–123, 2022.
https://doi.org/10.1007/978-3-031-16364-7_9

Coronavirus can enter to our body when the people are not maintaining social distance from the affected person. To avoid this type of disease, the best practice can be following the social distancing guidelines. Consider our own country India, which is the second most populated country after China. Due to more population, the public places are becoming more crowded. In turn, it is increasing the COVID-19 cases rapidly.

In this paper, a live social distancing tracker tool is proposed. It will detect the people who are not following the social distancing rules. This can be done by computing the distance between the nearby people.

The technique has been specially utilized for the more crowded area like markets, bus stands, and other public areas as the probability of spreading COVID-19 is more in these particular areas. In addition to that the proposed technique can also be used for any type of restricted area or property like a gas station where entry of any outsider is prohibited; this will detect the person and alert the security in that area.

The following sections discuss the related works, proposed work, results and discussions and the conclusions.

2   Research Background

Yibing ma et al. [3], have proposed a fire alarm with person detection and thermal camera. This system uses the YOLOv3-tiny algorithm for detecting a person. It also checks the temperature of the environment, if the temperature is high in the region, then it automatically starts detecting the people and alerting them.

In [4], the authors have proposed a technique for the detection of people from a distance of several miles using a high-quality camera lens. They have used a two-stage approach for detecting a person and distinguish between them using a distanced captured video. A convolution neural network (CNN) is used for distinguishing a person carrying a bundle or a person carrying a long arm. Several methods have been proposed for background subtraction to robustly detect the people in thermal video across different environmental conditions. They have used thermal cameras which can detect the images even at the night, this is done by capturing the thermal radiation reflected from objects which are captured by the camera, and hence the person is detected even in low light conditions. In the developed technique, the shadow of the person will not be considered in the detection process unless the person is stationary for a long time [5]. Video surveillance has become a necessary part in most of the places such as malls, railway station, bus station and no entry places. Detecting a person in these areas has a major role. In [6], a simple and fast real-time person tracking system was developed based on the SiamMask network for video surveillance. This system uses deep learning-based algorithms. Initially the person is detected using SiamMask technique then additional training is done using transfer learning for obtaining the accuracy of 95%. The technique [7] has proposed for multispectral pedestrian detection in the field of computer vision. The authors have detected the pedestrian by leveraging RGB-thermal pairs with convolutional neural networks. Further, they have studied the effect of noise on the output by creating a clean version of ground-truth annotations.

Current industries required a minimum separation of workers from the machine which is a good practice in industries for the safety of workers [8]. It was performed using a sensor-based approach and this system uses a background model for representing an aspect of the environment. Ideally, the safety zone surrounds by a green mark and the danger zone is surrounded by a red mark. However, people have not considered robots and if they enter in the red region then the system alerts the workers. Siezen et al. [9] have implemented a technique using magnetic fields for keeping the track of social distancing between people. They made a wearable device, it measures the distance between people by the strength of the coil present in. If the distance is less than two meters then the wearable device alerts the person to follow the social distancing guidelines. Nowadays many digital companies involve Bluetooth technology which they have declared the most effective technique for tracking persons in contact. Scott et al. [10], proposed a system that tracks the people based on Bluetooth in smartphone. It calculates the distance between the smartphones and then track the social distancing, but this study may not be feasible in all cases. Yadhav et al. [12], have conducted a mean accuracy and maximum validation performance comparison between different models for detecting the objects. Lin et al. [13], have performed an experiment on using the dataset of Common Object in Context (COCO) and the obtained results are shown in Table 1.

Table 1. Mean average Precision score results of different pre-trained model

Model	Minimum validation	Mean accuracy
MobileNet SSD	19.3	19
Resnet-101 Fast R-CNN	31	30.9
Resnet-101 Fast R-CNN	33.2	33
Inception Resnet Fast R-CNN	34.7	34.2
MobileNet R-FCN	13.8	13.4

Yadhav et al. [12] have discussed the different dataset for analyzing Execution time and number of object detected in each frame for the different dataset and the highest accuracy between the different model and datasets is shown in Table 2.

Table 2. Comparison work for different pre trained model

Model name	Execution time(s)	Highest accuracy	Object detected
SSD Mobilenet V1 COCO	219.58	94%	2
SSD Inception V2 COCO	298.22	97%	2
Inception V2 COCO Faster RCNN	420.71	99%	3
Resenet V2 Atrous Mask RCNN Inception	6008.02	99%	5

On the basis of the results obtained, the data show that the MobileNet SSD has a faster recognition rate and this model runs faster for object recognition. Hence, the proposed work choses the MobileNet SSD as a recognition model.

3 Proposed Work

In this work, a live spatial detection technique is proposed. Initially, the live video will be converted to frames. The people standing nearby will be detected in each and every frame by the proposed technique. When the people are not maintaining the social distancing, they will be surrounded by 'Red mark' or else they will be surrounded by 'Green mark'. The method will be achieved by calculating the center point of the people and a rectangular box type border will be created around them and if the box overlaps the other nearby box, then the alarm or warning will be given to the people. This algorithm can be divided into four functions where each function has its definitive algorithms and properties.

Check Function: In this function, the person class which is identified by the mobile net SSD is bounded by a box. This box may get overlapped to prevent the infinite bounding box which is due to the simultaneous change in the position of the object. After the boxes are bounded, the centroid of the bounded box is calculated and it is given a unique ID and the same is done with other objects after the centroid is calculated.

Image Process Function: It is the major and most significant function as this function is responsible for the generation of input and output files which are in the form of mp4 formats. This function takes the input file and splits it into many frames which are then processed individually and then the output frames are joined to create an output video in the same video format.

Input Parameters: Here, the algorithm is initialized, and input is given as 'filename' while the output file is generated and stored in the.mp4 format as 'opname'. The input video is in form of mp4, or live feed and the output video also can be downloaded in the same format. Centroid rotation is used for finding the distance between a people. When the separation distance is less than the set limit, then the person is shown with red mark. The person finder function is used for finding and identifying the person of the figure in the taken videos.

Main Function: This is the final part of the algorithm where the output processed frames generated by the algorithm are joined once again to create an output mp4 file. This output file contains the time stamp which tells the speed and duration of the video, the.time() functions calculate the duration of the video and process which can be used for finding the efficiency of the proposed algorithm. As the frames generation and object class identification go simultaneously, this process time depends greatly on the hardware used, speed of the processor, and also the duration of video like in my example – a 3-s video took almost 150 s to be processed with the social distancing tool.

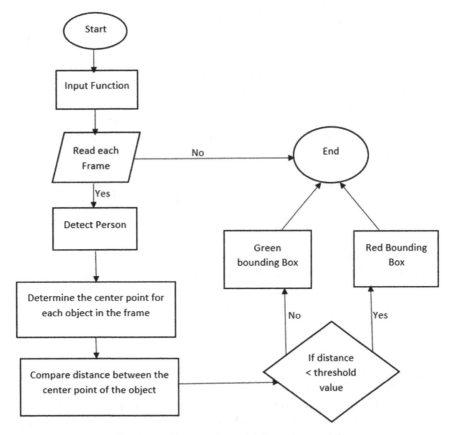

Fig. 1. Architecture for social distancing model

A. Object Detection Model

In this paper, the machine learning and person identification tool is used with deep learning algorithms. While there are many other algorithms with higher accuracy, the Mobile net SSD came out to be the fastest, with the least hardware requirements it has a detection rate of 87.7% with a ratio of 80:20, training to test data [11].

B. Determine person location

In determining the person and identification of the location of the object, some of the important factors are kept in mind. For example, the camera should be at a high point where it can capture large and stable video with minimum contact and disturbance. The camera quality should be above average so that at least the person in close proximity can be seen and identified. It should also have some kind of protection or covering for preventing the outer disturbance to be captured as video.

C. Calculate the center point in the bounding box

The person bounded by the bounding box is first given an id and then the centroid of the person is calculated by using the formulae:

$$C(x, y) = \frac{(xmin + xmax), (ymax + ymin)}{2} \tag{1}$$

Each of the minimum and maximum values for the corresponding width, xmin + xmax and height ymax + ymin, of the bounding box will be used to calculate the center point of the bounding box.

D. Distance between bounding box

After the bounding box has been calculated and the centroid has been determined, the distance between two centroids are calculated using the Eq. (2).

$$D = ((xmax - xmin)^2 + (ymax - ymin)^2)^{1/2} \tag{2}$$

This distance is the maximum distance between two objects or persons. It is compared with two meters which is the distance to be maintained for preventing the spread of COVID-19, thus the person can be then marked according to them as a green or red box.

3.1 Algorithm

Step 1: Give the video (mp4) or live feed as an input.

Step 2: The video is divided into frames.

Step 3: Each frame is processed, and the person can be detected in each frame.

Step 4: A bounding box is created and the distance between each bounding box is calculated.

Step 5: The value is compared with the minimum distance.

Step 6: The box is either displayed red or green based on the measured distance.

Step 7: Output processed frames are then converted to a video.

4 Results and Discussions

In this work, python 3 and openCV libraries are used for image processing techniques. For calculation, two videos are taken on bridge and a 4-way road. Camera recordings are processed, while some random frames are taken from the output video to create a proper dataset. We have taken four different benchmark dataset as Self video, Bridge video, 4-Lane video and PETS-2009 video whose accuracy has been listed in Table 3. This dataset has been analyzed based on the multitude i.e., if this system can detect the person.

To measure the accuracy of the system, values for true positive (TP) - number of persons correctly identified who are following the social distancing norms, true negative (TN) - number of persons identified who are not following the social distancing norms, false positive (FP) - number of persons correctly identified and not following the social distancing norms, false negative (FN) - number of persons not correctly identified and following the social distancing norms.

$$Accuracy = \frac{TP + TN}{TP + TN + FP + FN}$$

The accuracy is then taken from these environments to give the final results and comparison with the different models and algorithms.

Table 3. Accuracy of the proposed algorithm.

No	Video	TP	TN	FP	FN	Accuracy
1	SELF-video	2	2	0	0	100%
2	BRIDGE	6	9	2	1	83%
3	4-LANE	10	16	10	4	65%
4	PETS-2009	0	7	1	2	70%

Fig. 2. BRIDGE output

Based on the findings and results, it can be proven that the proposed technique yields better results when compared to other algorithms.

The obtained results for different dataset are tabulated in Table 3. For the self-video, the proposed technique achieves 100% accuracy. But when the other datasets are considered, the accuracy is getting decreased because of the increased density of crowd.

Figures 3 and 4 shows the model detection for the dataset PATS-2009 and 4-lane. The detection model accuracy is reduced when the person is too close because both the persons are identified as a single person. The technique achieves very high efficiency in tracking static or less crowded places. The algorithm is implemented in Intel I5, 7th generation. The computational complexity is reduced and accuracy can be improved by removing the redundant frames. After analyzing all four dataset we understood that in Figs. 1 and 2 has less crowd so the efficiency of the system is 100% but when coming to

Fig. 3. PATS-2009 output

Figs. 3 and 4 this dataset contains more crowd so the system is unable to detect person who are very close to each other it detects them as a single person so this can be a major issue of the existing system.

Fig. 4. Four lane output

5   Conclusion and Future Work

The major precaution to prevent COVID- 19 and its variant is maintaining social distance and avoiding physical contact of the person who is affected from the disease. Viral transmission rates will be increased because of non-compliance with these rules. To achieve the desired functionalities, proposed technique is implemented using Python and the OpenCV library. The first component detects social distancing infractions, while the second feature detects violations of entering restricted locations. The accuracy of both characteristics has been verified. The proposed technique has met all its objectives and it can be proven from the obtained results. In future, we are planned to remove the redundant frames in the dataset which will improve the accuracy and reduces the computational complexity.

References

1. World Health Organization: "Coronavirus Disease 2019," Coronavirus disease (COVID-19) pandemic (2021). https://www.who.int/health-topics/coronavirus#tab=tab_1
2. World Health Organization: Considerations for implementing and adjusting public health and social measures in the context of COVID-19, Interim guidance, 14 June 2021
3. Ma, Y., et al.: Smart fire alarm system with person detection and thermal camera. In: Krzhizhanovskaya, V.V., et al. (eds.) ICCS 2020. LNCS, vol. 12143, pp. 353–366. Springer, Cham (2020). https://doi.org/10.1007/978-3-030-50436-6_26
4. Wei, H., Laszewski, M., Kehtarnavaz, N.: Deep learning-based person detection and classification for far field video surveillance, Department of Electrical & Computer Engineering, University of Texas at Dallas, November 2018
5. Davis, J.W., Sharma, V.: Robust Background-Subtraction for Person Detection in Thermal Imagery, Department of Computer Science and Engineering, Ohio State University Columbus, OH 43210 USA
6. Ahmed, I., Jeon, G.: A real-time person tracking system based on SiamMask network for intelligent video surveillance. J. Real-Time Image Proc. **18**(5), 1803–1814 (2021). https://doi.org/10.1007/s11554-021-01144-5
7. Li, C., Song, D., Tong, R., Tang, M.: Multispectral Pedestrian Detection via Simultaneous Detection and Segmentation, State Key Lab of CAD&CG, Zhejiang University Hangzhou, China, 14 August 2018
8. Rybski, P., Anderson-Sprecher, P., Huber, D., Niess, C., Simmons, R.: Sensor fusion for human safety in industrial workcells. In: IEEE/RSJ International Conference on Intelligent Robots and Systems, 24 December 2012
9. Bian, S., Zhou, B., Bello, H., Lukowicz, P.: A Wearable Magnetic Field Based Proximity Sensing System for Monitoring COVID-19 Social Distancing. German Research Center for Artificial Intelligence Kaiserslautern, Germany (2020)
10. McLachlan, S., et al.: Bluetooth Smartphone Apps: Are they the most private and effective solution for COVID-19 contact tracing? (2020)
11. Kurdthongmee, W.: A comparative study of the effectiveness of using popular DNN object detection algorithms for pith detection in cross-sectional images of parawood, School of Engineering and Technology, Walailak University 222 Thaibury, Thasala, Nakornsithammarat, 80160, Thailand, February 2020
12. Yadav, N., Binay, U.: Comparative study of object detection algorithms. Int. Res. J. Eng. Technol. **4**, 586–591 (2017). www.irjet.net
13. Lin, T.-Y., et al.: Microsoft COCO: common objects in context. In: Fleet, D., Pajdla, T., Schiele, B., Tuytelaars, T. (eds.) ECCV 2014. LNCS, vol. 8693, pp. 740–755. Springer, Cham (2014). https://doi.org/10.1007/978-3-319-10602-1_48

Assessing Layer Normalization with BraTS MRI Data in a Convolution Neural Net

Akhilesh Rawat[✉] and Rajeev Kumar

Data to Knowledge (D2K) Lab, School of Computer and Systems Sciences,
Jawaharlal Nehru University, New Delhi 110 067, India
akhile62_scs@jnu.ac.in

Abstract. Deep learning-based Convolutional Neural Network (CNN) architectures are commonly used in medical imaging. Medical imaging data is highly imbalanced. A deep learning architecture on its own is prone to overfit. As a result, we need a generalized model to mitigate total risk. This paper assesses a layer normalization (LN) technique in a CNN-based 3D U-Net for faster training and better generalization in medical imaging. Layer Normalization (LN) is mostly used in Natural Language Processing (NLP) tasks such as question-answering, handwriting sequence generation, etc. along with Recurrent Neural Network (RNN). The usage of LN is yet to be studied in case of medical imaging. In this context, we use brain MRI segmentation and train our model with LN and without normalization. We compare both models and our LN-based model gives 32% less validation loss over without normalization-based model. We achieve validation dice scores of unseen input data passes to LN based model of 0.90 (7.5% higher than without normalization) for edema, 0.74 (12.5% higher than without normalization) for non-enhancing tumor and 0.95 (1.5% higher than without normalization) for enhancing tumor.

Keywords: Convolution neural net · 3D U-Net · Generalization · Layer normalization · Batch normalization · Group normalization · Instance normalization · BraTS · MRI

1 Introduction

Recent advances in the interpretation of medical images specially in brain tumor segmentation make substantial use of deep neural networks [5,10–12,16,18]. Convolutional Neural Network (CNN) is most often used as deep learning architecture for medical image analysis. We used BraTS 2016 and 2017 datasets [17] for brain MRI tumor segmentation and applying layer normalization on CNN based 3D U-Net to generalize the deep neural network. By default, traditional approaches employ Batch Normalizaion (BN) with a 3D U-Net architecture [3,12]. Many researchers apply Group Normalization (GN) [6,11] and Instance Normalization (IN) [13] in medical imaging. Layer Normalization (LN)

© IFIP International Federation for Information Processing 2022
Published by Springer Nature Switzerland AG 2022
L. Kalinathan et al. (Eds.): ICCIDS 2022, IFIP AICT 654, pp. 124–135, 2022.
https://doi.org/10.1007/978-3-031-16364-7_10

is mostly used in NLP tasks such as question-answering, handwriting sequence generation, etc. with Recurrent Neural Network (RNN) [1].CNN served as an effective tool for representing and presenting the input data. CNN-produced features were so effective that handcrafted features became obsolete. This is not to say that domain specialists are unnecessary. Domain knowledge aids in the selection of the CNN architecture, resulting in a more performant CNN model.

The backpropagation learning technique poses difficulties for deep learning networks. Numerous gradient descent algorithms may be accelerated using various optimization approaches, such as momentum optimization, Adagrad, and so on. Additionally, normalization approaches improve the stability and learning speed of DL. Many studies employed batch normalization [4], instance normalization [20], group normalization [21], layer normalization [1], and other normalizing approaches in DNN. When CNN is used for computer vision or medical image analysis, a problem called "covariate shift" occurs [15]. This issue confuses the training process, and the model provides biased output. Loffe and Szegedy [4] devised the method of Batch Normalization (BN) to solve this issue. They normalize the input data and output of the hidden layers and make learning faster and helps in faster convergence. The primary issue with BN is that performance is highly dependent on batch size. Also, we cannot normalize until the complete batch is operated on during batch mode training. This research assessing normalization with Layer Normalization (LN) approach in a CNN-based 3D U-Net [2] deep learning architecture. The same technique has been used in Recurrent Neural Network (RNN) [1]. We normalized the input layer data and hidden layer data through LN. There are several parameters and weights in the hidden layer, and each layer is a weighted sum of the activations of the preceding layers. Backpropagation updates all weights repeatedly. Hence, the distribution may not be consistent over epochs. As a result, we're facing the identical issue of covariate shift [15]. To see the impact of this issue on training, we train our model using layer normalization as well as without normalization. Finally, we assess training effects in CNN-based 3D U-Net on the brain tumor segmentation with BraTS MRI dataset.

The remainder of the paper is laid out as follows: the literature review is included in Sect. 2. The normalizing approach is explained in Sect. 3. Section 4 discusses the dataset and explains the functioning of the 3D U-Net architecture with Layer Normalization. Section 5 concludes this work.

2   Literature Review

In general, normalization improves the efficiency, speed, and accuracy of the learning process. When a deep neural network is trained and an image dataset is passed to network through batches, one of the problems encountered by Shimodaira [15] is referred to as covariate shift. It is defined as the shift in the distribution of network activations caused by training parameter changes. To address this issue, Ioffe and Szegedy [4] developed a technique called Batch Normalization (BN). This paper defines how BN works. They presented an algorithm for how to train a network with BN. They used the MNIST dataset to predict the digit class and the ImageNet dataset to perform classification. They employed a CNN architecture with BN with a batch size of 32 for their analysis. They found that BN required fewer iterations during training to achieve higher accuracy than a normal architectures. To address the shortcoming of batch normalization Yao et al. [22] have proposed a cross iteration batch normalization (CBN) in which numerous recent iterations are jointly utilizes to improve performance. They used benchmark dataset Image net and COCO for image classification and object detection respectively. With batch size 4, Naive CBN surpasses BN by 1.7% in accuracy.

Wu and He [21] introduced the technique of Group Normalization (GN). This technique is used to normalize the values. They demonstrated that GN is a straightforward representation of Batch Normalization (BN). They used ImageNet, COCO dataset for image segmentation and detection, and the Kinetics dataset for video classification. They trained our model using ResNet-50. As a result, they determined that GN had a 10.6% lower error rate than BN (batch size 2). When we reduce the batch size or use a tiny batch size during training, BN becomes worse. GN is independent of batch sizes, so it outperforms BN.

Ba et al. [1] have proposed another normalization technique called Layer Normalization (LN). They noticed that in BN, performance suffers when the batch size is decreased. The BN-related architecture can not normalize the input until the complete batch is processed. In LN, they computed the mean and variance used to normalize all the features extracted from a single example. For the experiment, they used the MNIST dataset for classification and RNN for image-sentence ranking, contextual language modeling, etc.

Shrestha et al. [16] have enhanced the batch normalization technique to make deep learning faster. They used this for segmentation in medical imaging, specifically prostate cancer segmentation. These images are magnetic resonance (MR) images. They found that DL had not accurately performed prostate segmentation on MR images, so they used BN in deep neural architecture. They used the U-Net architecture and modified the state-of-the-art BN through a modification in the loss function. Finally, their proposed method reduces the processing time and enhances accuracy. To compute the accuracy, they use the dice similarity coefficient.

Reinhold and Jacob [14] have investigated seven different techniques for normalizing the intensity distribution of MRI brain images over three contrasts (T1-w, T2-w, and FLAIR). They demonstrated with the Kirby-21 dataset. The experiments have shown that synthesis results are robust to the choice of the normalization technique. They recommended normalization as a pre-processing step before data is served as an input in patch-based DNN.

Numerous authors [6, 11–13] have applied data augmentation and normalization techniques on the BraTS dataset of MR images. Data normalization built new patches from the existing patches to overcome the large variation in brain tumor heterogeneity and scarcity of data. Andriy [11] has used group normalisation followed by ReLU and traditional CNN to iteratively decrease the dimension of images. Zeyu and Changxing have proposed a novel two stage U-Net architecture to segment the brain tumor from coarse to fine. They used group normalisation as pre-processing technique followed by ReLU on BraTS 2019 dataset. Kumar and Rockett [8] proposed a bootstrapping procedure to partition data across modules and thus improving generalization through training of multiple neural modules [9].

Zeyu et al. [6] have proposed a novel two stage U-Net architecture to segment the brain tumor from coarse to fine. They used group normalisation as pre-processing technique followed by ReLU on BraTS 2019 dataset. Hiba and Ines have proposed a pre-processing technique based on contrast enhancement and intensity normalization on MRI data. Automatic segmentation is a challenging due to the high spatial and structural heterogeneity of brain tumours. Sergio et al. [13] have presented a CNN based architecture for automatic segmentation on brain tumor (BraTS-2013 and 2015). MR images. They applied intensity normalization at pre-processing level to make the contrast and intensity ranges between patients and acquisitions more consistent. Experimentally, they shown that intensity normalization enhances the dice similarity coefficient matrix. Nosheen et al. [18] have proposed a smart segmentation technique for whole brain tumor based on standard dataset BraTS-2019 and BraTS-2020. They use 3D UNET architecture followed by batch normalization layer.

3   Normalization

Normalization is a technique used to normalize the heterogeneous data on the same scale. This approach is used in deep learning architectures to accelerate learning, maintain network stability, and improves the overall performance. Normalization procedures are classified into four categories. Normalization by Batch (BN) [4], Layer (LN) [1], Instance (IN) [20], and Group (GN) [21]. We employ layer normalization to train our CNN-based 3D U-Net. In LN, we consider the feature maps calculated by multiple kernels over a single training data. Then we calculate mean and standard deviation over these feature maps derived from one training sample. Another mean and s.d. derived from another training sample, and so on. The dimensionality of μ and σ vector is the same as the channel 'C.' Let us suppose we have 'N' number of training images and C number of channels and every channel is of width 'W' and height 'H' provided by CNN. x represents one feature element in channel C at point j, k in that particular channel feature.

$$\mu_N = \frac{1}{CWH} \sum_{i=1}^{C} \sum_{j=1}^{W} \sum_{k=1}^{H} x_{Nijk} \tag{1}$$

$$\sigma_N^2 = \frac{1}{CWH} \sum_{i=1}^{C} \sum_{j=1}^{W} \sum_{k=1}^{H} (x_{Nijk} - \mu_N)^2 \tag{2}$$

We calculate μ_N and $\sigma_N 2$, the mean and variance of the same data across the same layer, using the preceding Eqs. 1 and 2. Now, we'll normalize feature by using Eq. 3, where ϵ is a tiny positive constant and prevent the division by zero condition.

$$\hat{x} = \frac{x - \mu_N}{\sqrt{\epsilon + \sigma_N^2}} \tag{3}$$

$$Y_i = \gamma \hat{x}_i + \beta \tag{4}$$

After applying LN, we do reparameterization. We have to compute Y_i (Eq. 4), where γ and β are new parameters that define the distribution of the data. During the backpropagation, γ and β are tunable, and it can be tuned in the same gradient descent procedure along with the weight vector of the neural network.

In Batch Normalization (BN), we collect all identical channels from all training examples in the same batch and group them together. We make mini batches and compute their mean and standard deviation using Eqs. 5 and 6, where μ_C and σ_C^2 are mean and variance of a particular mini batche, respectively.

$$\mu_C = \frac{1}{NWH} \sum_{i=1}^{N} \sum_{j=1}^{W} \sum_{k=1}^{H} x_{iCjk} \tag{5}$$

$$\sigma_C^2 = \frac{1}{NWH} \sum_{i=1}^{N} \sum_{j=1}^{W} \sum_{k=1}^{H} (x_{iCjk} - \mu_C)^2 \tag{6}$$

After calculating the values of μ_C and σ_C2 from identical channels, we use Eq. 7 to normalize the values of mini batches. The fundamental issue with BN is that its performance is highly dependent on the batch size. The performance will decrease if we reduce the batch size.

$$\hat{x} = \frac{x - \mu_C}{\sqrt{\epsilon + \sigma_C^2}} \tag{7}$$

$$Y_i = \gamma \hat{x}_i + \beta \tag{8}$$

Finally, Eq. 8 does the reparameterization; during the backpropagation phase, γ and β values are tunable with the same gradient descent procedure as the weight vector of the neural network. In BN, we cannot normalize the data unless the entire batch is processed.

In Instance Normalization (IN), we consider every individual channel and normalize them with respect to that channel's mean μ and standard deviation σ. We do not consider features belonging to other channels for computing the mean and standard deviation. We normalize the features in that particular channel only. After applying IN, we do the reparameterization as in Eq. 8 and both γ and β are tunable during the backpropagation.

In Group Normalization (GN), the different channels are grouped into different groups. We compute the mean and standard deviation of each group. After computing μ and σ we normalize the features belonging to that group only with respect to the corresponding mean and standard deviation. Furthermore, we follow the same process of reparameterization that projecting back the normalized data as $y_i = \gamma \hat{x}_i + \beta$. The values of γ and β will be tuned during the backpropagation along with the parameters or weight vector of the neural network.

4 Experimental Setup and Results

4.1 Dataset Description

In this paper, we have used MRI multisequence data for brain tumor segmentation from the BraTS 2016 and BraTS 2017 datasets. Simpson [17] et al. combined these datasets and made them publicly available. This dataset was used in the challenge (brain tumor segmentation) organized by the Medical Segmentation Decathlon [17]. The segmentation task aims to segment three brain tumors called Edema, Enhancing tumor and Non-Enhancing tumor. The dataset contains 750 MRI scans of patients receiving High Grade Glioma (HGG) or Low Grade Glioma (LGG) treatment. All images are a 4-dimensional array of size $240 \times 240 \times 160 \times 4$. The MRI scans are multi-parametric and comprise native (T1) and post-Gadolinium (Gd) contrast T1-weighted (T1-Gd) volumes, as well as native T2-weighted (T2) and T2- Fluid Attenuated Inversion Recovery (FLAIR) volumes shown in Fig. 1.

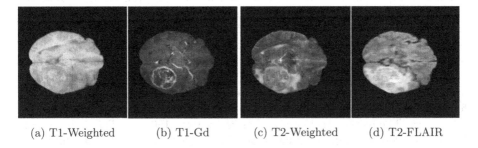

(a) T1-Weighted   (b) T1-Gd   (c) T2-Weighted   (d) T2-FLAIR

Fig. 1. Four Modalities of MRI Images in 2D space for visualization.

The BraTS dataset includes MRI scans from 19 medical institutes. These MRI scans were performed routinely on patients at specific hospitals. There are 484 data points in this dataset. For our experiments, we divided these data points into two partitions: training and testing. The 80% of all data points are utilized for training, while the rest 20% are used for prediction. As a result, our training set has 387 data points, whereas our test data contains 97. Every MRI scan is rescaled to a $1mm^3$ isotropic resolution, and the skull is eliminated. These scan is done in four modalities, we denoted these modalites with numbers (ranging from 0 to 3); 0→ FLAIR, 1→ T1-weighted, 2→ T1-Gd, and 3→ T2-weighted. Multidimensional MRI images (each shape $240 \times 240 \times 160 \times 4$) are given as input to the 3D U-Net architecture. The first three dimensions represent each point in 3D volume called a voxel. A voxel is a data point in a three dimensional volume. The fourth dimension aids in accurately detecting tumor locations by providing additional information beyond the voxel data. These images (represented in voxel) are associated with four labels. Various numbers denote these labels (ranging from 0 to 3), i.e., 0 denotes the background, 1 denotes edema, 2 denotes non-enhancing tumors, and 3 denotes enhancing tumors.

We extracted five subvolumes from each input image to efficiently train and test the model. Each volume size is $160 \times 160 \times 16 \times 4$. We have removed those subvolumes that contain 5% tumor regions. We used the HDFS-5 data format to store these subvolumes and made a large dataset for training. 3D U-Net architecture to learn with these subvolumes. For training, we have 1560 subvolumes, and 440 subvolumes are used for validation. Before training we standardize the input data with its mean (μ) and standard deviation (σ). The sub-volume obtained is standardized over each channel and Z plane.

4.2   3D U-Net Architecture

To learn the underlying pattern, a deep neural network requires a substantial amount of input samples. In medical applications, the training data is small, which is insufficient to quell the deep learning architecture's surge. As a result, we need a system capable of learning from a small sample size. For automated brain tumour image segmentation. Figure 2 depicts our suggested architecture. It includes an encoder and a decoder. Both contraction and expansion exist. These two portions are symmetric. 3D U-Net uses 3D volumetric images. Hmeed et al. [3] have used the U-Net architecture with the Batch Normalization (BN) for MRI brain tumor segmentation. The contraction component gets the input image first. The model accepts input of size $4 \times 160 \times 160 \times 16$. The input layer is followed by downsampling units with one 3D convolution with $3 \times 3 \times 3$ filter size and $1 \times 1 \times 1$ stride. Then normalization, activation, dropout, 3D convolution, and max pooling with a pool size of $2 \times 2 \times 2$.

The network with a 25% dropout rate has four downsampling units and symmetric upsampling till the output layer. Each upsampling layer has a concatenation layer. The network uses 'ReLU' activation function for hidden layers, whereas the output layer uses the 'sigmoid' activation function. We employ standardization, dropout, layer normalization, and data augmentation techniques to improve brain tumor image segmentation through this architecture. We also train the 3D U-Net architecture shown in Fig. 2 without normalization layer and evaluate both performances.

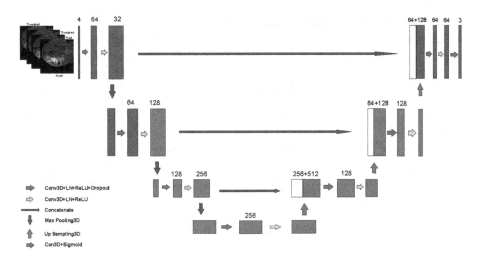

Fig. 2. Layer Normalization based 3D U-Net Architecture for MRI segmentation

4.3   Experimental Results

This part trains our model and compares the results with and without layer normalization. We utilized Adam Optimizer to optimize [7]. The experiment was conducted using a 6 GB NVIDIA GeForce RTX 3060/144 Hz GPU and 16 GB of RAM. We utilized Python as the platform and two open-source libraries, Keras and TensorFlow. We have worked with a batch size of one. Given the computational restrictions, we trained the model for 50 epochs with an initial learning rate of 10^{-4}. We performed MRI scans on 1560 sub volumes (defined in Sect. 4.1). It takes around one day to complete the training iterations. However, the behavior of the findings is almost the same. As a consequence of space constraints, we give just one set of findings.

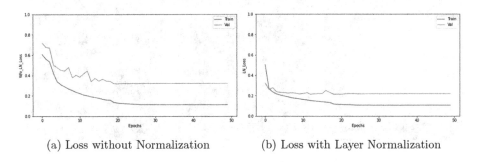

(a) Loss without Normalization   (b) Loss with Layer Normalization

Fig. 3. Loss without normalization and with Layer Normalization.

4.3.1 Generalization for Improved Performance

In Fig. 3, training and validation loss behavior keep on changing with their ups and downs in errors. However, both training and validation loss errors converge after some particular epochs. LN-based architecture (training loss 0.1115 and validation loss 0.3220) gives 32% less validation loss than without Normalization based architecture (training loss: 0.1054 and validation loss: 0.2182). This demonstrates that the LN-based model outperforms the competition and provide stability.

Additionally, we calculated performance in terms of the Dice Coefficient, as the majority of publications have done (i.e., [13, 19] etc.). Figure 4 illustrates that the validation dice coefficient of the LN-based design has increased by 17.5% when Normalization is used. Both of the training dice Coefficient after 50 epoch almost the same. Hence, it shows that the LN-based approach gives better prediction and generalizability than without the normalization-based approach.

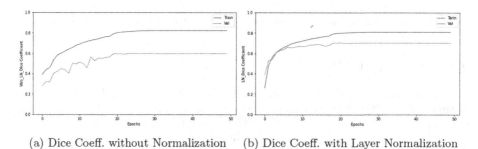

(a) Dice Coeff. without Normalization   (b) Dice Coeff. with Layer Normalization

Fig. 4. Dice Coefficient without normalization and with Layer Normalization

4.3.2 Segmentation for Abnormality

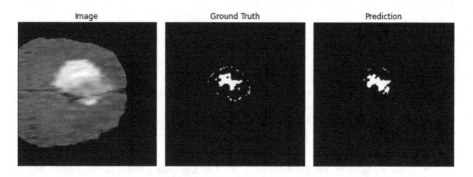

Fig. 5. Tumor prediction with layer Normalization deep learning architecture

This section demonstrates that how all LN-based empirical results are better than without normalization based model. In high-dimensional MRI data, the

LN-based 3D U-Net model is trained to correctly determine the shape, size, and location of the tumorous cell. We have also evaluated the segmentation of the subvolumes. The tumor region of ground truth and predicted data can be visualized in 2D images. Both architectures probabilistically predict cancer for each pixel. We have chosen 0.5 as a threshold for visualization to discretize the output.

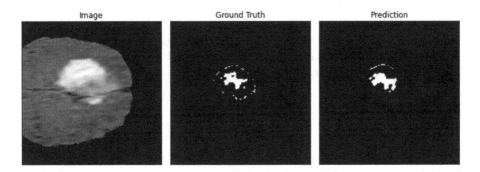

Fig. 6. Tumor prediction with out layer Normalization deep learning architecture

The segmentation results of unseen random data with corresponding ground truth are shown in Fig. 5 and Fig. 6 with LN-based and without normalization-based 3D U-Net architectures, respectively. For the evaluation of both model predictions, we used different evaluation measures [13], including sensitivity, specificity, FPR, and dice score.

Table 1. Performance measures with Layer Normalization 3D U-NET

	Edema	Non-enhancing-tumor	Enhancing tumor
Sensitivity	0.8785	0.7391	0.9231
Specificity	0.9973	0.9978	0.9989
FPR	0.0027	0.0022	0.0011
Dice_Score	0.9014	0.7419	0.9465

Table 1 and Table 2 show the empirical results of both architectures. Dice_score of the LN-based architecture gained 7.5% higher than without normalization-based architecture for Edema, 12.5% higher Dice_score for Non-Enhancing-Tumor, and 1.5% higher Dice_score for Enhancing Tumor. Specificity outperforms sensitivity in both architectures (Table 1 and Table 2).

Table 2. Performance measures without Normalization in 3D U-NET

	Edema	Non-enhancing-tumor	Enhancing tumor
Sensitivity	0.8355	0.7440	0.9117
Specificity	0.9940	0.9956	0.9982
FPR	0.0060	0.0044	0.0018
Dice_Score	0.8381	0.6591	0.9317

LN based model shows less FPR than, without normalization based model. For all three forms of cancer, the LN-based model is more generalized and exhibits significant outcomes on random input.

5   Conclusion

In summary, We assessed the LN technique on CNN based 3D U-Net architecture for brain MRI tumor segmentation. Additionally, we trained 3D U-Net architecture without using any normalizing approaches. During training phase with 50 epoch and batch size 1, LN based approach has given 32% reduction in validation loss compared to the without normalization based approach. Tumor segmentation has also done better in empirical studies using LN-based architecture. By and large, Layer Normalization minimizes covariate shift effects during training, resulting in faster learning and a more generic architecture.

References

1. Ba, J.L., Kiros, J.R., Hinton, G.E.: Layer normalization. arXiv preprint arXiv:1607.06450 (2016)
2. Çiçek, Ö., Abdulkadir, A., Lienkamp, S.S., Brox, T., Ronneberger, O.: 3D U-Net: learning dense volumetric segmentation from sparse annotation. In: Ourselin, S., Joskowicz, L., Sabuncu, M.R., Unal, G., Wells, W. (eds.) MICCAI 2016. LNCS, vol. 9901, pp. 424–432. Springer, Cham (2016). https://doi.org/10.1007/978-3-319-46723-8_49
3. Hmeed, A.R., Aliesawi, S.A., Jasim, W.M.: Enhancement of the U-net architecture for MRI brain tumor segmentation. In: Kumar, R., Mishra, B.K., Pattnaik, P.K. (eds.) Next Generation of Internet of Things. LNNS, vol. 201, pp. 353–367. Springer, Singapore (2021). https://doi.org/10.1007/978-981-16-0666-3_28
4. Ioffe, S., Szegedy, C.: Batch normalization: accelerating deep network training by reducing internal covariate shift. In: Proceedings of the International Conference on Machine Learning (ICML), pp. 448–456. PMLR (2015)
5. Jiang, X., Chang, L., Zhang, Y.D.: Classification of Alzheimer's disease via eight-layer convolutional neural network with batch normalization and dropout techniques. J. Med. Imaging Health Inf. **10**(5), 1040–1048 (2020)
6. Jiang, Z., Ding, C., Liu, M., Tao, D.: Two-stage cascaded U-Net: 1st place solution to BraTS challenge 2019 segmentation task. In: Crimi, A., Bakas, S. (eds.) BrainLes 2019. LNCS, vol. 11992, pp. 231–241. Springer, Cham (2020). https://doi.org/10.1007/978-3-030-46640-4_22

7. Kingma, D., Ba, J.: Adam: a method for stochastic optimization. arxiv (2014). arXiv preprint arXiv:1412.6980 (2017)
8. Kumar, R., Chen, W.C., Rockett, P.: Bayesian labelling of image corner features using a grey-level corner model with a bootstrapped modular neural network. In: Proceedings of the Fifth International Conference on Artificial Neural Networks (Conf. Publ. No. 440), pp. 82–87 (1997). https://doi.org/10.1049/cp:19970706
9. Kumar, R., Rockett, P.: Multiobjective genetic algorithm partitioning for hierarchical learning of high-dimensional pattern spaces: a learning-follows-decomposition strategy. IEEE Trans. Neural Netw. **9**(5), 822–830 (1998). https://doi.org/10.1109/72.712155
10. Lin, F., Wu, Q., Liu, J., Wang, D., Kong, X.: Path aggregation u-net model for brain tumor segmentation. Multim. Tools Appl. **80**(15), 22951–22964 (2021)
11. Myronenko, A.: 3D MRI brain tumor segmentation using autoencoder regularization. In: Crimi, A., Bakas, S., Kuijf, H., Keyvan, F., Reyes, M., van Walsum, T. (eds.) BrainLes 2018. LNCS, vol. 11384, pp. 311–320. Springer, Cham (2019). https://doi.org/10.1007/978-3-030-11726-9_28
12. Mzoughi, H., et al.: Deep multi-scale 3D convolutional neural network (CNN) for MRI gliomas brain tumor classification. J. Digital Imaging **33**, 903–915 (2020)
13. Pereira, S., Pinto, A., Alves, V., Silva, C.A.: Brain tumor segmentation using convolutional neural networks in MRI images. IEEE Trans. Med. Imaging **35**(5), 1240–1251 (2016)
14. Reinhold, J.C., Dewey, B.E., Carass, A., Prince, J.L.: Evaluating the impact of intensity normalization on MR image synthesis. In: . Proceedings of the International Society Optics and Photonics Medical Imaging: Image Processing. vol. 10949, p. 109493H (2019)
15. Shimodaira, H.: Improving predictive inference under covariate shift by weighting the log-likelihood function. J. Statist. Plan. Inference **90**(2), 227–244 (2000)
16. Shrestha, S., Alsadoon, A., Prasad, P., Seher, I., Alsadoon, O.H.: A novel solution of using deep learning for prostate cancer segmentation: enhanced batch normalization. Multim. Tools Appl. **80**(14), 21293–21313 (2021)
17. Simpson, A.L., et al.: A large annotated medical image dataset for the development and evaluation of segmentation algorithms. arXiv preprint arXiv:1902.09063 (2019)
18. Sohail, N., Anwar, S.M., Majeed, F., Sanin, C., Szczerbicki, E.: Smart approach for glioma segmentation in magnetic resonance imaging using modified convolutional network architecture (u-net). Cybern. Syst. **52**(5), 445–460 (2021)
19. Tseng, K.L., Lin, Y.L., Hsu, W., Huang, C.Y.: Joint sequence learning and cross-modality convolution for 3D biomedical segmentation. In: Proceedings of the IEEE Conference Computer Vision and Pattern Recognition (CVPR), pp. 6393–6400 (2017)
20. Ulyanov, D., Vedaldi, A., Lempitsky, V.: Instance normalization: the missing ingredient for fast stylization. arXiv preprint arXiv:1607.08022 (2016)
21. Wu, Y., He, K.: Group normalization. In: Proceedings of the European Conference Computer Vision (ECCV), pp. 3–19 (2018)
22. Yao, Z., Cao, Y., Zheng, S., Huang, G., Lin, S.: Cross-iteration batch normalization. In: Proceedings of the IEEE/CVF Conference Computer Vision and Pattern Recognition (CVPR), pp. 12331–12340 (2021)

Data Set Creation and Empirical Analysis for Detecting Signs of Depression from Social Media Postings

Kayalvizhi Sampath[✉] and Thenmozhi Durairaj

Sri Sivasubramaniya Nadar College of Engineering, Chennai, India
{kayalvizhis,theni_d}@ssn.edu.in

Abstract. Depression is a common mental illness that has to be detected and treated at an early stage to avoid serious consequences. There are many methods and modalities for detecting depression that involves physical examination of the individual. However, diagnosing mental health using their social media data is more effective as it avoids such physical examinations. Also, people express their emotions well in social media, it is desirable to diagnose their mental health using social media data. Though there are many existing systems that detects mental illness of a person by analysing their social media data, detecting the level of depression is also important for further treatment. Thus, in this research, we developed a gold standard data set that detects the levels of depression as 'not depressed', 'moderately depressed' and 'severely depressed' from the social media postings. Traditional learning algorithms were employed on this data set and an empirical analysis was presented in this paper. Data augmentation technique was applied to overcome the data imbalance. Among the several variations that are implemented, the model with Word2Vec vectorizer and Random Forest classifier on augmented data outperforms the other variations with a score of 0.877 for both accuracy and F1 measure.

Keywords: Depression · Data set · Data augmentation · Levels of depression · Random Forest

1 Introduction

Depression is a common and serious mental illness that negatively affects the way one feels, thinks and acts [1]. The rate of depression is rapidly increasing day by day. According to Global Health Data Exchange (GHDx), depression has affected 280 million people worldwide [3]. Detecting depression is important since it has to be observed and treated at an early stage to avoid severe consequences[1]. The depression was generally diagnosed by different methods modalities clinical interviews [5,12], analysing the behaviour [6], monitoring facial and

---

[1] https://www.healthline.com/health/depression/effects-on-body.

© IFIP International Federation for Information Processing 2022
Published by Springer Nature Switzerland AG 2022
L. Kalinathan et al. (Eds.): ICCIDS 2022, IFIP AICT 654, pp. 136–151, 2022.
https://doi.org/10.1007/978-3-031-16364-7_11

speech modulations [19], physical exams with Depression scales [14,28], videos and audios [18], etc. All these methods of diagnosing involves more involvement of an individual or discussion about their feeling in person.

On the other hand, social media is highly emerging into our lives with a considerable rate of increase in social media users according to the statistics of statista [4]. Slowly, the social media became a comfortable virtual platform to express our feelings. And so, social media platform can be considered as a source to analyse people's thoughts and so can also be used for analysing mental health of an individual. Thus, we aim to use social media texts for analysing the mental health of a person.

The existing works collect social media texts from open source platforms like Reddit [32], Facebook [13], Twitter [11,16,27,30], Live journals [20], blog posts [31], Instagram [26] etc. and used them to detect depression.

Research Gaps

All these research works concentrate on diagnosing depression from the social media texts. Although detecting depression has its own significance, detecting the level of depression also has its equal importance for further treatment. Generally, depression is classified into three stages namely mild, moderate and severe [2]. Each stage has its own symptoms and effects and so detecting the level of depression is also a crucial one. Thus, we propose a data set to detect the level of depression in addition to detection of depression from the social media texts. This paper explains the process of data set creation that detects the levels of depression along with some baseline models.

Our Contributions in this Research Include

1. Creating a new bench mark data set for detecting the depression levels from social media data at postings level.
2. Developing base line models with traditional learning classifiers.
3. Analysing the impact of data augmentation

2   Related Work

The main goal of our research study is to create a data set that identifies the sign of depression and detect the level of depression and thus, the existing works are analysed in terms of data collection, modalities and methodologies of detecting depression.

2.1   Modalities and Methodologies of Depression Detection

For detecting depression, the data was collected by various methods like clinical interviews [5,12], analysing the behaviour [6], monitoring facial and speech modulations [19], physical exams with Depression scales [14,28], videos and audios

[18], etc. Since, the social media users are rapidly increasing day by day, social media can also be considered as a main platform of source for detecting the mental health. This key idea gave rise to the most utilized data set E-Risk@CLEF-2017 pilot task data set [17] that was collected from Reddit. In addition to this data set, many other data sets such as DAIC corpus [5], AVEC [18], etc. also evolved that detects depression from the social media data. Though few benchmark data set exists to detect depression, more researchers tend to collect data from applications of social network and create their own data sets.

2.2   Data Collection from Applications of Social Network

The social media texts were collected from open source platforms like Reddit [29,32], Facebook [13], Twitter [11,16,27,30], Live journals [20], blog posts [31], Instagram [26] etc. The data from twitter was collected using API's and annotated into depressed and not depressed classes based on key words like "depressed, hopeless and suicide" [11], using a questionnaire [30], survey [27], etc. The data was also scrapped from groups of live journals [20], blog posts [31] and manually annotated into depressed and not depressed.

Among these social media platforms, Reddit possess large amount text discussion than the other platforms and so Reddit has become widely used platform to collect social media text data recently.

The data were collected from these platforms using Application Programming Interface (API) using hashtags, groups, communities, etc. The data from reddit was collected from Subreddits and annotated manually by two annotators into depressed and not depressed class [32]. The data was also from subreddits like "r/anxiety, r/depression and r/depression_help" and annotated into a data set [24]. A data set was created with classes depression, suicide_watch, opiates and controlled which was collected using subreddits such as "r/suicidewatch, r/depression", opioid related forums and other general forums [33]. A survey was also done on the studies of anxiety and depression from the Reddit data [8].

From the Table 1, it is clear that all these research works have collected the social media data only to detect the presence of depression. Although, diagnosing depression is important, detecting the level of depression is more crucial for further treatment. And thus, we propose a data set that detects the level of depression.

3   Proposed Work

We propose to develop a gold standard data set that detects the levels of depression as not depressed, moderately depressed and severely depressed. Initially, the data set was created by collecting the data from the social media platform, Reddit. For collecting the data from archives of Reddit, two way communication is needed, which requires app authentication. After getting proper authentication, the subreddits from which the data must be collected are chosen and the data was extracted. After extracting the data, it is pre-processed and exported

Table 1. Existing custom data sets comparison

Existing system	Social media platform	Class labels
Eichstaedt et al. [13]	Facebook	Depressed and not depressed
Nguyen et al. [20]	Live journal	Depressed and control
Tyshchenko et al. [31]	Blog post	Clinical and Control
Deshpande et al. [11]	Twitter	Neutral and negative
Lin et al. [16]	Twitter	Depressed and not depressed
Reece et al. [27]	Twitter	PTSD and Depression
Tsugawa et al. [30]	Twitter	Depressed and not depressed
Losada et al. [17]	Reddit	Depression and Not depression
Wolohan et al. [32]	Reddit	Depressed and not depressed
Tadesse et al. [29]	Reddit	Depression indicative and standard
Pirina et al. [24]	Reddit	Positive and negative
Yao et al. [33]	Reddit	depression, suicide watch, control and opiates
Proposed data set	**Reddit**	**Not depressed, moderately depressed & severely depressed**

in the required format which forms the data set. The data were then annotated into levels of depression by domain experts following the annotation guidelines. After annotation, the inter-rater agreement is calculated to analyze the quality of data and annotation. Then, the corpus is formed using the mutually annotated instances. Baseline models were also employed on the corpus to analyze the performance. To overcome the data imbalance problem, data augmentation technique was applied and their impact on performance was also analyzed.

3.1 Data Set Creation

For creating the data set, a suitable social media platform is chosen initially and data is scraped using suitable methods. After scraping the data, the data is processed and stored in a suitable format.

Data Collection: For creating the data set, Reddit[2], an open source social media platform is chosen since it has more textual data when compared to other social media platforms. This data will be of postings format which includes only one or more statements of an individual. The postings data are scraped from the Reddit archives using the API "pushshift".

App Authentication: For scraping the data from Reddit achieves, Python Reddit API Wrapper(PRAW) is used. The data can be only scraped after getting authentication from the Reddit platform. This authentication process involves creation of an application in their domain, for which a unique client secret key and client id will be assigned. Thus, PRAW allows a two way communication only with these credentials of user_agent (application name), client_id and client_secret to get data from Reddit.

---
[2] https://www.reddit.com.

Subreddit Selection Reddit is a collection of million groups or forums called subreddits. For collecting the confessions or discussion of people about their mental health, data was scraped from the archives of subreddits groups like "r/Mental Health, r/depression, r/loneliness, r/stress, r/anxiety".

Data Extraction: For each posting, the details such as post ID, title, URL, publish date, name of the subreddit, score of the post and total number of comments can be collected using PRAW. Among these data, PostID, title, text, URL, date and subreddit name are all collected in dictionary format.

Data Pre-processing and Exporting: After collecting these data, the text and title part are pre-processed by removing the non-ASCII characters and emoticons to get a clean data set. The processed data is dumped into a Comma Separated Values (.csv) format file with the five columns. The sample of the collected postings is shown in Table 2.

Table 2. Sample Reddit postings

Post ID	Title	Text	Url	Publish date	Subreddit
g69pqt	Don't want to get of bed	I'm done with me crying all day and thinking to myself that I can't do a thing and I don't what to get out of bed at all	https://www.reddit.com/r/depression/comments/g69pqt/dont_want_to_get_of_bed/	2020-04-23 02:51:32	Depression
gb9zei	Today is a day where I feel emptier than on other days	It's like I am alone with all my problems. I am sad about the fact I can't trust anyone and nobody could help me because I feel like nobody understand how I feel. Depression is holding me tight today...	https://www.reddit.com/r/depression/comments/gb9zei/today_is_a_day_where_i_feel_emptier_than_on_other/	2020-05-01 08:10:06	Depression

3.2   Data Annotation

After collecting the data, the data were annotated according to the signs of depression. Although all the postings were collected from subreddits that exhibit the characteristics of mental illness, there is a possibility of postings that do not confess or discuss depression. Thus, the collected postings data were annotated by two domain experts into three labels that denote the level of signs of depression namely "Not depressed, Moderate and Severe". Framing the annotation guidelines for postings data is difficult since the mental health of an individual has to be analyzed using his/her single postings. For annotating the data into three classes, the guidelines were formatted as follows:

Label 1 - Not Depressed: The postings data will be annotated as "Not Depressed", if the postings data reflect one of the following mannerism:

- If the statements have only one or two lines about irrelevant topics.
- If the statements reflect momentary feelings of present situation.
- If the statements are about asking questions about any or medication
- If the statement is about ask/seek help for friend's difficulties.

Example:

I struggled to count to 20 today : For some context I work in mcdonalds and I was on the line finishing the stuff and almost every 20 box I think I sent was wrong. I just couldn't focus and do it quick and it was awful, and I ended up doing almost 2 hours of overtime because I can't say no to people. I fucking hate it, the worst part is I was an A grade maths student less than 5 years ago. Now I can't count to 20, I can't do basic maths without a calculator, I can barely focus at times. I feel like I'm just regressing as a person in every way but still aging, I just wanna end it all before it gets worse and I become a fucking amoeba of a person.

Label 2 - Moderately Depressed: The postings data will be annotated as "moderately depressed", if the postings falls under these conditions:

- If the statements reflect change in feelings (feeling low for some time and feeling better for some time).
- If the statement shows that they aren't feeling completely immersed in any situations
- If the statements show that they have hope for life.

Example :

If I disappeared today, would it really matter?
I'm just too tired to go on, but at the same time I'm too tired to end it. I always thought about this but with the quarantine I just realised it is true. My friends never felt close to me, just like the only two relationships I have ever been in. They never cared about me, to the point where I even asked for help and they just turned a blind eye. And my family isn't any better. I don't know what to do, and I believe it won't matter if I do something or not. I'm sorry if my English isn't good, it isn't my first language.

Label - 3: Severely Depressed: The data will be annotated as "Severely depressed", if the postings have one of the following scenarios:

- If the statements express more than one disorder conditions.
- If the statements explain about history of suicide attempts.

> **Example:**
>
> Getting depressed again?
> So I'm 22F and I have taken antidepressants the last time 4 years ago. I've had ups and downs when I got off and with 19 I was having a rough time for two months - started drinking and smoking weed a lot. Kinda managed to get back on track then and haven't been feeling too bad until now. Lately I've been feeling kinda blue and started making mistakes or have to go through stuff multiple times to do it correctly or to be able to remember it. Currently I'm having a week off and have to go back to work on Monday. I just don't know I feel like I'm getting worse and want to sleep most of the time and at first I thought it's because I'm used to working a lot, but when I think about having to go back soon I feel like throwing up and at the same time doing nothing also doesn't sit well with me.I guess I'm kinda scared at the moment because I don't want to feel like I was feeling years ago and I still don't feel comfortable with my own mind and don't trust myself that I'm strong enough to pull through if depression hits me again.

3.3   Inter-rater Agreement

After annotating the data, inter-rater agreement was calculated between the decisions of two judges using kappa coefficient estimated using a per-annotator empirical prior over the class labels [7]. Inter-rater agreement[3] is the degree of agreement among independent observers who rate, code, or assess the same phenomenon. The inter rater agreement is measured using Cohen's kappa statistics [10].

Table 3. Landis and Koch measurement table of inter rater agreement

Kappa value (κ)	Strength of agreement
<0	Poor
0.01–0.20	Slight
0.21–0.40	Fair
0.41–0.60	Moderate
0.61–0.80	Substantial
0.81–0.99	Almost perfect agreement

The inter-rater agreement between the annotations was calculated using sklearn [22]. For our annotation, the kappa value (κ) is 0.686. According to Landis & Koch [15] in the Table 3, the κ value denotes substantial agreement between the annotators, which proves the consistency of labeling according to the annotation guidelines. Thus, the mutually annotated instances form the corpus.

3.4   Corpus Analysis

Initially 20,088 instances of postings data were annotated, out of which 16,613 instances were found to be mutually annotated instances by the two judges,

---

[3] https://en.wikipedia.org/wiki/Inter\discretionary-rater_reliability.

and thus they were considered as instances of data set with their corresponding labels. The complete statistics of the corpus is tabulated in Table 4.

Table 4. Postings data analysis

Category	Count
Count of instances annotated	20,088
Data set instances (*number of instances mutually annotated*)	16,632
Count of sentences	1,56,676
Count of words	26,59,938
Count of stop-words	12,47,016
Count of words other than stop-words	14,12,922
Count of unique words	28,415
Count of unique stop-words	150
Count of unique words other than stop-words	28,265
Range of sentences per instance	1–260
Range of words per instance	1–5065
Average number of sentences per posting instance	9.42
Average number of words per posting instance	159.92

The whole corpus has 1,56,676 sentences with 26,59,938 words which shows the size of the corpus created. In the corpus, each posting with its labels is considered as each instance in the corpus. An instance in the corpus will have an average of 9.42 sentences each that varies in the range of 1 to 260 sentences with an average of 159.92 words that lies between 1 to 5065 words.

The distribution of three class labels in the data set is represented as Fig. 1. As shown in figure, the data set is unbalanced with 10,494 instances of "moderately depressed" class, 1489 instances of "severely depressed" class and 4649 instances of "Not depressed" class which includes some duplicate entries.

3.5   Base Line Models

The data set has been evaluated using traditional models which are considered as baseline models. The data set has four columns namely id, title, text and class label. For implementation, the title data and text data are initially combined. The combined text data is pre-processed, extracted features, balanced, classified using traditional classifiers and evaluated by cross validation.

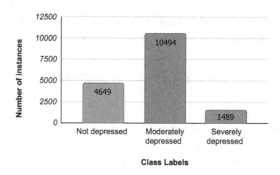

Fig. 1. Class wise distribution of dataset

Data Pre-processing: The title and text column are combined together as a single text data column by filling the "NA" instances of both title and text data. The combined text data is cleaned by converting the words to lower case letters and removing unwanted punctuation, "[removed]" tags, web links, HTML links, stop words and small words. After cleaning, the instances are tokenized using regexptokenizer [21], stemmed using porter stemmer [25] and lemmatized using wordnet lemmatizer.

Feature Extraction: The features were extracted using three vectorizers namely Word2Vec, Term Frequency - Inverse Document Frequency (TF-IDF) and Glove [23].

- **Word2Vec:** It produces a vector that exhibits the context of the word considering the occurrence of the word. The vectors are generated using Continuous Bag Of Words.
- **TF-IDF:** It produces a score considering the occurrence of the word in the document. The calculation is based on the importance of a topic in a particular document. The vectors are calculated using four grams considering a maximum of 2000 features.
- **Glove:** It produces the word embeddings considering the occurrence and co-occurrence of the words with reduced dimensionality. The words are mapped to a word embedding using 6 Billion pre-trained tokens with 100 features each.

Classifiers: Twelve different classifiers that include Ada Boost Classifier, Decision Tree, K-Nearest Neighbour, Linear Deterministic Analysis, Logistic Regression, Multi-layer Perceptron, Gaussian Naive Bayes, Quadratic Deterministic Analysis, Support Vector Machine (Radial Basis Function, linear) and Random Forest of Scikit-learn [21] were used for classification.

- **Ada Boost Classifier(ABC):** The Adaptive Boosting algorithm is a collection of N estimator models that assigns higher weights to the mis-classified samples in the next model. In our implementation, 100 estimator models with t0 random state at a learning rate of 0.1 were used to fine tune the model.

- **Decision Tree (DT):** The decision tree classifier predicts the target value based on the decision rules that was formed using features to identify the target variable. The decision rules are formed using gini index and entropy for information gain. For implementing the decision trees, the decision tree classifier was fine tuned with two splits of minimum samples of one leaf node each by calculating gini to choose the best split and random state as 0.
- **Gaussian Naive Bayes (GNB):** The Gaussian normal distribution variant of Naive Bayes classifier that relies on the Bayes theorem is Gaussian Naive Bayes.
- **K-Nearest Neighbour(KNN):** It classifies the data points by plotting them and finding the similarity between the data points. In implementation, number of neighbours were set as three with equal weights and euclidean distance as metric to calculate distance.
- **Logistic Regression (LR):** The probabilistic model that predicts the class label based on the sigmoid function for binary classification. As our data set are multi-class data sets, multi-nominal logistic regression was used to evaluate the data sets. For implementation, the classifier was trained with a tolerance of $1e{-}4$, 1.0 as inverse of regularization strength and intercept scaling as 1.
- **Multi-layer Perceptron (MLP):** The neural network trained to predict the class label along with back propagation of error. The multi-layer perceptron of two layers of 100 hidden nodes each was trained at learning rate of 0.001 with relu activation function for a maximum 300 iterations.
- **Discriminant Analysis:** The generative model that utilizes Gaussian distribution for classification by assuming each class has a different co-variance. For implementation, the co-variance is calculated with threshold of $1.0e{-}04$. Linear DA (LDA) and Quadratic DA (QDA) both were implemented.
- **Support Vector Machine:** The data is projected into higher dimensions and then classifies the data using hyper-planes in this model. The model was trained with RBF kernel (RBF-SVM) and linear kernel (L-SVM) function of three degree, 0.1 regularization parameter without any specifying any maximum iterations.
- **Random Forest (RF):** Random Forest combines many decision trees as in ensemble method to generate predictions. It overcomes the limitation of decision trees by bagging and bootstrap aggregation. It was implemented with 100 number of estimators.

4   Implementation and Results

The features extracted in Subsect. 3.5 are classified using the above classifiers in Subsect. 3.5 and evaluated using stratified k-fold sampling of Scikit-learn [21]. In this validation, data are split into 10 folds and the evaluation results with respect to weighted average F1-score is tabulated in Table 5. Table 5, shows that

Table 5. Performance of baseline models

F1 - score	TF- IDF	Glove	Word2Vec
ABC	0.263	0.496	0.451
DT	0.273	0.614	0.469
GNB	0.271	0.415	0.302
KNN	0.258	0.604	0.594
L-SVM	0.273	0.309	0.273
LDA	0.270	0.395	0.391
LR	0.270	0.329	0.395
MLP	0.269	**0.647**	0.625
QDA	0.276	0.459	0.368
RBF-SVM	0.273	0.560	0.452
RF	0.272	**0.647**	0.456

(a) F1 score

Accuracy	TF- IDF	Glove	Word2Vec
ABC	0.384	0.654	0.616
DT	0.388	0.697	0.579
GNB	0.351	0.464	0.351
KNN	0.379	0.717	0.694
L-SVM	0.388	0.646	0.623
LDA	0.388	0.659	0.619
LR	0.387	0.650	0.619
MLP	0.386	0.754	0.700
QDA	0.393	0.499	0.485
RBF-SVM	0.388	0.733	0.667
RF	0.388	**0.760**	0.695

(b) Accuracy

the model with Random Forest Classifier and Multi-Layer Perceptron (MLP) applied on features extracted using Glove performs equally well with an F1-score of 0.647. The performance of the models with accuracy as metric is depicted in Table 5. It is clear from the table that the model with Random Forest classifier and Glove vectorizer performs better with an accuracy of 0.760.

4.1 With Data Augmentation

The postings data is populated with more "moderately depressed" instances and thus, the data has to be balanced before classification for better performance. For balancing the data, Synthetic Minority Oversampling Technique (SMOTE) [9] was applied after vectorization. The effect of data augmentation can be observed in Fig. 2.

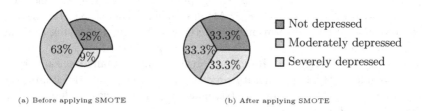

(a) Before applying SMOTE   (b) After applying SMOTE

Fig. 2. Effect of data augmentation in the data

The features extracted in Subsect. 3.5 are augmented using SMOTE and then classified using the classifiers in Subsect. 3.5. The performance of these models in terms of F1-score and accuracy after data augmentation are shown in Table 6

Table 6. Performance of baseline models after data augmentation

F1 - score	TF- IDF	Glove	Word2Vec
ABC	0.451	0.622	0.559
DT	0.469	0.772	0.721
GNB	0.290	0.449	0.389
KNN	0.549	0.814	0.834
L-SVM	0.273	0.570	0.642
LDA	0.391	0.550	0.540
LR	0.395	0.544	0.551
MLP	0.625	0.775	0.852
QDA	0.368	0.592	0.477
RBF-SVM	0.452	0.762	0.788
RF	0.449	0.854	**0.877**

(a) F1-score

Accuracy	TF- IDF	Glove	Word2Vec
ABC	0.616	0.628	0.562
DT	0.579	0.781	0.728
GNB	0.351	0.479	0.427
KNN	0.695	0.839	0.854
L-SVM	0.623	0.575	0.642
LDA	0.619	0.550	0.550
LR	0.619	0.547	0.559
MLP	0.700	0.780	0.857
QDA	0.485	0.615	0.497
RBF-SVM	0.667	0.769	0.792
RF	0.689	0.864	**0.877**

(b) Accuracy

and 6 respectively. From the tables, it is clear that the performance was improved and model with Random Forest classifier applied on the features extracted using Word2Vec performs well with a score of 0.877.

The significance of improvement in the performance by incorporating data augmentation in terms of F1 score was measured using Benefit Cost Ratio (BCR) value. In general, BCR value is computed by dividing the proposed total benefit cost by the proposed total cost. If the calculated value is greater than one, then the proposed cost is proven as significant one. In terms of performance, the metric is calculated by dividing the performance score of proposed model by the performance of existing model. Since, F1 score is considered to be the suitable performance metric, BCR value is calculated by dividing F1 score of model with data augmentation by that of model without data augmentation as shown below.

$$\text{BCR metric}_{(f1)} = \frac{\text{F1 score of proposed model (with data augmentation)}}{\text{F1 score of model without data augmentation}}$$

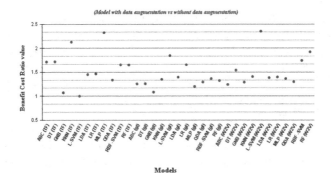

Fig. 3. BCR values of F1 scores

The BCR values calculated with F1 score are plotted in a graph and is shown in Fig. 3. In the figure, vertical axis represents the scores, horizontal axis represents the different baseline models built and the data points represent the distribution of BCR values. From Fig. 3, it is clear that the improvement in performance data augmentation is significant with respect to F1 scores since all the BCR values are greater than one.

5   Research Insights

The researchers can further extend this work by implementing the following methods:

– Extend the data set by considering the images along with text data.
– Implement deep learning models in the data set.
– Implement other methods of data augmentation to improve performance.

6   Conclusions

Depression is a common mental illness that has to be detected and treated early to avoid serious consequences. Among the other ways of detecting, diagnosing mental health using their social media data seems much more effective since it involves less involvement of the individual. All the existing systems are designed to detect depression from social media texts. Although detecting depression is more important, detecting the level of depression also has its equal significance. Thus, we propose a data set that not only detects depression from social media but also analyzes the level of depression. For creating the data set, the data was collected from subreddits and annotated by domain experts into three levels of depression, namely not depressed, moderately depressed and severely depressed.

An empirical analysis of traditional learning algorithms was also done for evaluating the data sets. Among the models, the model with Glove vectorizer and Random Forest classifier performs well with a F1 of 0.647 and accuracy of 0.760.

While analyzing the data set, "the moderately depressed" class seems to be highly populated than the classes and so, a data augmentation method named SMOTE was applied, and the performance is analyzed. Data augmentation improved the performance by 23% and 12% in terms of F1-score and accuracy respectively, with both F1-score and accuracy of 0.877. The significance of improvement in performance by incorporating data augmentation was also proved using BCR values.

The data set can also be extended by considering the images along with texts for more accurate detection. The work can be extended further by implementing other traditional learning and deep learning models. Other augmentation techniques can also be experimented with to improve the model's performance.

Data set availability

The data is available to public in a repository of a Github in the link: https://github.com/Kayal-Sampath/detecting-signs-of-depression-from-social-media-postings.

Acknowledgements. We would like to thank the Department of Science and Technology - Science and Engineering Research Board (DST-SERB) for providing funds to annotate the collected data.

References

1. American Psychiatric Association. https://www.psychiatry.org/patients-families/depression/what-is-depression. Accessed 17 Nov 2021
2. Healthline. https://www.healthline.com/health/depression/mild-depression. Accessed 17 Nov 2021
3. Institute of Health Metrics and Evaluation. Global Health Data Exchange (GHDx). http://ghdx.healthdata.org/gbd-results-tool?params=gbd-api-2019-permalink/d780dffbe8a381b25e1416884959e88b. Accessed 17 Nov 2021
4. Statista statistics. https://www.statista.com/statistics/278414/number-of-worldwide-social-network-users/. Accessed 17 Nov 2021
5. Al Hanai, T., Ghassemi, M.M., Glass, J.R.: Detecting depression with audio/text sequence modeling of interviews. In: Interspeech, pp. 1716–1720 (2018)
6. Alghowinem, S., et al.: Multimodal depression detection: fusion analysis of paralinguistic, head pose and eye gaze behaviors. IEEE Trans. Affect. Comput. **9**(4), 478–490 (2016)
7. Artstein, R., Poesio, M.: Inter-coder agreement for computational linguistics. Comput. Linguist. **34**(4), 555–596 (2008)
8. Boettcher, N., et al.: Studies of depression and anxiety using reddit as a data source: scoping review. JMIR Ment. Health **8**(11), e29487 (2021)
9. Chawla, N.V., Bowyer, K.W., Hall, L.O., Kegelmeyer, W.P.: SMOTE: synthetic minority over-sampling technique. J. Artif. Intell. Res. **16**, 321–357 (2002)
10. Cohen, J.: A coefficient of agreement for nominal scales. Educ. Psychol. Measur. **20**(1), 37–46 (1960)
11. Deshpande, M., Rao, V.: Depression detection using emotion artificial intelligence. In: 2017 International Conference on Intelligent Sustainable Systems (ICISS), pp. 858–862. IEEE (2017)
12. Dibeklioğlu, H., Hammal, Z., Yang, Y., Cohn, J.F.: Multimodal detection of depression in clinical interviews. In: Proceedings of the 2015 ACM on International Conference on Multimodal Interaction, pp. 307–310 (2015)
13. Eichstaedt, J.C., et al.: Facebook language predicts depression in medical records. Proc. Natl. Acad. Sci. **115**(44), 11203–11208 (2018)
14. Havigerová, J.M., Haviger, J., Kučera, D., Hoffmannová, P.: Text-based detection of the risk of depression. Front. Psychol. **10**, 513 (2019)

15. Landis, J.R., Koch, G.G.: The measurement of observer agreement for categorical data. Biometrics **33**, 159–174 (1977)
16. Lin, C., et al.: SenseMood: depression detection on social media. In: Proceedings of the 2020 International Conference on Multimedia Retrieval, pp. 407–411 (2020)
17. Losada, D.E., Crestani, F., Parapar, J.: eRISK 2017: CLEF lab on early risk prediction on the internet: experimental foundations. In: Jones, G.J.F., et al. (eds.) CLEF 2017. LNCS, vol. 10456, pp. 346–360. Springer, Cham (2017). https://doi.org/10.1007/978-3-319-65813-1_30
18. Morales, M.R., Levitan, R.: Speech vs. text: a comparative analysis of features for depression detection systems. In: 2016 IEEE Spoken Language Technology Workshop (SLT), pp. 136–143. IEEE (2016)
19. Nasir, M., Jati, A., Shivakumar, P.G., Nallan Chakravarthula, S., Georgiou, P.: Multimodal and multiresolution depression detection from speech and facial landmark features. In: Proceedings of the 6th International Workshop on Audio/Visual Emotion Challenge, pp. 43–50 (2016)
20. Nguyen, T., Phung, D., Dao, B., Venkatesh, S., Berk, M.: Affective and content analysis of online depression communities. IEEE Trans. Affect. Comput. **5**(3), 217–226 (2014)
21. Pedregosa, F., et al.: Scikit-learn: machine learning in Python. J. Mach. Learn. Res. **12**, 2825–2830 (2011)
22. Pedregosa, F., Varoquaux, G., Gramfort, A., et al.: Scikit-learn: machine learning in Python. J. Mach. Learn. Res. **12**, 2825–2830 (2011)
23. Pennington, J., Socher, R., Manning, C.D.: GloVe: global vectors for word representation. In: Proceedings of the 2014 Conference on Empirical Methods in Natural Language Processing (EMNLP), pp. 1532–1543 (2014)
24. Pirina, I., Çöltekin, Ç.: Identifying depression on Reddit: the effect of training data. In: Proceedings of the 2018 EMNLP Workshop SMM4H: The 3rd Social Media Mining for Health Applications Workshop & Shared Task, Brussels, Belgium, pp. 9–12. Association for Computational Linguistics, October 2018. https://doi.org/10.18653/v1/W18-5903, https://aclanthology.org/W18-5903
25. Porter, M.F.: An algorithm for suffix stripping. Program **14**(3), 130–137 (1980)
26. Reece, A.G., Danforth, C.M.: Instagram photos reveal predictive markers of depression. EPJ Data Sci. **6**, 1–12 (2017)
27. Reece, A.G., Reagan, A.J., Lix, K.L., Dodds, P.S., Danforth, C.M., Langer, E.J.: Forecasting the onset and course of mental illness with Twitter data. Sci. Rep. **7**(1), 1–11 (2017)
28. Stankevich, M., Latyshev, A., Kuminskaya, E., Smirnov, I., Grigoriev, O.: Depression detection from social media texts. In: Data Analytics and Management in Data Intensive Domains: XXI International Conference DAMDID/RDCL 2019, p. 352 (2019)
29. Tadesse, M.M., Lin, H., Xu, B., Yang, L.: Detection of depression-related posts in Reddit social media forum. IEEE Access **7**, 44883–44893 (2019). https://doi.org/10.1109/ACCESS.2019.2909180
30. Tsugawa, S., Kikuchi, Y., Kishino, F., Nakajima, K., Itoh, Y., Ohsaki, H.: Recognizing depression from Twitter activity. In: Proceedings of the 33rd Annual ACM Conference on Human Factors in Computing Systems, pp. 3187–3196 (2015)
31. Tyshchenko, Y.: Depression and anxiety detection from blog posts data. Nature Precis. Sci., Institute of Computer Science, University of Tartu, Tartu, Estonia (2018)

32. Wolohan, J., Hiraga, M., Mukherjee, A., Sayyed, Z.A., Millard, M.: Detecting linguistic traces of depression in topic-restricted text: attending to self-stigmatized depression with NLP. In: Proceedings of the 1st International Workshop on Language Cognition and Computational Models, pp. 11–21 (2018)
33. Yao, H., Rashidian, S., Dong, X., Duanmu, H., Rosenthal, R.N., Wang, F.: Detection of suicidality among opioid users on Reddit: machine learning-based approach. J. Med. Internet Res. **22**(11), e15293 (2020)

Classification and Prediction of Lung Cancer with Histopathological Images Using VGG-19 Architecture

N. Saranya$^{(\boxtimes)}$, N. Kanthimathi$^{(\boxtimes)}$, S. Boomika$^{(\boxtimes)}$, S. Bavatharani$^{(\boxtimes)}$, and R. Karthick raja$^{(\boxtimes)}$

ECE, Bannari Amman Institute of Technology, Sathyamangalam, India
{saranyan,kanthimathi,boomika.ec18,bavatharani.ec18,
karthickrajar.ec18}@bitsathy.ac.in

Abstract. In recent times, for the diagnosis of several diseases, many Computer-Aided Diagnosis (CAD) systems have been designed. Among many life-threatening diseases, lung cancer is one of the leading causes of cancer-related deaths in humans worldwide. It is a malignant lung tumor distinguished by the uncontrolled growth of cells in the tissues of the lungs. Diagnosis of cancer is a challenging task due to the structure of cancer cells. Predicting lung cancer at its initial stage plays a vital role in the diagnosis process and also increases the survival rate of patients. People with lung cancer have an average survival rate ranging from 14 to 49% if they are diagnosed in time. The current study focuses on lung cancer classification and prediction based on Histopathological images by using effective deep learning techniques to attain better accuracy. For the classification of lung cancer as Benign, Adenocarcinoma, or Squamous Cell Carcinoma, some pre trained deep neural networks such as VGG-19 were used. A database of 15000 histopathological images was used in which 5000 benign tissue images and 10000 malignant lung cancer-related images to train and test the classifier. The experimental results show that the VGG-19 architecture can achieve an accuracy of 95%.

Keywords: Lung cancer · VGG-19 · Histopathological images · Convolutional neural network

1 Introduction

Abnormal growth of cells in the body causes cancer. Lung cancer is a dreadful disease amongst the most widely recognized malignant tumors, also known as lung carcinoma. It affects the estimation of 2.3 million people every year all over the world. About 85% of patients affected by lung cancer are by smoking and tobacco consumption. 10–15% of cancer arises in a person who never smoke is because of their exposure to secondhand smoke, air pollution, or any other chemical exhaust. Two different types of Lung cancer are present, Small Cell Lung Cancer (SCLC) and Non-Small Cell Lung Cancer (NSCLC).

© IFIP International Federation for Information Processing 2022
Published by Springer Nature Switzerland AG 2022
L. Kalinathan et al. (Eds.): ICCIDS 2022, IFIP AICT 654, pp. 152–161, 2022.
https://doi.org/10.1007/978-3-031-16364-7_12

NSCLC grows out of control when there is a change in the healthy cells of the lung, which in turn form a mass called a tumor, or a nodule. These cells can grow anywhere in the lung and the tumor can be cancerous or benign. NSCLC is of three different types. Adenocarcinoma- It is more often found in former smokers or non-smokers. This type of cancer usually grows in the outer edges of the lungs, Squamous Cell carcinoma- This cancer type will develop more in smokers or former smokers. It will start growing in the middle part of the lungs near the bronchi, Large Cell carcinoma- It is very hard to treat. This tumor grows rapidly and effectively spreads to different organs. The tumor size will determine the stages of cancer in the nodes of the lung. The biopsy report is used to confirm the disease. By utilizing traditional techniques that are widely used by physicians, radiologists around the world, lung cancer can generally be detected at the mature stage and after it has spread to a great extent. A patient's chances of survival when lung cancer is discovered at that stage are very low. In the previously mentioned issue, the problem of misdiagnosis is also a cause for concern. Sometimes it is possible for doctors to diagnose benign conditions as malignant and vice versa. The life of the patients will also be put in very high-risk situations. Computer-based analysis techniques can be used as support tools for radiologists and physicians to overcome this concern. Transfer learning reuse already acquired knowledge gained from the pre-trained model in previous tasks and use it to improve generalization with new data. The idea is to use a model which is already been trained on a larger dataset for a long time and has proven to work well. By providing an input Histopathological image and possibly adding appropriate infected person metadata, to deliver a measurable result linked to lung cancer risk. According to the framework, there are two objectives to be taken into account. The first step was to minimize the inconsistency in the evaluation and observation of lung cancer by inferring with the diverse clinicians. Inevitably, Computer-based strategies have been shown to enhance a physician's ability to work across diverse medical backgrounds.

2   Literature Survey

Many authors and researchers were continuing to detect the early stage of lung cancer with high accuracy by using a suitable algorithm.

Authors Acucena R. S. Soares, et al. [1], this method uses a 3D U net and 3D V net, which takes the 888 images which contain 1159 nodules of CT images from LIDCIDRI. After preprocessing the samples obtained are 1664. This model achieved an accuracy of 74% for 3D U-Net and 99% accuracy for the 3D V-net model accordingly. In this work, they attain the best result in 3D V-net. Aishwarya Kalra [2] the dataset was collected from LUNA and the prediction of nodules was done in Kaggle. Next the detection and segmentation were achieved by using the LUNA dataset to provide us with cancer in the lungs using the CNN model which obtains the accuracy of AOU is 97% recall 74% precision 87% and it has the specificity of 97%. Author BardhRushiti [3], the lung cancer classification was done using Artificial Intelligence, the model used are VGG16, VGG19, ResNet 34, and ResNet 50. At last ResNet 50 obtains an accuracy of 88.93%, precision is 95.83%, and F1 score is 88.46%. For this, the dataset was get from LIDC-IDRI and the training was done in Google Colab.

Authors Siddharth Bhatia, et al. [4] used UNet and ResNet models for detection. The dataset of CT images from LIDC-IDRI, to highlight the lung regions. This model achieves an accuracy of 84%. For classifying the image the architecture used are XGBoost and Random forest. Authors Muhammad ZiarurRehman, et al. [5] has used a model CAD system for detection. Here they used SVM for classification and this model achieves an accuracy of 88.97% while the CAD system shows higher sensitivity of 89.9%. Ashnil Kumar, et al. [6] used CT images to diagnose lung cancer by using computer-aided diagnosis applications. They compared their working model to baseline techniques for the purpose of multi-modality image fusion, multi-branch techniques, and multichannel and segmentation. This detection shows that the CNN model shows a maximum accuracy of 99%.

SajjaTulasi Krishna [7], used Convolution Neural Network (CNN) to analyze and classify both malignant and benign tissues from computed topology images. In this work neural network is designed on GoogleNet with a high dropout ratio to minimize the time of processing.25 epochs were used for training. GoogleNet shows 95.42% accuracy. Authors AbdarahmaneTraore, et al. [8], have used deep learning detectors such as Faster-RCNN, YOLO, SSD, RetinaNet, and EfficientNet in the critical task for the detection of the pulmonary nodule. This Faster RCNNmodel achieved a precision score of 35.73% and RetinaNet achieved a precision score of 34.15%. Wafaa Alakwaa, et al. [9], the network can be trained end to end from crude image patches. Its principle necessity is the accessibility of training database, yet otherwise no suspicions are made about the objects of interest or basic image methodology. Later on, it very well may be feasible to stretch out our present model to not just decide if the patient has disease, yet additionally decide the specific location of the cancerous knobs.

Authors Muhammad Attique Khan, et al. [10], started their work in the field of lung cancer identification with the assist of certain CT images. In this work, they approached the VGG-SegNet for segmenting the nodule in the lungs. Then they used the VGG19 algorithm with the SVM-RBF classifier for the purpose of the classification part. Finally, they acquired a maximum accuracy of about 94.83%. Mohammad Ali Abbas [11], used the Histopathological images in the diagnosis of cancer cells that are present in the lungs. In this paper, they used 3 CNN architectures and compared the results between them. VGG-19, AlexNet, ResNet-50 are the three methods used in this work. By comparing F1 score VGG-19 attains 0.997, Alex net attains 0.973 and ResNet-50 attains 0.999. Authors Imam Ali, et al. [12], diagnosed the malignant lung tissue with the assist of CT medical images. In this work, they approached the Transferrable texture CNN architecture for classification, this model mainly consists of three convolution layer networks and only one Energy layer for replacing the pooling layer. Totally this CNN model comprises nine layers for extracting the features and then goes for the classification part. Finally, this texture model acquired a higher accuracy of 96.69%.

The classification of medical images is one of the primary techniques of Computer-Aided Diagnosis (CAD) systems. Traditionally, image classification is based on shape, color, and texture features, and also their combinations. These features are probably problem-specific and have ended up being integral in clinical images diagnosis. Therefore, we have a framework that is unequipped for catching undeniable level issue area ideas and has restricted capacities in summing up models. As of late, deep learning techniques have ended up being a powerful method for developing a general model for registering the name for clinical images got from raw pixels. Not-withstanding, deep learning models are costly and have restricted layers and channels because of the resolution of the clinical images and the small dataset size [13–18].

3 Methodology

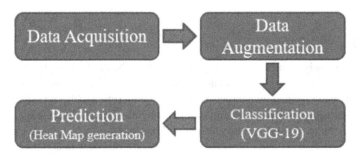

3.1 Database Collection

Artificial Intelligence relies heavily on information. Computers find it difficult to learn without data. It is the most critical factor that makes the teaching of algorithms possible. The images will be in DICOM format, which means that the added data is useful as a means of visualizing and classifying. The dataset consists of 15000 histopathological images with three different classes namely Adenocarcinoma, Squamous Cell Carcinoma, and lung normal each with 5000 images. Images are in the size of 768 * 768 pixels and are in .jpeg file format. The images are converted to the .jpeg format with the same labels for effective classification. The converted images are stacked in separate directories based on the type of cancer

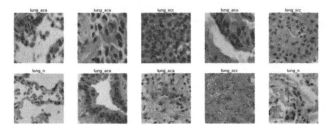

Fig. 1. Histopathological images

3.2 Data Augmentation

As our model is based on deep learning, the number of samples will have a significant impact on its performance. Augmentation is a technique largely used to enhance the training and testing accuracy of convolutional neural networks, and further increases the performance of deep neural networks. A model can take into account what the image subject looks like in different situations by randomly adjusting the rotation, brightness, or scaling of an input image. It helps to expose our classifier to a wider variety of situations to make our classifier more robust. It is a technique in computer vision includes adding noise, cropping, flipping, rotation, scaling, translation, brightness, contrast, color augmentation, and saturation, and for the Natural language processing EDA, Back translation, and text generation. Data augmentation algorithm increases the accuracy of machine learning models.

3.3 Transfer Learning

Transfer learning reuses already acquired knowledge gained from the pre-trained model in previous tasks and use it to improve generalization with new data. The idea is to use a model which is already been trained on a larger dataset for a long time and has proven to work well. Transfer learning improves learning new tasks by transferring knowledge and skills of tasks already learned solving other problems. Transfer learning is a method of the optimization training process with pre-trained models in a different task. Once the model is trained it can be reused as a base for a different tasks.

3.4 VGG-19 Architecture

VGG-19 is an optimized VGG model, which consists of 19 layers (16 convolution layers, three fully connected layers, five Maxpool layers, and one Softmax layer). There are different variants in VGG namely VGG11, VGG16, and so on. Unlike many other networks, the VGG-19 is a much simpler one with fewer hyper parameters. The classification is performed using 3 fully connected layers, consisting of two layers with 4096 neurons each, and one last layer with 1000 neurons.

ReLu activation functions are used in all layers except the last one, while for the last layer Softmax is used to distribute probabilities between classes. More than a million images from the ImageNet database were used to train VGG19. Using the network, images can be classified into 1000 categories, each with 19 layers. Using this network, images can be classified into 1000 object categories, each with 19 layers Fig. 1 and Fig. 2.

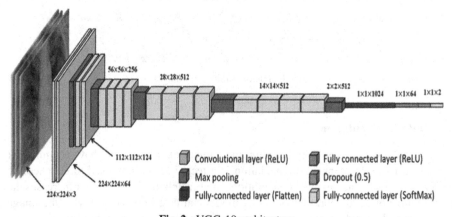

Fig. 2. VGG-19 architecture

4 Results and Discussion

Below are the results of VGG-19 model for classifying the lung cancer.

Validation Accuracy	95%
Training Accuracy	95.80%
Validation Loss	30%
Training Loss	29%

For a deep learning model to achieve better while training the neural networks, loss function consider as a main key for adjusting the weights of the neural networks. While training the datasets during backpropagation, loss function penalizes the model assuming that there is any deviation between the label predicted by model and actual target label. Hence the use of loss function is hyper-critical to attain efficient model performance. Triplet loss is used as loss function in this work. This was first developed for recognition of face to find the similarities that occur in face. This method can be used when the images are blurred with the help of the distances between faces of similar and different identities. The triplet loss guides the model to minimize the distance between images of the similar category and increases the distance between images that belong to dissimilar categories. The use of triplet loss method resulted better accuracy in binary classification while the use of base model shows the less accuracy.

In Fig. 3, ten epochs were given to attain an expected accuracy based on training and validation datasets. The epochs indicate the number of passes that the learning algorithm will work through the entire training dataset comprised of one or more batches. VGG-19 model does not overfit. This model worked efficiently and gives the accuracy and loss by using all layers in VGG-19.

Starting training using base model VGG19 training all layers

Epoch	Loss	Accuracy	V_loss	V_acc	LR	Next LR	Monitor	Duration
1 /10	2.981	80.431	1.82777	85.667	0.00100	0.00100	accuracy	87.72
2 /10	1.193	90.667	1.11538	88.111	0.00100	0.00100	val_loss	87.09
3 /10	0.702	92.417	1.20456	54.222	0.00100	0.00050	val_loss	88.16
4 /10	0.740	93.306	1.11964	64.333	0.00050	0.00025	val_loss	87.58
5 /10	0.775	93.000	0.85363	90.333	0.00025	0.00025	val_loss	88.03
6 /10	0.642	93.931	0.71352	85.778	0.00025	0.00025	val_loss	85.67
7 /10	0.541	94.208	0.54909	93.000	0.00025	0.00025	val_loss	87.91
8 /10	0.453	95.083	0.46861	92.444	0.00025	0.00025	val_loss	85.53
9 /10	0.402	94.833	0.49259	89.778	0.00025	0.00013	val_loss	86.42
10 /10	0.398	95.403	0.50074	92.111	0.00013	0.00006	val_loss	86.14

Fig. 3. Accuracy analysis for each epoch

4.1 Plots of Accuracy and Loss

Fig. 4 and 5 indicates the plots of accuracy and loss on the training and validation datasets over training epochs. The training and validation accuracy of the fine-tuned VGG-19 model is observed to have a similar convergence rate in both training and validation outcomes.

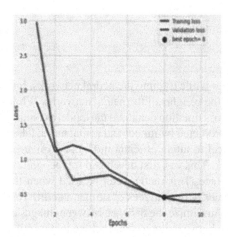

Fig. 4. Model loss   **Fig. 5.** Model accuracy

4.2 Heat Map Generation Using Vgg-19

Heat map generation is the process of analyzing and reviewing heat map data to gather insights about the performance of the model during the training process. In Fig. 4.4, The Heat Map Generation method represents the accuracy of on testing process. Prediction is based on the following parameters such as,

TRUE POSITIVE (TP) - As a test result, it correctly indicates the presence of a condition.

TRUE NEGATIVE (TN) - As a test result, it correctly indicates the absence of a condition.

FALSE POSITIVE (FP) - As a test result, it wrongly indicates that a certain condition is present.

FALSE NEGATIVE (FN) - As a test result, it wrongly indicates that a certain condition is absent.

The classification report in Table 1 shows the evaluation parameters such as Precision, Recall, F1-Score for our proposed model. These values are calculated based on the below mathematical expressions,

$$\text{Accuracy(ACC)ACC} = (TP + TN)/(TP + TN + FP + FN)$$

$$\text{F1 Score F1} = 2TP/(2TP + FP + FN)$$

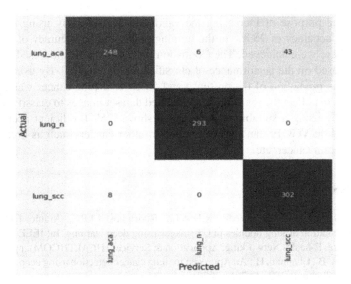

Fig. 6. Heat map generation

Precision, recall, and F1 score are more helpful than accuracy performance metrics for predicting outcomes. As both indicates the accuracy of the model. These are some of the important metrics for evaluating models.

Table 1. Evaluation parameters

	PRECISION	RECALL	F1 - SCORE	SUPPORT
Lung_aca	0.97	0.84	0.90	297
Lung_n	0.98	1.00	0.99	293
Lung_scc	0.88		0.92	310
Accuracy			0.94	900
Macro avg	0.94	0.94	0.94	900
Weighted avg	0.94	0.94	0.94	900

5 Conclusion

In this project, the deep learning approach is used for classifying and predicting lung cancer using histopathological images. Early detection of lung cancer surely increases the possible chances of survival rate but it is more difficult to identify the lung cancer in the beginning stage itself. The proposed technique is based on Convolution neural networks model and transfer learning like VGG19. Here, the transfer learning method is used for the purpose of increasing the validation accuracy. By using the VGG19 algorithm, an accuracy of 95.8% in the training phase and an accuracy of 95% in the validation phase were achieved. The Confusion matrix is the method used to evaluate parameters based on the performance of classification by VGG19. By using heat map generation, the prediction of the presence and absence of lung cancer was done. The future work is to collect the real-time lung affected dataset images to classify and detect lung cancer by using a suitable algorithm that shows526570 efficient accuracy. The proposed method VGG19 can also be applied to other cancers such as breast cancer, skin cancer, brain cancer, etc.

References

Soares, A.R.S., Lima, T.J.B., Rabelo, R.D.A.L., Rodrigues, J.J.P.C., Araujo, F.H.D.: Automatic segmentation of lung nodules in CT images using deep learning. In: IEEE International Conference on E-health Networking, Application & Services (HEALTHCOM), pp. 1–6 (2020)

Kalra, A., Singh, B., Chauhan, H.: An Approach for lung cancer detection using deep learning. Int. Res. J. Eng. Technol. (IRJET) 7(9) (2020)

Rushiti, B.: Automatic lung cancer detection using artificial intelligence university of business and technology Kosovo (2019)

Bhatia, S., Sinha, Y., Goel, L.: Lung Cancer Detection: A Deep Learning Approach. In: Bansal, J., Das, K., Nagar, A., Deep, K., Ojha, A. (eds) Soft Computing for Problem Solving. Advances in Intelligent Systems and Computing, vol 817, Springer, Singapore. (2019). https://doi.org/10.1007/978-981-13-1595-4_55

ur Rehman, M.Z., Javaid, M., Shah, S.I.A., Gilani, S.O., Jamil, M., Butt, S.I.: An appraisal of nodules detection techniques for lung cancer in CT images. Biomed. Sig. Process. Cont. 41, 140–151 (2018)

Kumar, A. Fulham, M., Feng, D., Kim, J.: Co-learning feature fusion maps from PET-CT images of lung cancer. IEEE Trans. Med. Imag. 39(1) 204–217 (2020)

Krishna, T.K., Devarapalli, Hemantha, R.M., Kalluri, H.K.: Lung cancer detection based on CT scan images by using deep transfer learning. Traitement du Sig. 36 339–344 (2019)

Traore, A., Ly, A.O., Akhloufi, M.A.: Evaluating Deep Learning Algorithms in Pulmonary Nodule Detection, New Brunswick Health Research Foundation (NBHRF) (2020)

Alakwaa, W., Nassef, M., Badr, A.: Lung cancer detection and classification with 3D convolutional neural network (3D-CNN). Int. J. Adv. Comput. Sci. Appl. 08 (2017)

Khan, M.A.: VGG19 network assisted joint segmentation and classification of lung nodules in CT images. Diagnostics 11(12), 2208 (2021)

Abbas, M.A., Bukhari, S.U.K., Syed, A., Shah, S.S.H.: The histopathological diagnosis of adenocarcinoma & squamous cells carcinoma of lungs by artificial intelligence: a comparative study of convolutional neural networks (2020)

Salaken, S.M., Khosravi, A., Khatami, A., Nahavandi, S., Hosen, M.A.: Lung cancer classification using deep learned features on low population dataset. In: IEEE 30th Canadian Conference on Electrical and Computer Engineering (CCECE) (2017)

Song, Q.Z., Zhao, L., Luo, X.K., Dou, X.C.: Using deep learning for classification of lung nodules on computed tomography images. J. Healthcare Eng. 04 (2017)

Lakshmanaprabu, S.K., Mohanty, S.N., Shankar, K., Arunkumar N., González, G.R.: Optimal deep learning model for classification of lung cancer on CT images. Future Gen. Comput. Syst. 92 374–382 (2019)

Ali, I., Muzammil, M., Haq, I.U., Khaliq, A.A., Abdullah, S.: Efficient lung nodule classification using transferable texture convolutional neural network. IEEE Access 8 175859–175870 (2020)

Traoré, A. Ly, A.O., Akhloufi, M.A.: Evaluating deep learning algorithms in pulmonary nodule detection. In: 2020 42nd Annual International Conference of the IEEE Engineering in Medicine & Biology Society (EMBC) (2020)

Kumar, A., Fulham, M., Feng, D., Kim, J.: Co-learning feature fusion maps from PET-CT images of lung cancer. IEEE Trans. Med. Imag. **39**(1), 204–217 (2020)

Singh, G., Gupta, G.K.: Performance analysis of various machine learning-based approaches for detection and classification of lung cancer in humans. Neural Comput. Appl. **31**, 6863–6877 (2019)

Analysis of the Impact of White Box Adversarial Attacks in ResNet While Classifying Retinal Fundus Images

D. P. Bharath Kumar, Nanda Kumar, Snofy D. Dunston,
and V. Mary Anita Rajam$^{(\boxtimes)}$

Department of Computer Science and Engineering, College of Engineering Guindy,
Anna University, Chennai, India
anitav@annauniv.edu

Abstract. Medical image analysis with deep learning techniques has been widely recognized to provide support in medical diagnosis. Among the several attacks on the deep learning (DL) models that aim to decrease the reliability of the models, this paper deals with the adversarial attacks. Adversarial attacks and the ways to defend the attacks or make the DL models robust towards these attacks have been an increasingly important research topic with a surge of work carried out on both sides. The adversarial attacks of the white box category, namely Fast Gradient Sign Method (FGSM), the Box-constrained Limited Memory Broyden-Fletcher-Goldfarb-Shanno (L-BFGS-B) attack and a variant of the L-BFGS-B attack are studied in this paper. In this work, we have used two defense mechanisms, namely, Adversarial Training and Defensive distillation-Gradient masking. The reliability of these defense mechanisms against the attacks are studied. The effect of noise in FGSM is studied in detail. Retinal fundus images for the diabetic retinopathy disease are used in the experimentation. The effect of the attack reveals the vulnerability of the Resnet model for these attacks.

1 Introduction

Deep learning models have been used in medical diagnostic systems for a long time now. The deep learning models are well accepted for their performance and their aid in decision making for the medical professionals. Some of the convolutional neural network models commonly used for this purpose are AlexNet, GoogleNet, ResNet and Inception Net. The other deep learning models commonly used are auto encoders, Recurrent Neural Network, Long Short Term Memory networks, Generative Adversarial neural network and so on. However, recent studies reveal that deep learning models could be tampered by attacks known as adversarial attacks. These attacks aim to reduce the reliability and performance of the deep learning system by the introduction of data samples which mislead the system to wrong predictions. These wrong predictions may also have higher confidence value in certain cases.

© IFIP International Federation for Information Processing 2022
Published by Springer Nature Switzerland AG 2022
L. Kalinathan et al. (Eds.): ICCIDS 2022, IFIP AICT 654, pp. 162–175, 2022.
https://doi.org/10.1007/978-3-031-16364-7_13

The adversarial attacks create images with slight variation from the original images. The difference between the newly created images and the original image is imperceptible to human vision. However, to the learning model, the variation leads to misprediction. Some of the methods for generating these adversarial images listed in the literature are Limited Memory Broyden-Fletcher-Goldfarb-Shanno attack (L-BFGS) attack [6], Fast gradient sign method (FGSM) [9], iterativative least-likely class, Jacobian based Saliency Map attack, Deep Fool, Carlini and Wagner attack (C&W), compositional pattern-producing network-encoded evolutionary algorithm, Zeroth order optimization based attack, Universal Perturbation, One-Pixel Attack, Feature Adversary, general generative adversarial networks, Hot/Cold, Model-Based Ensembling Attack and Ground-Truth Attack [14].

The adversarial attacks perturbate the original input, then present them to the model and lead to misclassification. Based on the intention of misclassification, the attacks are classified as targeted attacks and untargeted attacks. In targeted attacks, the intention of the adversaries is to misclassify the records to a particular class. In untargeted attacks, the intention of the adversaries is to misclassify the records to any other class other than its true class.

Based on the knowledge of the model being targeted, the attacks are classified into White box attack, Black box attack and Grey box attack. In case of white box attacks, the structure and parameters of the model are known to the adversaries. The structure of the model alone is known to the adversaries in the case of grey box attacks. In case of black box attacks, these details are not known to the adversaries.

Defense mechanisms are used to improve the robustness of the DL models against adversarial attacks. The defense mechanisms to countermeasure adversarial attacks can be proactive or reactive. Various defense methods have been proposed in literature. Adversarial detecting, input reconstruction and network verification are some of the reactive methods, and adversarial training and network distillation are some of the proactive methods.

In this paper, the vulnerability of Resnet model towards the optimization based L-BFGS-B, and the gradient based Fast Gradient Sign method (FGSM) white-box adversarial attacks is studied. These attacks are studied on a Diabetic Retinopathy (DR) dataset comprising of retinal fundus images and their effects are discussed. The success rate of the attacks and the trade-off between producing less distorted and more confident adversaries is studied. We have used two defense mechanisms, namely, Adversarial Training and Defensive distillation-Gradient masking. The reliability of these defense mechanisms against the attacks are tested.

The rest of the paper is organized as Sect. 2 on the related work, Sect. 3 on the different attacks used in this work, Sect. 4 on results and discussion and Sect. 5 concludes the paper.

2   Literature Survey

The study of universal perturbation as an adversarial attack was done by Guohua Cheng and Hongli Ji on U-Net. It was found that, among the four modalities in the brain MRI images, if all the modalities were altered the perturbation has an effect on the classifier [4]. Hassan et al. demonstrated the FGSM attack and Basic Iterative Model of FGSM on a ResNet model with 85 public datasets which contained time series data [10].

Gerda Bortsova et al. experimented the FGSM and projected gradient descent (PGD) attacks on Inception-v3 and Densenet121. The datasets used were diabetic retinopathy dataset with 88,702 color fundus images, ChestX-Ray14 dataset with 112,120 frontal-view X-rays and patchCamelyon (PCam) with 327,680 patches extracted from histopathology whole-slide images of lymph node sections [2].

Yupeng Cheng et al. examined the camera exposure to the Retinal Fundus images and proposed an exposure based adversarial attack on the retinal fundus images. Their experiments were carried out in three stages as multiplicative-perturbation based attack, adversarial bracketed exposure fusion and convolutional bracketed exposure fusion. The convolutional bracketed exposure fusion was found to be more effective in Resnet50, MobileNet and EfficientNet [5].

Saeid Asgari et al. have demonstrated 10 attacks belonging to gradient based, score based and decision based adversarial attacks on chest X-ray images with Inception-ResNet-V2 and Nasnet-Large neural network models. The gradient based attacks were superior in fooling the network in comparison to the other attacks. Secondly, by modifying the pooling layer of the network, the effect of adversarial attacks were reduced [1]. A susceptibility score has been proposed by Mengying Sun et al. using the global maximum perturbation (GMP), global average perturbation (GAP) and the probability of being perturbed across all records (GPP). This score provides the efficiency of an attack by uncovering the locations which are susceptible for an attack. A perturbation distance has been defined by the authors and is used to find the optimal adversarial records [13].

Rida El-Allami et al. have performed a study on the Spiking neural network's robustness against adversarial attacks. The study reports that the parameters of the network namely Spiking Voltage threshold, spiking time window and attack's noise budget play a major role in the robustness of the network [7].

3   Materials and Methods

3.1   Attacks Studied

The attacks studied in this work are explained in this section.

FGSM Attack
The goal of FGSM attack is to find the optimal direction in the input space where the loss function ascends/descends steeply and so pushing the inputs in

that direction will find an input that is close to the given image and also is misclassified.

The perturbed input x' is given by

$$x' = x + \epsilon \cdot sign[\nabla x J(\Theta, x, y)] \tag{1}$$

where

x is the input value, ϵ is the epsilon value that determines the magnitude of the perturbation allowed on a pixel value, $J(\Theta, x, y)$ is the loss function and ∇x is the gradient of the loss function.

L-BFGS-B Attack

In this type of attack, finding adversarial examples is modeled as an optimization problem. The optimization aims at finding an adversarial example with minimum perturbation and classifies into non-true class.

The perturbed input x' is found by solving the below minimization (optimization) equation

$$f(x) = c \cdot L2norm(x - x') + loss(x') \tag{2}$$

where x' is chosen between 0 to 1.

3.2 Deep Neural Network Classifier Used

The basic deep learning model used in this work is Resnet50, a variant of ResNet model with 48 Convolution layers, one MaxPool layer and one Average Pool layer. It has been widely used for transfer learning in the classification of medical images. The final network architecture used in this work consists of additional layers to the Resnet50 namely, a global average pooling layer, dropout layer and the final classifying dense layer. The model is initialized with pretrained weights from ImageNet and then trained on the retinal fundus dataset specified in Sect. 4.1, that is, transfer learning from ImageNet is done. The ImageNet dataset contains 14,197,122 annotated images.

3.3 Defense Mechanisms Studied

This section describes the defense mechanisms studied in this work.

3.3.1 Adversarial Training

In adversarial training, adversarial samples that are misclassified by the classifier are created. The neural network classifier is then trained with the created samples and their correct labels.

3.3.2 Defensive Distillation

Hinton et al. [8], introduced knowledge distillation, a process to extract knowledge from one model and apply it to a compressed model and make it perform with the same accuracy. Papernot et al. [12] exploited this process and modified it to act as a defense. Distillation, is training a teacher model by introducing a temperature parameter T in the final softmax activation. The softmax output of the teacher model (soft labels) is used for training another model with the same temperature T.

The intuition behind this is that, whatever the model learns during training is encoded not only in its weights, but also in the output probability vector, which holds the relative information about classes. Informally, in case of hard labels, the model only gets to know that it belongs to one particular class and no information about its similarity with the remaining classes. All samples belonging to a class are treated with the same weights, even though some samples might be less similar to that class. Soft labels use a vector of probabilities and the similarities of the given sample, with all classes, are learned by the model. This helps the model understand the relative difference between the classes.

3.4   Details of the Study on the Impact and Working of Adversarial Attacks

A binary classifier is first built using the deep learning model specified in Sect. 3.2, which classifies retinal images as 'DR affected' or 'Unaffected'. The FGSM and the L-BFGS-B attacks are performed on 100 image samples which were correctly classified by the Resnet classifier model used as 'DR affected' or 'Unaffected'. The number of samples in each class is chosen to be approximately the same. Different metrics are studied for the attacks performed. The robustness of the Resnet model against the FGSM and the L-BFGS-B attacks using adversarial training and distillation defense mechanisms is studied. Various studies are also done on the perturbations generated using the attacks.

4   Results and Discussion

4.1   Dataset

The dataset used for training the Resnet classifier model has 35,638 images, out of which 35,126 retina scan images are taken from the diabetic retinopathy (DR) Kaggle dataset and 512 retina scan images are taken from IDRiD (Indian Diabetic Retinopathy Image Dataset) dataset.

The classes in the dataset are grouped into two classes as affected by DR and unaffected by DR. The dataset is first balanced and the images are resized to 224×224 pixels. The dataset is divided into train set, validation set and test set. The details of the dataset used for training are given in Table 1.

Table 1. Details of training dataset

Dataset name	Number of images	Unaffected by DR	Affected by DR	Number of images in		
				Training set	Validation set	Testing set
Kaggle	18966	9556	9400	14956	3720	290
IDRiD	554	184	370	444	–	110

4.2   Hyper Parameters Used for Tuning the Resnet Model

Batch Size   A batch size of 16 is used for training since it provides better generalization and is small enough to fit in memory.

Learning Rate   A learning rate of 0.0001 is found to give the best results based on experimentation.

Loss function   Categorical cross entropy loss function is used in the training process since the model is a multi-class classifier.

Optimizer   Several optimizers were tested for stochastic gradient descent parameter updates namely Adam, RMSprop and SGD. No optimizer showed significant increase in performance over the other two. It is found that the Adam optimizer performed the best overall.

4.3   Attack Framework

100 original retina image samples that are correctly classified by the Resnet model explained in Sect. 3.2 are used for implementing the attacks. As it is a binary classification problem, only targeted attacks have been tested, and we believe untargeted attacks should also provide similar results. The norm or distance metric is taken as L2 norm because it is widely used in the literature for the implementation of FGSM and L-BFGS-B attacks.

4.4   Sample Images After the Adversarial Attacks

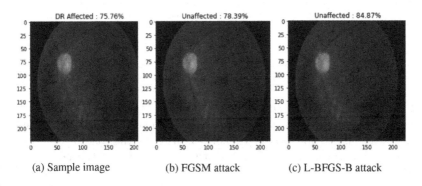

(a) Sample image   (b) FGSM attack   (c) L-BFGS-B attack

Fig. 1. Effect of adversarial attacks

Figure 1 depicts the effects of the different adversarial attacks considered on a sample image from the dataset. While the sample image is classified as DR affected with a confidence of 75.76%, after the FGSM and the L-BFGS-B attacks, the image is classified as unaffected with confidence values of 78.39% and 84.87% respectively. Thus, it is seen that the attacks were successful in misleading the classifier.

4.5   Metrics Used for Evaluation

Success Rate The number of adversarial examples created successfully.

Confidence Confidence value for the classification of the adversarial image.

Strong adversaries The number of adversarial examples which were classified with more confidence than their original counterparts (here, out of the 100 samples considered).

Norm L2 distance between the original image and perturbed image.

4.6   Analysis of the Effect of the FGSM Attack

Experimentation was done with the FGSM attack with different epsilon values namely 0.1, 0.25, 0.5, 1, 2 and the results are tabulated in Table 2. It is seen that, when the epsilon value is 0.25, FGSM produces adversarial images with minimum average norm value and achieved a success rate of 95%.

The step size in the direction of gradient is controlled by epsilon, with smaller values of epsilon giving less distortion and adversaries with low confidence. It is observed that, very high epsilon values also did not help, as it prevented the gradient descent to converge to an optimal perturbation and so the success rate decreased for very high values of epsilon. (Started decreasing from epsilon $= 3$)

Table 2. FGSM results

Epsilon	Success rate (%)	Avg. conf (%)	Max. conf (%)	Avg. norm	Max. norm	Strong adversaries
0.1	85	69.78	94.37	38.77	38.78	64
0.25	95	83.54	98.78	96.9	97	71
0.5	99	91.62	99.9	193.98	193.98	86
1	99	94.77	99.98	387.97	388	92
2	99	95.12	99.97	775.95	776	92

4.6.1 Analysis of the Effect of the L-BFGS-B Attack

Experimentation is done with the Box constrained L-BFGS attack and the results are tabulated in Table 3. L2-norm is used as the distance metric, and the constant c, which measures the relative importance of the norm and loss in the objective function is computed using binary search. To control the perturbation, a parameter, epsilon is used. This value controls the lower and the

Table 3. L-BFGS-B attack

Epsilon	Success rate (%)	Avg. conf (%)	Max. conf (%)	Avg. norm	Max. norm	Strong adversaries
0.07	61	66.891	92.01	29.1	31.4	36
0.2	90	76.71	96.23	70.7	75.1	71
0.5	100	99.43	99.86	175	189.76	99
1	100	99.43	99.86	346.35	360.99	100

upper thresholds of the box. High values of epsilon give large perturbation and the confidence of adversaries is increased.

We have also tried a variant of the L-BFGS-B attack by removing the norm constraint in the objective function and optimizing only the loss function. The motive is to observe the extent of optimization without the norm constraint and whether this could produce images with acceptable perturbations with a better confidence level. The results are tabulated in Table 4.

Table 4. Modified L-BFGS-B attack

Epsilon	Success rate (%)	Avg. conf (%)	Max. conf (%)	Avg. norm	Max. norm	Strong adversaries
0.07	100	96.2	99.92	171.56	421.79	100
0.2	100	99.21	99.93	146.56	301.57	100
0.5	100	99.65	99.94	158.65	279.88	100
1	100	99.66	99.87	250.52	1067.82	100

The perturbation increased as expected, but the confidence remained almost similar. The same maximum confidence was observed and the average confidence increased slightly. The effect of epsilon was minimal and the results stayed almost constant. So, it is concluded that the cost given for high perturbation is not balanced in the confidence level and so the previous method (L-BFGS-B) produces better adversaries. Though the confidence value is higher compared to the unmodified attack, the norm value is also higher. The higher norm values indicate that the difference in the images may be much visible.

4.6.2 Experiments Using Perturbation

For the creation of adversaries, the attack models generate perturbation that are added to the original image. We wanted to analyse the impact of the noise/perturbations generated by the attacks.

Figure 2 shows the perturbations/noises created by FGSM to transform a 'DR affected' sample to 'Unaffected' for different values of epsilon. It is seen that the perturbation is high when the value of epsilon is larger.

We first try to find how the Resnet classifier classifies, when just the perturbation is given as input to the classifier. Figure 3 shows a perturbation image created by FGSM to push the classification towards 'Unaffected' class. Interestingly, this noise (just the perturbation) shown in Fig. 3 is classified by the classifier as 'Unaffected' with 81.68% confidence.

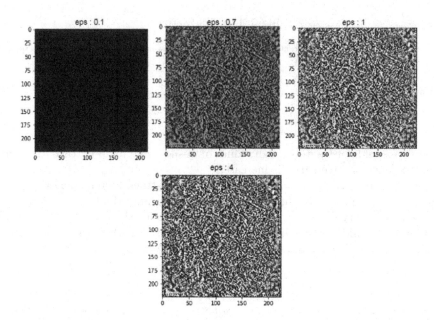

Fig. 2. Perturbation images created by FGSM for different epsilon values

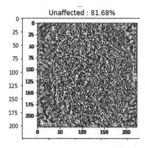

Fig. 3. Perturbation created by FGSM to push towards Unaffected class

Now, if this perturbation in Fig. 3 is given as input further to FGSM, it successfully completes the attack on the perturbation image and the resultant image (perturbation using FGSM + FGSM) is classified with the Resnet classifier as DR affected with a confidence of 55.42% (Fig. 4). Though the attack is successful, the low confidence value signifies again that the original perturbation given as input had the properties of unaffected retina image strongly, as expected.

When the perturbation in Fig. 3 is given as input further to L-BFGS-B (perturbation using FGSM + L-BFGS-B), the resultant image is misclassified as DR affected with 61.3% confidence and a high norm of 350.5 (Fig. 5). So, now we have created an artificial image sample of an unaffected retinal fundus image.

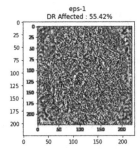

Fig. 4. FGSM on perturbation created by FGSM

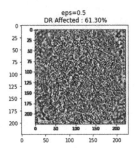

Fig. 5. L-BFGS-B on perturbation created by FGSM

We have also tested if the perturbation created from one input image is enough to convert another image to an adversary. For this, we added the perturbation obtained from one image of the dataset on a different image of the dataset and tested if the second image behaved as an adversary. When we add the perturbation shown in Fig. 3 to a different DR affected image (Fig. 6a), the result shown in Fig. 6b is got.

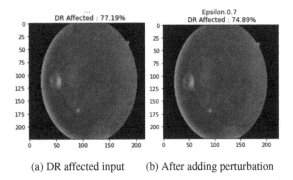

(a) DR affected input   (b) After adding perturbation

Fig. 6. Image before and after adding the perturbation shown in Fig. 3

The confidence is observed to have reduced. Even though the added noise didn't misclassify the retinal image as unaffected, it has pushed the sample towards 'Unaffected' direction and we believe a stronger noise sample will perform the misclassification too, proving the claim by Goodfellow et al. [9] that the adversarial directions stay the same for all samples within a training set.

To examine whether any random noise can cause a misclassification, a random noise is generated (Fig. 7) and given as input to the Resnet classifier. This noise is unrelated to the DR image but the classifier classifies the image as 'Unaffected' with 73.24% confidence.

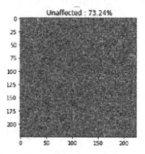

Fig. 7. Random noise

When the L-BFGS-B attack is applied on this random noise shown in Fig. 7, the image was easier to attack and it produced an adversarial image with less epsilon value of 0.39 and with a less norm of 130.5 (Fig. 8).

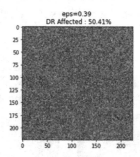

Fig. 8. Adversary created from random noise

This observation signifies that the perturbation noise added in FGSM is not just a random noise but carefully structured one to push a benign sample towards target direction and so it captured the features of Unaffected image, which made it hard to misclassify it as seen above.

4.7   Defenses

4.7.1 Adversarial Training

We created Adversarial examples with FGSM and retrained the Resnet model along with the adversarial examples with the correct labels.

We tested the adversarial training defense mechanism against the FGSM attack and the results are summarized in Table 5. It is observed that training with the adversarial examples decreased the success rate of the attacks and decreased the confidence of the adversaries while the norm remained the same. However, it couldn't still prevent attacks and misclassifications. The number of strong adversaries reduced to 0, showing that the adversaries produced had low confidence values. The more the adversarial samples we used for training, the less the confidence of the adversaries.

Table 5. FGSM results after adversarial training

Epsilon	Success rate (%)	Avg. conf (%)	Max. conf (%)	Avg. norm	Max. norm	Strong adversaries
0.1	38	57	94	38.77	38.78	0
0.25	41	56	98.78	96.9	97	0
0.5	44	56	99.9	193.98	193.98	0
1	44	55	99.98	387.97	388	0

The results of the adversarial training defense mechanism against the L-BFGS-B attack are summarized in Table 6. It is seen that norm values decreased more than confidence values. The number of strong adversaries reduced to 0 again, similar to FGSM attack. The success rate didn't reduce but as the number of strong adversaries is 0, it is inferred that the produced adversaries are of low confidence as expected.

Table 6. Modified L-BFGS-B results after adversarial training

Epsilon	Success rate (%)	Avg. conf (%)	Max. conf (%)	Avg. norm	Max. norm	Strong adversaries
0.2	100	76.71	96.23	78.7	163.7	0
0.5	100	96	99.86	115	148.44	0
1	100	97.6	99.86	205.46	246.36	0

4.7.2 Defensive Distillation

We trained the teacher network with a softmax output at temperature 100. The teacher network has the same architecture as the previously discussed ResNet model. We utilized the same hyperparameters that were previously shown to be the most ideal. We used a custom loss function that computes the softmax cross entropy between logits and labels.

A new training set was formed with soft labels obtained from the teacher model. We trained the student model with the same network architecture as the teacher model using the new training set and the temperature of the softmax layer was kept at 100. This new model is referred to as the distilled model.

We tested the distilled model against the two adversarial attacks and the results are summarized in Table 7.

Table 7. Success rate of attack after defensive distillation

Adversarial attack	Success rate against original model	Success rate against distilled model
FGSM	99%	60%
L-BFGS-B	100%	82%

5   Conclusion

The paper aims to demonstrate the applicability of adversarial attacks and their implications on classifying retinal images. This enforces the importance of making neural networks robust and the need to combat these attacks. The experiments on the FGSM and the L-BFGS-B attacks have revealed that the performance of the ResNet model could be reduced by the adversarial attacks. Defense mechanisms can reduce the effect of the attacks. It is also observed that a random noise may also pose the risk of reducing the confidence of the classifier in its true prediction.

Acknowledgements. This research is funded by Science & Engineering Research Board (SERB).

References

1. Asgari Taghanaki, S., Das, A., Hamarneh, G.: Vulnerability analysis of chest X-ray image classification against adversarial attacks. In: Stoyanov, D., et al. (eds.) MLCN/DLF/IMIMIC -2018. LNCS, vol. 11038, pp. 87–94. Springer, Cham (2018). https://doi.org/10.1007/978-3-030-02628-8_10
2. Bortsova, G., et al.: Adversarial attack vulnerability of medical image analysis systems: unexplored factors. Med. Image Anal. **73**, 102141 (2021). https://doi.org/10.1016/j.media.2021.102141
3. Carlini, N., Wagner, D.: Towards evaluating the robustness of neural networks. In: 2017 IEEE Symposium on Security and Privacy (SP) (2017)
4. Cheng, G., Ji, H.: Adversarial perturbation on MRI modalities in brain tumor segmentation. IEEE Access **8**, 206009–206015 (2020). https://doi.org/10.1109/ACCESS.2020.3030235
5. Cheng, Y., et al.: Adversarial exposure attack on diabetic retinopathy imagery. arXiv arXiv:2009.09231 (2020)
6. Christian, S., et al.: Intriguing properties of neural networks. arXiv preprint arXiv:1312.6199 (2013)

7. El-Allami, R., Marchisio, A., Shafique, M., Alouani, I.: Securing deep spiking neural networks against adversarial attacks through inherent structural parameters. In: 2021 Design, Automation & Test in Europe Conference & Exhibition (DATE) (2021). https://doi.org/10.23919/DATE51398.2021.9473981
8. Hinton, G., Vinyals, O., Dean, J.: Distilling the knowledge in a neural network. In: Deep Learning and Representation Learning Workshop, NIPS 2014 (2014). arXiv preprint arXiv:1503.02531
9. Goodfellow, I.J., Shlens, J., Szegedy, C.: Explaining and harnessing adversarial examples. arXiv preprint arXiv:1412.6572 (2014)
10. Ismail Fawaz, H., Forestier, G., Weber, J., Idoumghar, L., Muller, P.: Adversarial attacks on deep neural networks for time series classification. In: 2019 International Joint Conference on Neural Networks (IJCNN) (2019). https://doi.org/10.1109/IJCNN.2019.8851936
11. Newaz, A., Haque, N., Sikder, A., Rahman, M., Uluagac, A.: Adversarial attacks to machine learning-based smart healthcare systems. In: 2020 IEEE Global Communications Conference, GLOBECOM 2020 (2020). https://doi.org/10.1109/GLOBECOM42002.2020.9322472
12. Papernot, N., McDaniel, P., Wu, X., Jha, S., Swami, A.: Distillation as a defense to adversarial perturbations against deep neural networks. In: 2016 IEEE Symposium on Security and Privacy (SP), pp. 582–597 (2016). https://doi.org/10.1109/SP.2016.41
13. Sun, M., Tang, F., Yi, J., Wang, F., Zhou, J.: Identify susceptible locations in medical records via adversarial attacks on deep predictive models. In: Proceedings of the 24th ACM SIGKDD International Conference on Knowledge Discovery & Data Mining (2018). https://doi.org/10.1145/3219819.3219909
14. Yuan, X., He, P., Zhu, Q., Li, X.: Adversarial examples: attacks and defenses for deep learning. IEEE Trans. Neural Netw. Learn. Syst. **30**, 2805–2824 (2019). https://doi.org/10.1109/TNNLS.2018.2886017

Factors Influencing the Helpfulness of Online Consumer Reviews

Sudarsan Jayasingh[(⊠)] [iD] and Thiruvenkadam Thiagarajan [iD]

SSN School of Management, Kalavakkam 603110, Tamil Nadu, India
sudarsanj@ssn.edu.in

Abstract. Online product review and rating is considered one of the most important factors influencing the purchasing behaviour. Today ecommerce sites received massive number of reviews and it becomes difficult for customers to read all the reviews. Online product review is one of widely researched area and many researchers have proposed models to predict review helpfulness in ecommerce sites. This main objective of this research is to find the determinants of online review helpfulness. 12,389 online reviews related to 25 brands of mobile phones were collected from amazon.in. The results shows that word count is positively associated with helpfulness of the review. Star rating, neutral opinion and negative opinion are negatively associated with helpfulness of the review.

Keywords: Online review · Helpfulness · Star rating · Tendency · Sentiment analysis · Text mining · Amazon

1 Introduction

Nowadays, most of the consumers are increasingly relying on opinions and experience of other consumers. Most of the online shopping websites in India like Amazon, Flipkart, Snapdeal etc., provides product review option for consumers to post their opinions and experiences of the product they have purchased recently. The ecommerce website like Amazon, Flipkart etc. receive thousands of reviews constantly for their popular products. Online product reviews are usually written by consumers to express their opinion and experience about their purchases. It has become common for consumers to search and read online product reviews before any online or offline purchase. Research survey shows that 93% of customers read online review before purchasing any product [1]. According to a survey, 62% of respondents in US mentioned that online customer reviews were very helpful in choosing the product [2]. A survey conducted in India about online reviews shows that 57% of consumers don't trust product reviews on eCommerce sites [3].

Most ecommerce sites give an option to consumers to grade or rate customer experience and satisfaction of using a product. Online product ratings and review is considered as one of the major factors that influences online users purchase behavior. Research shows that not all online consumer reviews are found to be helpful and have an influence on purchase decisions. Online reviews which are considered to be helpful by consumers found

© IFIP International Federation for Information Processing 2022

Published by Springer Nature Switzerland AG 2022
L. Kalinathan et al. (Eds.): ICCIDS 2022, IFIP AICT 654, pp. 176–183, 2022.
https://doi.org/10.1007/978-3-031-16364-7_14

to have stronger effect on consumer purchase decisions than other reviews [4, 5]. Review helpfulness shows if the review is giving useful product assessment and help in buying decisions. Hence it is important to identify the determinants which can make the review more helpful to the consumers. The purpose of this study is to find the determinants of online review helpfulness based on theoretical framework and empirical model.

2   Literature Review and Hypothesis

Good number of researchers have demonstrated that online reviews provide consumers relevant information about a product and influence consumers buying decisions [6, 7]. The previous research studies can be grouped into two categories based on research method. One of the categories is based on survey and experiments method of measuring review helpfulness [8]. The second category is based on collecting data from the ecommerce websites. Some previous research studies analyzed factors influencing the consumers to click helpfulness of the review. Baek, Shn and Choi researched using online consumer reviews collected from Amazon.com using web data crawler and found that review rating, reviewer's credibility and content of review influenced the helpfulness of reviews [5]. Biswas, Sengupta and Ganguly research findings shows the review title, review sentiments, star rating and temporal features predicts the count of helpful votes [7]. Lee, Trimi and Yang research shows that high star ratings and long review postings were found to be more helpful to consumers [9].

Some researchers studied the impact of online reviews on the purchase decisions [9]. Studies conduct by Li, Wu and Mai shows that star rating and sentiment is positively related to sales [10]. Mariani and Borghi research findings shows the online review helpfulness affect financial performance of a company [11]. Akbarabadi and Hosseini research studies shows that reviewer, and readability features found to strongly associated with helpfulness of review [12]. Lian, Schuckert and Law studies show that extreme rating and informative review are more likely to get the helpful votes [8]. Some of the widely studied factors influencing consumers helpfulness are total number of reviews, average rating score (star rating score), tendency and word count of the review. Statistical techniques used by previous research comes under, multiple regression, Tobit regression, multiple logit regression, vector regression, negative binomial regression, and partial least squares SEM. The most widely used platform to collect review was Amazon, Yelp, and Trip Advisor.

The present study focuses on studying the determinants of helpfulness. The independent variables used for this study are star rating, word count, tendency (positive, neutral, negative). The star rating is one of the widely used format to evaluate the product reviews by consumers and it is displayed near to the review. The tendency included in a review is defined as subjective attitudes expressed through affective word choice based upon a consumer's opinion or feeling as a result of using product. The conceptual model was developed to test the proposed hypotheses. The following hypothesis is proposed based on literature review.

H_1: Review Star rating positively affects the review helpfulness.

H_2: Review word count positively affects the review helpfulness.

H_3: Positive opinion expressed in a review positively affects the review helpfulness.

H4: Neutral opinion expressed in a review positively affects the review helpfulness.
H5: Negative opinion expressed in a review positively affects the review helpfulness.

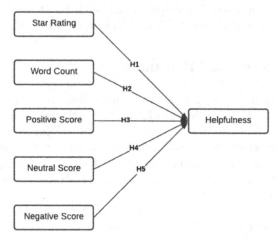

Fig. 1. Conceptual model

3   Research Methodology

The aim of this research is to find the determinants of online review helpfulness. 73,151 number of online reviews of 25 smart phones brands were collected from Amazon.in. During August 2020 and only 12,389 reviews have at least one helpfulness vote. The analysis is only done to 12,389 online reviews as high number of zero counts in the overall sample. This approach was supported by many previous researchers [7]. This shows that 16.9% of total online consumer reviews have at least one helpful vote.

Five brands in each price range are randomly selected. Python application was used to extract the reviews, creating a worksheet of review database, cleaning worksheet data, evaluating word books against WordNet to count the number of words in each review, determine the tendency, performing sentiment analysis and calculating the helpfulness rate a product or service. The sentiment analysis is one of the popular natural language processing (NLP) that automatically identifies and extract options in each text. The process of extracting opinion from online review is a form of text mining. This study uses Vader NLP algorithm to extract opinions found in an online review. These sentiments are categorized into three levels as positive, negative and neutral.

Many previous researchers have adopted linear regression for analyzing helpfulness [13]. This research adopted linear regression for predicting review helpfulness. The reviews with zero vote for helpfulness are excluded in the sample to investigate the determinants of helpfulness of the online review.

4 Results and Discussion

The descriptive data of all the variable is presented in the Table 1. The results shows that the average word count of reviews in the collected data set is 248.49 and average star rating is 3.73 for the 25 mobile phones. The average helpfulness vote from the online review is 9.71 and maximum review was given to one review for Apple mobile phone.

Table 1. Descriptive statistics of the variables

Variable	Mean	SD	Min	Max
Helpfulness Vote	9.71	102.27	1	5551
Star Rating	3.73	1.5937	1	5
Negative	0.08	0.1362	0	1
Neutral	0.67	0.2256	0	1
Positive	0.25	0.2431	0	1
No. of Characters	43.5	87.945	1	3271
No. of Words	248.49	501.87	2	18327

Table 2 lists the reviewer score which is also called as star rating. The reviewers provide a qualitative evaluation of a product by posting their subjective opinions on the review board. The average star rating is high for Oneplus and low for Mi. Overall mean of star rating is 3.73. Out of 12,389 reviews 2519 reviews was rated 1 star and 8.

Table 2. Distribution of star rating

Model	1	2	3	4	5	Average
Apple	292	41	59	157	855	3.88
Honor	158	49	81	228	605	3.96
Mi	74	14	10	21	63	2.92
Oneplus	489	142	262	569	2352	4.09
Oppo	161	32	70	148	371	3.69
Redmi	522	102	216	298	879	3.45
Samsung	649	174	210	294	527	2.93
Vivo	174	39	94	208	700	4.00
Total	2519	593	1002	1923	6352	3.73

This study selected reviews which has at least one vote for helpfulness of the review. Table 3 shows that certain reviews have maximum number of helpfulness vote. The

Table 3. Number of reviews and helpfulness

Model	Count of Reviews	Max number of Helpfulness	Average
Apple	292	41	3.88
Honor	158	49	3.96
Mi	74	14	2.92
Oneplus	489	142	4.09
Oppo	161	32	3.69
Redmi	522	102	3.45
Samsung	649	174	2.93
Vivo	174	39	4.00
Total	2519	593	3.73

average review helpfulness is found to be high for Apple mobile phones and low for Oneplus mobile phones.

Table 4 shows the review length and helpfulness. Interestingly helpfulness votes are higher when the length of review is less than 500 words. This study shows that when length of the review is less than 1000 word count then helpful votes are found to be higher. The length of the review has more impact on helpfulness. When the length of reviews is more than 1000 words, there is a possibility of getting avoided as peoples don't spend much time reading full reviews.

Table 4. Review length and helpfuness

Number of Words	Number of helpful votes	Percentage
1–500	98713	82.09%
501–1000	17237	14.33%
1001–1500	4167	3.47%
1501–2000	125	0.104%
2001–2500	2	0.002%
2501–3000	0	0
3000–3500	6	0.005%
Total	2519	593

The sentiment score of reviews is classified as Positive, Neutral, and Negative. The negative opinions are higher percentage for Samsung and less for Oneplus. The sentiment score for the eight brands is presented in Table 5. Sentiment levels are indeed an extremely influential component in determining a reviews helpfulness vote count. Customers tend to give higher star ratings when they also provide Negative or Critical reviews.

Table 5. Sentiment scores of reviews

Model	Positive	Neutral	Negative
Apple	64	30	7
Honor	64	29	7
Mi	68	19	12
Oneplus	68	26	6
Oppo	66	26	8
Redmi	68	23	9
Samsung	69	19	13
Vivo	63	31	7

The linear regression was conducted, and r-square is found to be only 0.040 and ANOVA F value is 130.217 with Sig 0.0001. The regression results are presented in Table 6. Word count and negative opinion in the review shows the positively associated with helpfulness of the review. This finding is consistent with previous research findings [14]. Star rating and neutral opinion in the review shows negatively associated with helpfulness of the review. The findings of the research show the customers tend to click helpfulness of the review when the star rating given to the mobile phone is low and also when the word count in the review is high. The research also shows that the negative opinions presented in the review more likely to be clicked by customer as helpful review.

Table 6. Results of regression analysis

Variable	Standardized coefficients	t value	Sig
Constant		3.550	0.001
Star rating	−0.028	−2.549	0.011
Word count	0.204	22.579	0.001
Neutral opinion	−0.022	−2.212	0.027
Negative opinion	0.20	−2.181	0.031

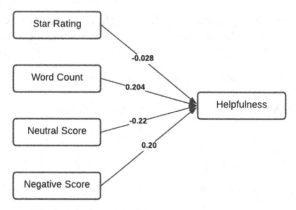

Fig. 2. Final model with standardized coefficients

5   Conclusion and Future Research

The present study investigated the determinants of helpfulness of the online reviews. The findings of this study shows that the review tendency, neutral and negative opinions affect the consumer to select the helpfulness of the review, but positive opinions did not show any effect on the consumer to select helpfulness of the review. The word count or the length of the review has more impact on helpfulness and has a stronger relationship. At the same time, the length of reviews should be less than 1000 words and when the review length is more than 1000 words, there is a possibility of getting avoided as peoples don't spend much time reading full reviews. It is necessary for the consumers to write reviews less than 1000 words as it encourages consumers to read and valuate the review as helpful. This study also found the review's tendency is perceived as more helpful for check the mobile phones in ecommerce site.

The present study investigated the determinants of helpfulness of the online reviews. The findings of this study shows that the review tendency, neutral and negative opinions affect the consumer to select the helpfulness of the review, but positive opinions did not show any effect on the consumer to select helpfulness of the review. The word count found to be positively associate to helpfulness of the review. The research finding could help the ecommerce companies to display the helpful reviews on the top which will help the company to increase their sales.

This study also has few limitations like the sample used for this research was selected based on condition that the review should have at least one star rating. This study collected data related to 25 brands of mobile phones. Future research can be extended to other products and comparative studies on online reviews can be conducted. Previous studies on online consumer review focused more on quantitative review contents like word count and star rating. Future research can include other review characteristics like emotions, verb counts, type of word, type of product etc. The consumer involvement in evaluating the review may also depend on the reviewer. Future research can be conducted to do a detailed analysis on the effect of reviewer profile.

References

1. Kaemingk, D.: Online reviews statistics to know in 2021. XM Blog. https://www.qualtrics. com/blog/online-review-stats/. Accessed 18 Nov 2020
2. Statista. Online reviews - statistics and facts. https://www.statista.com/topics/4381/online-rev iews/. Accessed 28 Aug 2021
3. Localcircles. Fake Product Review Ratings on ecommerce sites. https://www.localcirc les.com/a/press/page/fake-product-review-ratings-on-ecommerce-sites-survey#.YWwfNd lBzrA Accessed 08 Sep 2017
4. Chen, P.-Y., Dhanasobhon, S., Smith, M.D.: All reviews are not created equal: the disaggregate impact of reviews and reviewers at amazon. Com. SSRN Electron. J. , 31 (2011). https://doi. org/10.2139/ssrn.918083
5. Baek, H., Ahn, J., Choi, Y.: Helpfulness of online consumer reviews: readers objectives and review cues. Int. J. Electron. Commer 17, 99–126 (2012)
6. Liu, Y., Hu, H.f.: Online review helpfulness: the moderating effects of review comprehensiveness. Int. J. Contemp. Hosp. Manag. 33, 534–556 (2021)
7. Biswas, B., Sengupta, P., Ganguly, B.: Your reviews or mine? Exploring the determinants of perceived helpfulness of online reviews: a cross-cultural study. Electron. Mark (2021).https:// doi.org/10.1007/s12525-020-00452-1
8. Liang, S., Schuckert, M., Law, R.: How to improve the stated helpfulness of hotel reviews? A multilevel approach. Int. J. Contemp. Hosp. Manag. 31, 953–977 (2019)
9. Lee, S.G., Trimi, S., Yang, C.G.: Perceived usefulness factors of online reviews: a study of amazon.com. J. Comput. Inf. Syst. 58, 1–9 (2018)
10. Li, X., Wu, C., Mai, F.: The effect of online reviews on product sales: a joint sentiment-topic analysis. Inf. Manag. 56, 172–184 (2019)
11. Mariani, M.M., Borghi, M.: Online review helpfulness and firms financial performance: an empirical study in a service industry. Int. J. Electron. Commer. 24, 421–449 (2020)
12. Akbarabadi, M., Hosseini, M.: Predicting the helpfulness of online customer reviews: the role of title features. Int. J. Mark. Res. 62(3), 272–287 (2020)
13. Park, S., Chung, S., Lee, S.: The effects of online product reviews on sales performance: focusing on number, extremity, and length. J. Distrib. Sci. 17, 85–94 (2019)
14. Choi, H.S., Leon, S.: An empirical investigation of online review helpfulness: a big data perspective. Decis. Support Syst. 139, 113403 (2020)

Perspective Review on Deep Learning Models to Medical Image Segmentation

H. Heartlin Maria[✉], A. Maria Jossy, and S. Malarvizhi

Department of Electronics and Communication Engineering, SRM Institute of Science and Technology, Chennai, India
hm8472@srmist.edu.in

Abstract. In recent days, deep learning is on rage and is gaining a huge amount of popularity due to its supremacy in terms of accuracy. Deep learning is being used for a vast number of applications out of which healthcare is an important category. In this paper, we discuss the role of deep learning in medical image segmentation. It is also known as the automated or semi-automated detection of edges within various medical image modalities so as to identify the region of interest. Furthermore, we also explore the various deep learning networks that are widely preferred for medical image segmentation along with the architecture and overview of each network. This paper covers the most recent and widely preferred deep learning networks such as Convolutional Neural Network (CNN) and other related networks such as Alexnet, Resnet, U-net and V-net. The challenges and limitations of the emerging DL networks is also studied.

Keywords: Deep learning · Image segmentation · CNN · Alexnet · Resnet · U-net · V-net

1 Introduction

Deep learning is a process in which machines learn to process data and derive a conclusion using neural networks that are comprised of different levels, arranged according to hierarchy. Deep learning is used in various applications such as speech and image recognition, bio-informatics, military and most importantly medical image analysis. It is capable enough to transform the entire landscape of healthcare. The application of deep learning in healthcare is expected to grow in the time ahead. Deep learning is used alongside medical imaging for health check and monitoring, diagnosis and treatment of diseases, injuries etc. Medical image segmentation is yet another application of deep learning that is used to identify organs or lesions from different modalities of medical images such as Computed Tomography (CT), Magnetic Resonance Imaging (MRI), ultrasound etc.

Initially, edge detection filters and mathematical methods were being used, after which deep learning was brought into use predominantly alongside transfer learning. Later, 2.5 dimensional CNN was introduced and this produced a remarkable balance

© IFIP International Federation for Information Processing 2022
Published by Springer Nature Switzerland AG 2022
L. Kalinathan et al. (Eds.): ICCIDS 2022, IFIP AICT 654, pp. 184–206, 2022.
https://doi.org/10.1007/978-3-031-16364-7_15

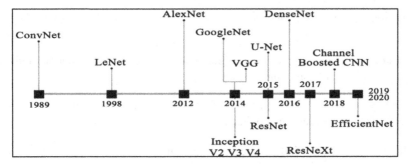

Fig. 1. Evolution of CNN architectures

between the performance and computational costs. After this, 3 dimensional CNN came into use and proved to be superior to 2.5D in terms of performance. Over time, various types of CNN architectures have evolved as shown in Fig. 1.

2 Deep Learning Network Architectures and Related Work

2.1 Basic CNN (1989)

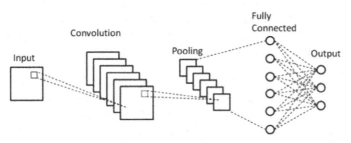

Fig. 2. Basic architecture of CNN [1]

CNN is a well-known class of deep learning networks. It is widely preferred in image segmentation, image classification, object detection etc. CNN is also known as ConvNet and is shift-invariant. They are regularized versions of the Multi-Layer Perceptron (MLP). MLP networks are often fully connected and hence result in overfitting of data. The basic architecture of CNN comprises of 5 layers as shown in Fig. 2

Input Layer: This layer is made up of artificial neurons that allows the initial data into the network for further processing.

Convolutional Layer: The convolutional layer is comprised of weights that should be trained as per the application that the network is being used [2]. This layer is also responsible for feature extraction which includes edges, objects, textures and scenes [3].

Pooling Layer: The feature map dimensions obtained from the convolutional layer are reduced in the pooling layer.

Fully Connected Layer: In this layer, each input from the previous layer is connected to each activation function in the next layer.

Output Layer: This layer is responsible for producing the final result of the segmentation, classification or relevant application.

The following authors have implemented various variants of CNN for segmentation of tumors in the brain. R. Thillaikkarasi et al. [4] presented a novel kernel-based CNN combined with a modified Support Vector Machine(SVM) for efficient and automatic segmentation of brain tumors. In this work, spectrum mixing was included along with the kernel to elevate the flexibility during segmentation. Sidra Sajid et al. [5] introduced a patch-based hybrid CNN approach to detect tumors in brain and considers local and contextual information. This method addressed the overfitting problem by making use of the dropout regulariser alongside with batch normalization procedure. This work also provided a solution for data imbalance using a two-phase training procedure and resulted in a DSC (Dice Score Co-efficient) of 86%. Farheen Ramzan et al. [6] proposed a 3D CNN network associated with residual learning and dilated convolutional operations to accurately analyze the end to end mapping from MRI volumes to the brain segments at the voxel level. DSC of $87 \pm 3\%$ was achieved for different datasets. Kumar et al. [7] also implemented a 3D CNN network for brain tumor segmentation.3D CNN was preferred over 2D CNN as it suffered a loss of quality in the input image due to compressed 2D image processing. The proposed 3D CNN consisted of five max-pooling layers, two fully connected layers and a soft max layer. Mostefa Ben naceur et al. [8] proposed a CNN network inspired by occipito temporal pathway which includes a special function known as selective attention that operates based on the receptive field sizes to identify the crucial objects in the scene.This method was used for segmentation of brain tumors and yielded a DSC of 90% for tumor, 83% for tumor core and 83% for enhanced tumor. Nai Qin Feng et al. [9] proposed a deep CNN framework in a cascaded structure with CRF (Conditional Random Field) for post processing which efficiently eradicates the contradiction between the accuracy of segmentation, depth of the network and the number of pooling layers in conventional CNN. This method was used to segment tumors in the brain and yielded a DSC of 86%. Zhaohan Xiong et al. [10] proposed a Dual FCN with 16 layers called AtriaNet for segmentation of the left atrium. This method yielded a DSC o 94%. Mamta Mittal et al. [11] proposed a combination of GCNN (Growing CNN) and SVM for segmentation of brain tumors. This GCNN permitted to encode properties of the inputs to improvise the next step and reduce the parameters. This method yielded a PSNR of 96%.W.V.Deng et al. [12] proposed a fusion of a Heterogeneous CNN (HCNN) and a CRF (Conditional Random Fields).The CRF has been developed as a Recurrent Regression NN (RRNN). This method could divide the brain images into several slices and yielded a precision and recall of 96.5% and 97.8% respectively which is by far the highest among the discussed methods.

Segmentation of Breast Tumors

Variants of CNN networks have been used in successful segmentation of breast masses as well. Ademola Enitan Ilesanmi et al. [13] proposed a DL based segmentation technique exclusively for breast tumors using VEU-Net. (Variant Enhanced Block).This method yielded a DSC varying from 74% to 91% for various datasets. Later, Mughad A. Al-Antari et al. [14] proposed a DL based segmentation technique for mammogram using full resolution CNN which yielded a DSC of 92% and accuracy of 92.97%.

Segmentation of Thyroid Nodules

Researchers have paved their way using DL into segmentation of thyroid nodules as welp in the recent days.Jeremy M. Webb et al. [15] implemented a combination of recurrent FCNN and DeepLab V3 for segmenting thyroid nodules from ultrasound images. This method yielded an Intersection over Union (IoU) of 42% for cycts, 53% for nodules and 73% for thyroid nodules. Viksit Kumar et al. [16] proposed a segmentation technique based on prong CNN for segmenting thyroid nodule, gland and cystic components. Prong is nothing but the network shape caused by splitting the architecture for generating multiple outputs. This method yielded a detection rate of 82% and 44% for thyroid nodules and cystic components respectively. Ngoc-Quang N Guyen et al. [17] proposed a Deep CNN network for segmenting boundaries in medical images. This DL network aids in identifying boundaries using multiscale effective decoders. This method yielded an accuracy of 95 ± 3% for segmenting boundaries in different datasets thus, proving to be superior to the other variants.

Segmentation of Parenchyma

Researches have been carried out using CNN variants on segmentation of various other diseases such as the implementation of a 3D patch-based CNN network for parenchymal segmentation from MRI images of the brain by Al-Louzi et al. [18]. This network not only resulted in robust and accurate outcome of brain atrophy and segmentation of lesions in PML but also proved to be valuable clinically and towards including standard forms of quantitative MRI measures in clinical therapies. This variant of the CNN network used was made up of a network architecture consisting of Multiview feature pyramid networks and hierarchical residual blocks consisting of embedded batch normalization and non-linear activation functions.Ying Chen et al. [19] introduced a dense deep CNN that includes popular optimization methods that include dense block, batch normalization and drop-out. This method was implemented to segment lung parenchyma and yielded an accuracy of 95%. J. Ramya et al. [20] introduced a technique for segmentation of optic cup combining DNN and hybrid particle swarm optimization technique which achieved superior performance with a DSC of 98%.

Segmentation of Prostate

The following authors have preferred CNN models for segmentation of prostate carcinoma.Davood Karimi et al. [21] proposed a variant of CNN which involved two strategies to segment prostate. The first strategy is to apply adaptive sampling strategy and the next is to use the disagreement of the CNN ensemble to identify the uncertain segmentations and estimate segmentation uncertainty map. Ke Yan et al. [22] proposed a P-DNN (Propagation DNN) for prostate segmentation. This method incorporates optimal combination of multi-level feature extraction on a single model. This method yielded a DSC of 89.9 ± 2%. Massimo Salvi et al. [23] proposed a CNN with rings (Rapid Identification of Glandular structures) for effective detection and segmentation of Gland segmentation in prostate histopathological images. This method yielded a DSC of 90%.

Segmentation of Cardiac Tissues

CNN networks have proved to be successful in segmentation of cardiac images as well.Huaifei Hu et al. [24] proposed a combination of FCN and 3D ASM (Active Shape Model) to segment right and left ventricles in cardiac MRI. The method yielded a JI of 89%. Hisham Abdeltawab et al. [25] proposed a segmentation technique for the left and right ventricle using dual FCN. FC1 and FC2 were concatenated at the final output. This method received a DSC of 88% to. 96% for different datasets.

Other Related ROI Segmentations

Futhermore, Tang et a. [26] implemented a multi-scale CNN network for Selective Internal Radiation Therapy (SIRT) patients. The trained model was not efficient enough on SIRT data which had low contrast due to reduced dosage as well as lesions having vast difference in density from their surroundings, abnormal liver shape or positioning. Ryu et al. [27] introduced a CNN network made-up of an encoder and inference branches which was combinedly for segmentation as well as classification purposes. This network takes the combination of an input image and its corresponding Euclidean distance maps of the foreground and background as the input data stream. However, several drawbacks were reported as it does not incorporate all kinds of machines available in clinical trials and hence results cannot be obtained for the left out machines. This resulted in a low Jaccard score index (JSI) of 68% and did not prove to be very efficient on heptic lesions. Terapap Apiparakoon et al. [28] proposed a modified CNN model with FPN for bone lesion segmentation.A ladder FPN was introduced to the top-down pathway to semi-supervise the network training and an additional layer was included to extract global features. This method yielded an F1 score of 84% and a precision of 85%. Kurnianingsih et al. [29] proposed a Mask R-CNN based technique for segmentation of cervical cells. This method used Resnet-10 as a backbone and yielded a precision of 92% and recall of 91%. Lee et al. [30] also used the 3D CNN network for the detection of plaque in major calcifications and obtained a decent F1 score of 92%. Nudrat Nida et al. [31] proposed a Region Based CNN (RCNN) in combination with fuzzy C-means clustering (FCM) technique for detecting and segmenting lesions in melanoma. In this method, CNN resolved the insufficient sample problem and FCM extracts affected patches with variable boundaries that aid in disease recognition. This method achieved a F1 score of 95% and accuracy of 94%.

Tariq Mahmood Khan et al. [32] proposed a CNN based network for segmentation of retinal vessel segmentation. This network was a Residual Connection based Encoder Decoder network. This architecture has is capable of retaining and exploiting low-level semantic edge information for robust vessel segmentation. The method yielded an accuracy on 96 ± 1% for different datasets. Veena et al. [33] introduced an optic disc and cup segmentation technique for the diagnosis of glaucoma which yielded an accuracy of 97% using a modified CNN network consisting of 39 layers including nineteen convolutional layers, four max-pooling layers, eleven drop out layers and a single merger layer.

Each of the above discussed CNN variants have their respective advantages and disadvantages. The most common advantage of CNN is that it does not require intense human supervision for feature detection unlike its predecessors but it also suffered certain drawbacks such as its requirement of a large training dataset and its inability to encode the position and orientation of the object.

2.2 Alexnet (2012)

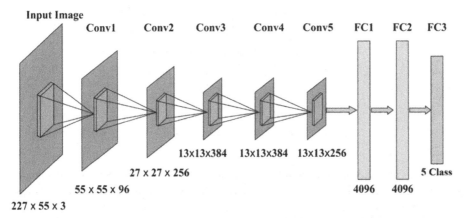

Fig. 3. Architecture of Alexnet [34]

The architecture of alexnet consists of eight layers including five convolutional layers and three fully connected layers as displayed in Fig. 3. But it isn't the layers that make the alexnet special. Instead, alexnet holds a series of features that ended up acting as new approaches to CNN frameworks which made alexnet stand out from the rest. AlexNet replaced the tanh function which was standard at the time with Rectified Linear Units (ReLU). Alexnet was designed to allow multi-GPUs which in turn enabled the training of bigger models at a reduced training time. Also, data augmentation and dropout techniques were deployed in alexnet as a means to the overfitting issue.

Lu et al. [35] proposed an improved alexnet network for detection and segmentation of abnormality in the brain from magnetic resonance images. The last few layers in this improved AlexNet were replaced with an exceptional learning machine which in turn was enhanced by a modified chaotic bat algorithm to attain improved generalization. Chen et al. [36] implemented a 3D framework of alexnet based on classic AlexNet to segment and reconstruct of prostate tumor medical images with adaptive improvement. This method yielded an accuracy of 92%. Also, in comparison with the conventional segmentation as well as depth segmentation methods, the efficacy of the proposed network was exemplary with respect to the time consumption during training, the amount of parameters considered, or evaluation of network performance. Alexnet was most preferred for feature extraction purposes rather than segmentation applications which yielded significant feature extraction results [37]. The disadvantage of alexnet is that because the model isn't particularly deep, it faces difficulty scannning for all attributes, resulting in models that aren't very good.

2.3 Resnet (2015)

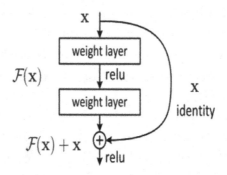

Fig. 4. Residual block [38]

Resnets are made up of residual blocks as shown in Fig. 4. The resnet was originally introduced to rectify the vanishing gradient issue faced by the previous neural networks. The ResNet uses a 34-layered basic network architecture which was influenced by VGG-19 in which a time saving alternate connection is introduced as in Fig. 5. These alternate connections thereby form a residual network. The alternate shortcut connections in resnet are called skip connections and these are the core of the residual. Also, these skip connections are padded with an extra zero in order to increase their dimensions. The skip connections in ResNet help to rectify the setback of vanishing gradients by permitting this new path through which the gradient is allowed to flow. These residual deep learning networks are widely preferred in classification applications but have been used in certain segmentation applications as well.

A resnet50 based mask r-cnn was implemented by Jeevakala et al. [39] as well to segment internal auditory canals and their nerves. The localization results yielded an accuracy of 79%. Song Guo et al. [40] proposed a combination of Resnet-101 and VGG-16 for segmentation of retinal vessels. This network was a multi-scale network and yielded a F1 score of $82 \pm 2\%$. Similarly, Resnet-101 was selected to be the foundation of Mask R-CNN proposed by Zhao et al. [41] where identity mapping block was used as a means to rectify degradation issues faced and facilitate training of the deeper network.

Comparatively, resnet based deep learning models proved to produce better results when used for classification applications rather than segmentation applications in terms of performance measures. Liu et al. [43] proposed a feature pyramid mask r-cnn network based on resnet to segment the nuclei present along the cervical. In this method, pixel-level information was used before hand to dispense supervisory information to train the mask r-cnn. The precision and recall yielded were 96%. One significant disadvantage of resnet is that deeper networks usually necessitate more training time.

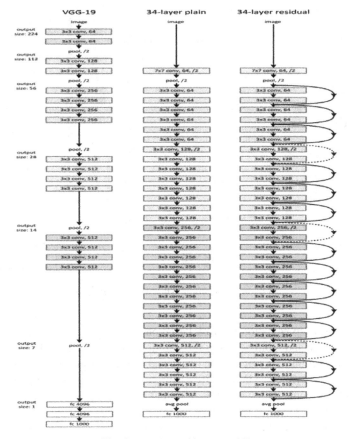

Fig. 5. Resnet architecture [38]

2.4 U-net (2015)

The U-net architecture first evolved from the traditional CNN in the year 2015 for bio-medical images. The u-net network is symmetric along both sides as represented in Fig. 6. The two major parts of the network architecture include the expansive 2D convolutional layers on the right and the contracting path comprising of the general convolutional process along the left. The pooling operations are replaced by upsampling operators consisting of multiple feature channels to amplify the resolution of the output. Different variants of u-net have been deployed for various medical segmentation tasks by researchers around the world as discussed below.

Segmentation of Optic Regions

U-net has been used for segmentation of various regions of the eye. Pan et al. [44] introduced a modified u-net based network for segmentation of retinal vessel segmentation in fundus images. As the traditional u-net was not deep enough, this network was proposed to bind the outcome of the convolutional layer with that of the deep CNN in the residual network under extreme depth conditions. Zheng qiang Jiang et al. [45] proposed

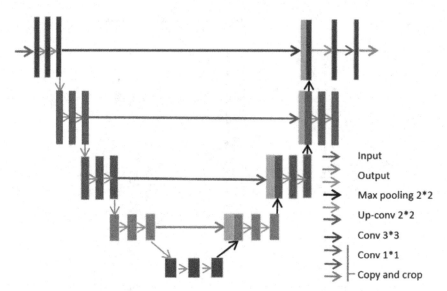

Fig. 6. Basic architecture of U-net [43]

a coronary vessel segmentation network based on U-net. This network was comprised of multi-resolution and multi-scales as the traditional U-net comprised of only a single convolution operation for a single scale image and hence does provide accurate segmentation. This method yielded a DSC of 79%. Xioming Liu et al. [46] proposed a modified U-net model for segmentation of fluids in retinal optic CT images. This variant has an automated attention mechanism to locate the fluid region to avoid the problem of excessive calculation in multi-stage methods. Also, the dense skip connections combined the high-level and low-level features thus making the results of segmentation more precise. This method achieved a DSC of $80 \pm 2\%$ for images from different devices. Sang Yoon Han et al. [47] proposed a segmentation technique inspired by the U-net for detection of pupil centreline this model the complexity of the U-net is reduced by decreasing the number of channels and floors in the U-net. This network achieved a detection rate of 87.3%. Bilel Daoud et al. [48] proposed a segmentation technique for nasopharyngeal carcinoma which was inspired by U-net. The proposed method consisted of 2 CNN based systems with overlapping patches with fixed sizes and with different sizes, thus yielding a DSC of 85% to 91% for axial, coronal and sagittal sections. Khaled Alsaih et al. [49] proposed a method for segmentation of retinal fluid segmentation using Seg-Net which resembled the U-net architecture except that the encoder path is replaced by VGG-16 network. This method yielded a DSC of 92%. Mangipudi et al. [50] proposed an improved u-net based network to segment optic disc and cup in glaucomatic images. Contrast to the original u-net architecture, this network consisted of only half the number of filters in each convolutional layer. Also, the input size was kept low so that the number of parameters used during training could reduce. By doing so, the computational time required for training was reduced to a significant extent and this method yielded a DSC of 93% and 95% for cup and disc segmentation respectively. Bhargav. J.Bhatkalkar et al. [51] proposed a segmentation technique for optic disc in fundus images. This method is a combination of DeepLab V3 + and U-net by incorporating an attention module between the encoder and de-coder to improve the accuracy. This method achieved a DSC of 95 ± 2% for different datasets. Monsumi et al. [52] implemented a iris segmentation method using an interactive variant of U-net that includes modules to squeeze and expand with an aim to reduce the training time and improve storage by reducing the parameters. This method resulted in a DSC of 98%. Shuang Yu et al. proposed a robust optic disc segmentation network based on U-net with Resnet-34 encoding layers. This method yielded an accuracy between 84% to 97% for different datasets.

Segmentation of Various Tumours
Researchers have also used U-net for segmentation of various tumours. A deeper 14 layer U-net model consisting of 26 blocks of VGG19 encoders with ImageNet was implemented by Lu et al. [53]. This method resulted in a DSC of 86% for segmenting the tumor mask, 76% for segmenting contour of the tumor and 66% for segmenting the contour of tumor after gaussian smoothening. Yong Zhou Lu et al. [54] implemented a U-net based DL model with VGG-16 encoder pretrained with ImageNet to segment tumours in PET images. This network yielded a DSC of 86% for mask of tumour and 76% for contour of tumour. Manhoor Ali et al. [55] introduced a model combining 3D CNN and U-net to segment brain tumour from MRI images. This method also replaced Relu activation function by leaky Relu and produced a DSC of 75%, 90% and 84% for enhancing tumour, whole tumour and tumour core respectively. In this method, 3D asymmetric kernels were used for convolution and flat stride was used for pooling to tackle anisotropic spacing. Mohamed A. Naser et al. [56] implemented a U-net model with 1 convolutional transpose layer instead of max-pooling in the de-coding part for segmentation of brain tumour yielded a DSC of 92%. Tran et al. [57] proposed the combined use of U-net and U^n-net for segmenting liver tumours. The U^n-net was designed in such a way that the skip connection path, pooling path and the up-convolutional path are replaced in the node structure. In the re-designed structure, the features in the node of the output layer are conjoined with the next node as well as the encoder node at the same level. This method yielded a DSC of 96.5% and 73% for liver and liver tumour segmentation respectively. Zhenxi Zhang et al. [58] proposed a U-net like model for segmenting 3D MRI images of the brain. Later, Tao Lei et al. [59] proposed an enhanced U-net model named as Def ED-Net (Deformable Encoder Decoder Network for liver and liver tumour segmentation. This method avoids loss of spatial contextual information of images by employing deformable convolution with residual structures to generate feature maps. This model yielded a DSC of 96% which is exemplary comparatively.

Segmentation of Blood Vessels
U-net has also proved to be efficient in segmentation of blood vessels. Manual E Gegundez-Arias et al. [60] proposed a simplified U-net with a combination of residual blocks and batch normalisation at the up and down scaling phases. This model was used to segment blood vessels in retinal images based and achieved an accuracy of 95% ± 1% on different datasets. Enda Boudegga et al. [61] proposed a DL network for segmenting blood vessels by extending the well-known U-net. In this method, the standard

convolutional layers were replaced using LCM (Light weight Convolutional Modules) in order to reduce the computations. This method yielded a superior accuracy of 97%.

Segmentation of Cardiac Diseases

Various parts and diseases of the heart have been effectively detected and segmented using U-net. Gurpreet Sing et al. [62] proposed the use of the traditional U-net for automatic segmentation of cardiac CT images and achieved an overall accuracy of 73%. Can Xiao et al. [63] implemented an improved 3D U-net based on FCN for heart coronary artery segmentation. The upper part of the FCN was modified to enable propagation of information to higher resolution layers. This method yielded a DSC of 82%. Lohendran Baskaran et al. [64] proposed a U-net based segmentation of cardiovascular structures from Cardiac CT images and achieved a DSC of 82% as well-lined-Lu et al. [65] proposed a ringed residual U-net for pancreatic segmentation. With the use of the ring residual module this method yielded exemplary results via deep convolution and can consolidate the characteristics of traditional deep learning networks. This network yielded a DSC of $88.32 \pm 2.84\%$. Tao Liu et al. [66] proposed a U-net based RCNN to efficiently segment heart diseases from cardiac images and obtained a DSC of 86% to 95% for different sections of the heart.

Segmentation of Brain Tissues and Tumours

Researches include the use of U-net in brain tissue segmentation tasks which have resulted to be successful. Sil C Van De Leemput et al. [67] proposed a FCNN based U-net model for brain tissue segmentation. In addition to the traditional U-net, shortcuts were added over every two convolutional layers as they speed up convergence and increase the overall performance, achieving a DSC of 87%. Nagaraj Yamanakumar et al. [68] proposed a brain-tissue segmentation model known as M-net which was inspired by U-net. M-net consisted of two side paths and two main encoding and de-coding paths which aids better feature learning. This method produced an accuracy of $94 \pm 2\%$. Fan Zhang et al. [69] proposed a brain tissue segmentation technique using 2D U-net with a novel augmented target loss function to increase accuracy in tissue boundaries. This method yielded a high accuracy of $95 \pm 2\%$.

Other Related ROI Segmentations

Variants of U-net have been employed in several other medical oriented segmentation tasks effectively.

The combination of U-net and U-net++ variant was proposed by Jonmohamadi et al. [70] to automatically segment multiple structures from knee arthroscopy. In U-net++, the skip connections are compensated using nested, dense skip connections as a means to develop a more efficient architecture. This modification was done to supress the semantic gap of the feature maps that lie between encoder and decoder operators. The U-net++ model yielded a DSC of 0.79%, 0.50%, 0.51% and 0.48% during segmentation of femur, tibia, anterio and meniscus respectively. Yuli Sun Hariyani et al. [71] proposed a dual attention based U-net variant for nailfold capillary segmentation. This model was named as the DA-CapNet and it improvised the U-net architecture by including a dual-attention module that captured feature maps more efficiently yielding an IoU of 64% and precision of 77%. Chen et al. [72] introduced the U-net plus variant which was used to segment esophagus and esophageal cancer. In this variant, 2 blocks were introduced to optimise feature extraction of tediously complex and abstract information and as a means to resolve irregular, vague boundaries with ease. The DSC obtained using the U-net plus was 79%. A dense U-net model was introduced by Li et al. [73] for segmentation of mammogramic masses. This variant of u-net combines densely connected CNN with attention gates. The encoder end is densely connected to the CNN whereas the attention gates are connected at the decoder end. This network produced an F1 score of 82.24%.

Sebastin Stenman et al. [74] introduced a U-net and ImageNet combination with Resnet backbone to segment leukocytes which yielded an IoU of 82%. Shyam Lal et al. [75] proposed a nuclei segmentation method for liver cancer detection using a modified U-net. This model was known as NucleiSeg Net and it included a residual block comprising of convolutional layers aiming to obtain high-level semantic features. This method yielded a F1 score of 83% and JSI of 72%. Yesenia Gonzalez et al. [76] proposed a sigmoid colon segmentation network based on U-net. The proposed network combined the use of 2D and 3D operationist DSC obtained using this network was 82% ± 6%. Xieli Li et al. [77] proposed a dual U-net based network for segmentation of overlapping nuclei. This method has a multi-task learning network in which the boundary and region information helps to improve the segmentation accuracy of glaucoma nuclei, especially overlapping ones yielded a F1 score of 82%. Junlong Chen et al. [78] proposed a variant of U-net for aortic dissection which when compared to the traditional encoder block consisted of an enhanced feature representation capability. This method achieved an accuracy of 85%.

Chanbo Huang et al. [79] employed a modified U-net for segmentation of cell images. This variant combined the advantages of U-net and resnet into one module and yielded an accuracy of 97% and IoU of 84%. Bing Bing Zheng et al. [80] introduced the Multi-scale Discriminative Net (MSD-Net) inspired by the U-net model. This variant was used to segment lung infections through four stages of operation. The four stages include feature map scale, a global average pooling layer to extract semantic consistence from the encoder and a pyramid convolutional block to achieve multi scale information. This method achieved a sensitivity of 82% to 86% for three different infections. Amine Amyar et al. [81] proposed a variant of U-net that comprised of convolutional layers with stride = 2 to replace pooling and maintain spatial information. Also, the number of filters were increased from 64 to 1024. This method yielded a DSC of 88%.

Catherine P. Jeyapandian et al. [82] proposed a network for segmenting histologic structures in the kidney cotex. This network was inspired by the conventional U-net with slightly tweaked parameters. The F1 scores obtained, varied from 81%–91% for various structures. Duo Wang et al. [83] implemented the 3D U-net for segmentation of pulmonary nodules and achieved an accuracy varying between 72% to 91% for different tasks. Van-Truong Pham et al. [84] proposed a DL network for segmentation of tympanic membranes from otoscopic images. This network was known as Ear U-net and was based on three paradigms. Firstly, efficientnet was used as encoder. The second paradigm is that, attention gate was used for skin connections and thirdly, residual blocks were used for decoder. The DSC achieved was 92%.

Zhang et al. [85] segmented epicardial fat using dual U-nets and a morphological processing layer. The function of the morphological layer is to accurately identify the pericardium. The first U-net network focuses solely on the detection of the pericardium and the second U-net network was used to locate and segment the epicardial fat. This dual network-based design has yielded a DSC of 91.19%.

Qi-Zhang et al. [86] proposed an Epicardial Fat segmentation network using dual U-nets including a morphological processing layer. The first U-net was for refining and obtaining the inside region of the pericardium and the second layer acted as a backbone for segmentation. This method yielded a DSC of 91%. Francesco Marzola et al. [87] proposed a segmentation technique for transverse musculoskeletal ultrasound images. This DL network was an ensemble NN that combined the predictions of U-net, U-net++, FPN and AttentionNet. This method yielded a precision of 88% and recall of 92%. Lian Ding et al. [88] proposed a light weight U-net variant for segmentation of pediatric hand bones. This model contained a reduced number of up sampling and down sampling operators as well as kernels. This method yielded a DSC of 92.9%.

A lightweight U-net model was introduced by Ding et al. [89] for segmentation of pediatric hand bones. Multiple filters with different kernel sizes were deployed along with two down-sampling operators, two up-sampling operators. This network frame yielded a DSC of 93.1% in the segmentation of pediatric bones. Javier Civit Mascot et al. [90] proposed a TPU (Tensor Processing Unit) cloud based U-net model for segmentation of eye fundus images and achieved a DSC of 94%. Tawsifur Rahman et al. [91] proposed a DL network for detection and segmentation of tuberculosis in chest x-ray with the use of two U-net models. The modified U-net includes a bi-directional convolutional long short term memory that combines feature maps. This method yielded an accuracy and F1 score of 96%. Guodong Zeng et al. [92] proposed a LP-Unet for segmentation of hip-joints in MRI images. In this network, the listic decomposition, convolution and dense up-sampling convolution were applied at the beginning of the 3D U-net. The main advantage of LP-net is that, it reduced the GPU memory. This method obtained a DSC of $97 \pm 2\%$. Al-Kofahi et al. [93] used a combination of the U-net and MXNet library to quantify pixel-level predictions of a number of classes.

U-net networks not only proved to be efficient in medical image segmentation but also generated significant results in image reconstruction and pixel regression as well. The disadvantage of U-Net topologies is that learning may slow down in the middle layers of deeper models, putting the network at danger of ignoring the layers that represent abstract characteristics.

2.5   Volumetric Convolution Network (V-net, 2016)

Although, the V-net was inspired by the U-net, both architectures have their differences. The left portion consists of the compression path and the right portion consists of the de-compression path which is responsible for reverting the original size of the signal. Each portion is divided into various stages that govern different resolutions. Pooling operations are replaced by convolution layers that vary between one to three and a residual function is familiarised at each stage. The convolutional layers are made up of volumetric kernels of $5 \times 5 \times 5$ voxels as displayed in Fig. 7. The Prelu non-linear activation function is present on the left portion and down-sampling is performed to increase the receptive field. On the other hand, the right portion performs a deconvolution operation to increase the size of the input. Few features are similar along both the portions such as the number of convolutional layers provided that the last convolutional layer is responsible for producing the same output size as the input. There are very few implementations of v-net by researchers which are discussed below and more works are to be expected in the latter days.

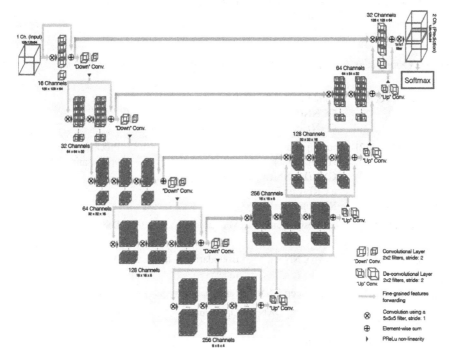

Fig. 7. V-net architecture [94]

Gibson et al. [95] introduced a dense v-net for segmentation of 8 organs in the abdominal region such as the stomach, duodenum, left kidney, liver, spleen, gallbladder and pancreas. Dense V-Net differs in certain ways. The down-sampler consisted of three dense feature stacks connected by down-sampling stridden convolutions. Every skip connection was a convolution of the associated stack output, and the up-sampler comprises bilinear up-sampling. Memory dependencies of the feature stack and spatial dropout enable deep networks at high resolutions, which is an advantage while segmenting smaller structures. Caixia Dong et al. [96] proposed a V-net based 3D DL network known as Di-Vnet for segmenting coronary arteries. It functions as two stages namely, cardiac segmentation, followed by a second stage of CAS(coronary arteries segmentation). This method achieved a DSC of 90 ± 1% for different datasets. Zeng et al. [97] implemented v-net architecture for image fetal segmentation. A combination of v-net and multi-scale loss function was used where v-net was used for the attention mechanism and the multi-scale loss function is used for deep supervision. The combination of these two functions induced significant results and helped to yield a DSC of 97.93%.

3   Discussion and Conclusion

Table 1. Summary of high-performance DL networks

Authors	Network	ROI	Performance
Mostefa Bennaceur et al. [8]	CNN network inspired by occipito temporal pathway	Brain tumour	DSC-90%
Zhaohan Xiong et al. [10]	Dual FCN	Left ventricle	DSC-94%
Mamta Mittal et al. [11]	combination of GCNN(Growing CNN) and SVM	Brain tumour	PSNR-96%
W.V.Deng et al. [12]	fusion of a Heterogeneous CNN(HCNN) and a CRF(Conditional Random Fields)	Brain tumour	Precision and recall of 96.5% and 97.8%
Al-Antari et al. [14]	full resolution CNN	Mammogram	DSC of 92% and accuracy of 92.97%
Ngoc-Quang N Guyen et al. [17]	Deep CNN	Boundaries in medical images	Accuracy of $95 \pm 3\%$
Ying Chen et al. [19]	dense deep CNN	Lung parenchyma	Accuracy of 95%
J.Ramya et al. [20]	combining DNN and hybrid particle swarm optimization technique	Optic cup	DSC of 98%
Massimo Salvi et al. [23]	CNN with rings(Rapid Identification of Glandular structures)	Prostate	DSC of 90%
Kurnianingsih et al. [29]	Mask R-CNN	Cervical cells	Precision of 92% and recall of 91%
Lee et al. [30]	3D CNN network	Detection of plaque in major calcifications	F1 score of 92%
Nudrat Nida et al. [31]	Region Based CNN(RCNN) in combination with fuzzy C-means clustering(FCM)	Lesions in melanoma	F1 score of 95% and accuracy of 94%

(continued)

Table 1. (*continued*)

Authors	Network	ROI	Performance
Tariq Mahmood Khan et al. [32]	CNN based network	Retinal vessel	Accuracy on 96 ± 1%
Veena et al. [33]	modified CNN	Optic disc and cup	Accuracy of 97%
Chen et al. [36]	Alexnet	Prostate tumour	Accuracy of 92%
Liu et al. [42]	resnet	Nuclei	Precision and recall yielded were 96%
Khaled Alsaih et al. [49]	U-net	Retinal fluid	DSC of 92%
Mangipudi et al. [50]	U-net	Optic disc and cup	DSC of 93% and 95% for cup and disc segmentation
Bhargav.J.Bhatkalkar et al. [51]	combination of DeepLab V3 + and U-net	Optic disc	DSC of 95 ± 2%
Monsumi et al. [52]	U-net	Iris	DSC of 98%
Mohamed A. Naser et al. [56]	U-net	Brain tumour	DSC of 92%
Tran et al. [57]	U-net and U^n-net	Liver tumour	DSC of 96.5%
Tao Lei et al. [59]	U-net	Liver tumour	DSC of 96%
Arias et al. [60]	U-net	Blood vessels	DSC of 95% ± 1%
Boudegga et al. [61]	U-net	Blood vessels	Accuracy of 97%
Nagaraj Yamanakumar et al. [68]	U-net	Brain-tissue	Accuracy of 94 ± 2%
Fan Zhang et al. [69]	U-net	Brain tissue	Accuracy of 95 ± 2%
Van-Truong Pham et al. [84]	U-net	Tympanic membranes	DSC achieved was 92%
Zhang et al. [85]	dual U-nets	Epicardial fat	DSC of 91.19%
Qi-Zhang et al. [86]	dual U-nets	Epicardial Fat	DSC of 91%
Lian Ding et al. [88]	U-net	Pediatric hand bones	DSC of 92.9%
Ding et al. [89]	U-net	Pediatric hand bones	DSC of 93.1%
Javier Civit Mascot et al. [90]	U-net	Eye	DSC of 94%
Tawsifur Rahman et al. [91]	U-net	Tuberculosis	F1 score of 96%

(*continued*)

Table 1. (*continued*)

Authors	Network	ROI	Performance
Guodong Zeng et al. [92]	U-net	Hip-joints	DSC of 97 ± 2%
Caixia Dong et al. [96]	V-net	Coronary arteries	DSC of 90 ± 1%
Zeng et al. [97]	V-net	Fetal segmentation	DSC of 97.93%

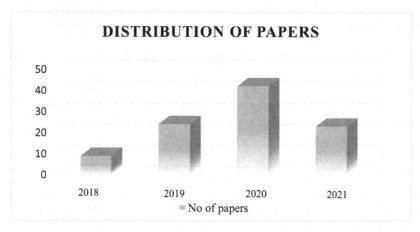

Fig. 8. Distribution of papers

Medical image processing using deep learning is a vast, interesting and challenging research area that conjoins the medical field and the computer field. This survey covers the recent works involving the widely used deep learning networks in medical image segmentation as per the distribution in Fig. 8. Researchers all around the world have been introducing and implementing several variants of DL networks that are derived from the standard DL architectures, geared towards amplifying the performance and rectifying the drawbacks faced by the existing network performances. Such research works are performed to contribute towards the advancement of the healthcare field and assist radiologists in precise diagnosis. This paper summarises the standard network architectures of CNN, Alexnet, Resnet, U-net, V-net and the related works that cover the implementation of its variants along with performance study and a comparison chart. From the study, we have understood that U-net is most preferred and widely used for segmentation of medical images due its high performance measures. We would like to conclude stating that from this survey it is understood that through collaborative research between computer vision techniques and DL techniques the medical field can draw huge benefits Table 1.

References

1. Van Hiep Phung, E.J.: A high-accuracy model average ensemble of convolutional neural networks for classification of cloud image patches on small datasets. Appl. Sci. **9**, 4500 (2019)
2. Ke, Q., Boussaid, F.: Computer vision for human–machine interaction. Comput. Vis. Assist. Heathcare (2018)
3. Yang, B., Guo, H.: Design of cyber-physical-social systems with forensic-awareness based on deep learning. Adv. Comput. **120**, 39–79 (2020)
4. Thillaikkarasi, R., Saravanan, S.: An enhancement of deep learning algorithm for brain tumor segmentation using kernel based CNN with M-SVM. J. Med. Syst. **43**, 1–7 (2019)
5. Sajid, S., Hussain, S.: Brain tumor detection and segmentation in MR images using deep learning. Arab. J. Sci. Eng. **44**, 9249–9261 (2019)
6. Ramzan, F., Khan, M.U.G., Iqbal, S., Saba, T., Rehman, A.: Volumetric segmentation of brain regions from MRI scans using 3D convolutional neural networks. IEEE Access **8**, 103697–103709 (2020). https://doi.org/10.1109/ACCESS.2020.2998901
7. Anand Kumar, G., Sridevi, P.V.: 3D deep learning for automatic brain MR tumor segmentation with T-spline intensity inhomogeneity correction. Autom. Control Comput. Sci. **52**(5), 439–450 (2018). https://doi.org/10.3103/S0146411618050048
8. Ben Naceur, M., Akil, M., Saouli, R., Kachouri, R.: Fully automatic brain tumour segmentation with deep learning-based selective attention using overlapping patches and multiclass weighted cross-entropy. Med. Image Anal. **63**, 101692 (2020). https://doi.org/10.1016/j.media.2020.101692. Epub 29 Apr 2020. PMID: 32417714
9. Feng, N., Geng, X., Qin, L.: Study on MRI medical image segmentation technology based on CNN-CRF model. IEEE Access **8**, 60505–60514 (2020). https://doi.org/10.1109/ACCESS.2020.2982197
10. Xiong, Z., Fedorov, V.V., Fu, X., Cheng, E., Macleod, R., Zhao, J.: Fully automatic left atrium segmentation from late gadolinium enhanced magnetic resonance imaging using a dual fully convolutional neural network. IEEE Trans. Med. Imaging **38**(2), 515–524 (2019). https://doi.org/10.1109/TMI.2018.2866845. PMID: 30716023; PMCID: PMC6364320
11. Mittal, M., Goyal, L.M., Kaur, S., Kaur, I., Amit Verma, D., Hemanth, J.: Deep learning based enhanced tumour segmentation approach for MR brain images. Appl. Soft Comput. **78**, 346–354 (2019)
12. Deng, W., Shi, Q., Wang, M., Zheng, B., Ning, N.: Deep learning-based HCNN and CRF-RRNN model for brain tumor segmentation. IEEE Access **8**, 26665–26675 (2020). https://doi.org/10.1109/ACCESS.2020.2966879
13. Ilesanmi, A.E., Chaumrattanakul, U., Makhanov, S.S.: A method for segmentation of tumours in breast ultrasound images using the variant enhanced deep learning. Biocybern. Biomed. Eng. **41**, 802–818 (2021)
14. Al-antari, M.A., Al-masni, M.A., Choi, M.-T., Han, S.-M., Kim, T.-S.: A fully integrated computer-aided diagnosis system for digital X-ray mammograms via deep learning detection, segmentation, and classification. Int. J. Med. Inform. **117**, 44–54 (2018)
15. Webb, J.M., Meixner, D.D., Adusei, S.A., Polley, E.C., Fatemi, M., Alizad, A.: Automatic deep learning semantic segmentation of ultrasound thyroid cineclips using recurrent fully convolutional networks. IEEE Access **9**, 5119–5127 (2021). https://doi.org/10.1109/ACCESS.2020.3045906
16. Kumar, V., et al.: Automated segmentation of thyroid nodule, gland, and cystic components from ultrasound images using deep learning. IEEE Access **8**, 63482–63496 (2020). https://doi.org/10.1109/ACCESS.2020.2982390

17. Nguyen, N., Lee, S.: Robust boundary segmentation in medical images using a consecutive deep encoder-decoder network. IEEE Access **7**, 33795–33808 (2019). https://doi.org/10.1109/ACCESS.2019.2904094

18. Al-Louzi, O.: Progressive multifocal leukoencephalopathy lesion and brain parenchymal segmentation from MRI using serial deep convolutional neural networks. NeuroImage Clin. **28**, 102499 (2020)

19. . Chen, Y, Wang, Y., Hu, F., Wang, D.: A lung dense deep convolution neural network for robust lung parenchyma segmentation. IEEE Access **8**, 93527–93547 (2020). https://doi.org/10.1109/ACCESS.2020.2993953

20. Ramya, J., Rajakumar, M.P., Uma Maheswari, B.: HPWO-LS-based deep learning approach with S-ROA-optimized optic cup segmentation for fundus image classification. Neural Comput. Appl. **33**(15), 9677–9690 (2021). https://doi.org/10.1007/s00521-021-05732-1

21. Karimi, D., et al.: Accurate and robust deep learning-based segmentation of the prostate clinical target volume in ultrasound images. Med. Image Anal. **57**, 186–196 (2019). https://doi.org/10.1016/j.media.2019.07.005

22. Yan, K., Wang, X., Kim, J., Khadra, M., Fulham, M., Feng, D.: A propagation-DNN: deep combination learning of multi-level features for MR prostate segmentation. Comput. Methods Programs Biomed. **170**, 11–21 (2019)

23. Salvi, M., et al.: A hybrid deep learning approach for gland segmentation in prostate histopathological images. Artif. Intell. Med. **115**, 102076 (2021)

24. Hu, H., et al.: Automatic segmentation of left and right ventricles in cardiac MRI using 3D-ASM and deep learning. Signal Process. Image Commun. **96**, 116303, 101902 (2021)

25. Abdeltawab, H., et al.: A deep learning-based approach for automatic segmentation and quantification of the left ventricle from cardiac cine MR images. Comput. Med. Imaging Graph. **81**, 101717 (2021)

26. Tang, X., et al.: Whole liver segmentation based on deep learning and manual adjustment for clinical use in SIRT. Eur. J. Nucl. Med. Mol. Imaging **47**(12), 2742–2752 (2020). https://doi.org/10.1007/s00259-020-04800-3

27. Ryu, H., Shin, S.Y., Lee, J.Y., Lee, K.M., Kang, H.-J., Yi, J.: Joint segmentation and classification of hepatic lesions in ultrasound images using deep learning. Eur. Radiol. **31**(11), 8733–8742 (2021). https://doi.org/10.1007/s00330-021-07850-9

28. Apiparakoon, T., et al.: MaligNet: semisupervised learning for bone lesion instance segmentation using bone scintigraphy. IEEE Access **8**, 27047–27066 (2020). https://doi.org/10.1109/ACCESS.2020.2971391

29. Allehaibi, K.H.S., et al.: Segmentation and classification of cervical cells using deep learning. IEEE Access **7**, 116925–116941 (2019). https://doi.org/10.1109/ACCESS.2019.2936017

30. Lee, J.: Segmentation of coronary calcified plaque in intravascular OCT images using a two-step deep learning approach. IEEE Access **8**, 225581–225593 (2020)

31. Nida, N., Irtaza, A., Javed, A., Yousaf, M.H., Mahmood, M.T.: Melanoma lesion detection and segmentation using deep region based convolutional neural network and fuzzy C-means clustering. Int. J. Med. Inform. **124**, 37–48 (2019)

32. Khan, T.M., Alhussein, M., Aurangzeb, K., Arsalan, M., Naqvi, S.S., Nawaz, S.J.: Residual connection-based encoder decoder network (RCED-Net) for retinal vessel segmentation. IEEE Access **8**, 131257–131272 (2020). https://doi.org/10.1109/ACCESS.2020.3008899

33. Veena, H.: A novel optic disc and optic cup segmentation technique to diagnose glaucoma using deep learning convolutional neural network over retinal fundus images. J. King Saud Univ. (2021)

34. Vaishnavi, J.: An efficient adaptive histogram based segmentation and extraction model for the classification of severities on diabetic retinopathy. Multimedia Tools Appl. **79**, 30439–30452 (2020)

35. Lu, S., Wang, S.-H., Zhang, Y.-D.: Detection of abnormal brain in MRI via improved AlexNet and ELM optimized by chaotic bat algorithm. Neural Comput. Appl. **33**(17), 10799–10811 (2020). https://doi.org/10.1007/s00521-020-05082-4

36. Chen, J.: Medical image segmentation and reconstruction of prostate tumor based on 3D AlexNet. Comput. Methods Programs Biomed. **200**, 105878 (2021)

37. Mansour, R.F.: Deep-learning-based automatic computer-aided diagnosis system for diabetic retinopathy. Biomed. Eng. Lett. **8**, 41–57 (2018)

38. He, K., Zhang, X.: Deep residual learning for image recognition. arXiv (2015)

39. Jeevakala, S., Sreelakshmi, C., Ram, K., Rangasami, R., Sivaprakasam, M.: Artificial intelligence in detection and segmentation of internal auditory canal and its nerves using deep learning techniques. Int. J. Comput. Assist. Radiol. Surg. **15**(11), 1859–1867 (2020). https://doi.org/10.1007/s11548-020-02237-5

40. Guo, S., Wang, K., Kang, H., Zhang, Y., Gao, Y., Li, T.: BTS-DSN: deeply supervised neural network with short connections for retinal vessel segmentation. Int. J. Med. Inform. **126**, 105–113 (2019)

41. Zhao, X.: EBioMedicine (2020)

42. Liu, Y.: Automatic segmentation of cervical nuclei based on deep learning and a conditional random field. IEEE Access **6**, 53709–53721 (2018)

43. Ding, L.: A lightweight U-Net architecture multi-scale convolutional network for pediatric hand bone segmentation in X-ray image. IEEE Access **7**, 68436–68445 (2019)

44. Pan, X.: A fundus retinal vessels segmentation scheme based on the improved deep learning U-Net model. IEEE Access **7**, 122634–122643 (2019)

45. Jiang, Z., Ou, C., Qian, Y., Rehan, R., Yong, A.: Coronary vessel segmentation using multiresolution and multiscale deep learning. Inform. Med. Unlocked **24**, 100602 (2021)

46. Xiong, Z., Fedorov, V.V., Fu, X., Cheng, E., Macleod, R., Zhao, J.: Fully automatic left atrium segmentation from late gadolinium enhanced magnetic resonance imaging using a dual fully convolutional neural network. IEEE Trans. Med Imaging **38**(2), 515–524 (2019). https://doi.org/10.1109/TMI.2018.2866845

47. Han, S.Y., Kwon, H.J., Kim, Y., Cho, N.I.: Noise-robust pupil center detection through CNN-based segmentation with shape-prior loss. IEEE Access **8**, 64739–64749 (2020). https://doi.org/10.1109/ACCESS.2020.2985095

48. Daoud, B., Morooka, K., Kurazume, R., Leila, F., Mnejja, W., Daoud, J.: 3D segmentation of nasopharyngeal carcinoma from CT images using cascade deep learning. Comput. Med. Imaging Graph. **77**, 101644 (2019)

49. Alsaih, K., Yusoff, M.Z., Faye, I., Tang, T.B., Meriaudeau, F.: Retinal fluid segmentation using ensembled 2-dimensionally and 2.5-dimensionally deep learning networks. IEEE Access **8**, 152452–152464 (2020). https://doi.org/10.1109/ACCESS.2020.3017449

50. Mangipudi, P.S., Pandey, H.M., Choudhary, A.: Improved optic disc and cup segmentation in Glaucomatic images using deep learning architecture. Multimedia Tools Appl. **80**(20), 30143–30163 (2021). https://doi.org/10.1007/s11042-020-10430-6

51. Bhatkalkar, B.J., Reddy, D.R., Prabhu, S., Bhandary, S.V.: Improving the performance of convolutional neural network for the segmentation of optic disc in fundus images using attention gates and conditional random fields. IEEE Access **8**, 29299–29310 (2020). https://doi.org/10.1109/ACCESS.2020.2972318

52. Sardar, M., Banerjee, S., Mitra, S.: Iris segmentation using interactive deep learning. IEEE Access **8**, 219322–219330 (2020). https://doi.org/10.1109/ACCESS.2020.3041519

53. Lu, Y.: Automatic tumor segmentation by means of deep convolutional U-Net with pre-trained encoder in PET images. IEEE Access **8**, 113636–113648 (2020)

54. Lu, Y., Lin, J., Chen, S., He, H., Cai, Y.: Automatic tumor segmentation by means of deep convolutional U-Net with pre-trained encoder in PET images. IEEE Access **8**, 113636–113648 (2020). https://doi.org/10.1109/ACCESS.2020.3003138

55. Ali, M., Gilani, S.O., Waris, A., Zafar, K., Jamil, M.: Brain tumour image segmentation using deep networks. IEEE Access **8**, 153589–153598 (2020). https://doi.org/10.1109/ACCESS.2020.3018160

56. Naser, M.A., Jamal Deen, M.: Brain tumour segmentation and grading of lower-grade glioma using deep learning in MRI images. Comput. Biol. Med. **121**, 103758 (2020)

57. Tran, S.-T.: A multiple layer U-Net, Un-Net, for liver and liver tumor segmentation in CT. IEEE Access **9**, 3752–3764 (2020)

58. Zhang, Z., Li, J., Tian, C., Zhong, Z., Jiao, Z., Gao, X.: Quality-driven deep active learning method for 3D brain MRI segmentation. Neurocomputing **446**, 106–117 (2021)

59. Lei, T., Wang, R., Zhang, Y., Wan, Y., Liu, C., Nandi, A.K.: DefED-Net: deformable encoder-decoder network for liver and liver tumor segmentation. IEEE Trans. Radiat. Plasma Med. Sci. (2021). https://doi.org/10.1109/TRPMS.2021.3059780

60. Gegundez-Arias, M.E., Marin-Santos, D., Perez-Borrero, I., Vasallo-Vazquez, M.J.: A new deep learning method for blood vessel segmentation in retinal images based on convolutional kernels and modified U-Net model. Comput. Methods Programs Biomed. **205**, 106081 (2021)

61. Boudegga, H., Elloumi, Y., Akil, M., Bedoui, M.H., Kachouri, R., Abdallah, A.B.: Fast and efficient retinal blood vessel segmentation method based on deep learning network. Comput. Med. Imaging Graph. **90**, 101902 (2021)

62. Gurpreet, S., et al.: Deep learning based automatic segmentation of cardiac computed tomography. J. Am. Coll. Cardiol. **73**, 1643–1643 (2019)

63. Xiao, C., Li, Y., Jiang, Y.: Heart coronary artery segmentation and disease risk warning based on a deep learning algorithm. IEEE Access **8**, 140108–140121 (2020). https://doi.org/10.1109/ACCESS.2020.3010800

64. Baskaran, L., et al.: Automatic segmentation of multiple cardiovascular structures from cardiac computed tomography angiography images using deep learning (2020). https://doi.org/10.1371/journal.pone.0232573

65. Lu, L., Jian, L., Luo, J., Xiao, B.: Pancreatic segmentation via ringed residual U-Net. IEEE Access **7**, 172871–172878 (2019). https://doi.org/10.1109/ACCESS.2019.2956550

66. Liu, T., Tian, Y., Zhao, S., Huang, X., Wang, Q.: Residual convolutional neural network for cardiac image segmentation and heart disease diagnosis. IEEE Access **8**, 82153–82161 (2020). https://doi.org/10.1109/ACCESS.2020.2991424

67. Van De Leemput, S.C., Meijs, M., Patel, A., Meijer, F.J.A., Van Ginneken, B., Manniesing, R.: Multiclass brain tissue segmentation in 4D CT using convolutional neural networks. IEEE Access **7**, 51557–51569 (2019). https://doi.org/10.1109/ACCESS.2019.2910348

68. Yamanakkanavar, N., Lee, B.: Using a patch-wise M-Net convolutional neural network for tissue segmentation in brain MRI images. IEEE Access **8**, 120946–120958 (2020). https://doi.org/10.1109/ACCESS.2020.3006317

69. Zhang, F., et al.: Deep learning based segmentation of brain tissue from diffusion MRI. Neuroimage **233**, 117934 (2021)

70. Jonmohamadi, Y.: Automatic segmentation of multiple structures in knee arthroscopy using deep learning. IEEE Access **8**, 51853–51861 (2020)

71. Hariyani, Y.S., Eom, H., Park, C.: DA-CapNet: dual attention deep learning based on U-Net for nailfold capillary segmentation. IEEE Access **8**, 10543–10553 (2020). https://doi.org/10.1109/ACCESS.2020.2965651

72. Chen, S.: U-Net plus: deep semantic segmentation for esophagus and esophageal cancer in computed tomography images. IEEE Access **7**, 82867–82877 (2019)

73. Li, S.: Attention dense-U-net for automatic breast mass segmentation in digital mammogram. IEEE Access **7**, 59037–59047 (2019)

74. Stenman, S., et al.: Antibody supervised training of a deep learning based algorithm for leukocyte segmentation in papillary thyroid carcinoma. IEEE J. Biomed. Health Inform. **25**(2), 422–428 (2021). https://doi.org/10.1109/JBHI.2020.2994970

75. Lal, S., Das, D., Alabhya, K., Kanfade, A., Kumar, A., Kini, J.: NucleiSegNet: robust deep learning architecture for the nuclei segmentation of liver cancer histopathology images. Comput. Biol. Med. **128**, 104075 (2021)
76. Gonzalez, Y., et al.: Semi-automatic sigmoid colon segmentation in CT for radiation therapy treatment planning via an iterative 2.5-D deep learning approach. Med. Image Anal. **68**, 101896 (2021)
77. Li, X., Wang, Y., Tang, Q., Fan, Z., Yu, J.: Dual U-Net for the segmentation of overlapping glioma nuclei. IEEE Access **7**, 84040–84052 (2019). https://doi.org/10.1109/ACCESS.2019.2924744
78. Cheng, J., Tian, S., Yu, L., Ma, X., Xing, Y.: A deep learning algorithm using contrast-enhanced computed tomography (CT) images for segmentation and rapid automatic detection of aortic dissection. Biomed. Signal Process. Control **62**, 102145 (2020)
79. Huang, C., Ding, H., Liu, C.: Segmentation of cell images based on improved deep learning approach. IEEE Access **8**, 110189–110202 (2020). https://doi.org/10.1109/ACCESS.2020.3001571
80. Zheng, B., et al.: MSD-Net: multi-scale discriminative network for COVID-19 lung infection segmentation on CT. IEEE Access **8**, 185786–185795 (2020). https://doi.org/10.1109/ACCESS.2020.3027738
81. Amyar, A., Modzelewski, R., Li, H., Ruan, S.: Multi-task deep learning based CT imaging analysis for COVID-19 pneumonia: classification and segmentation. Comput. Biol. Med **126**, 104037 (2020). https://doi.org/10.1016/j.compbiomed.2020.104037
82. Jayapandian, C.P., Chen, Y., Janowczyk, A.R., Palmer, M.B.: Development and evaluation of deep learning–based segmentation of histologic structures in the kidney cortex with multiple histologic stains. Kidney Int. **99**(1), 86–101 (2021)
83. Wang, D., Zhang, T., Li, M., Bueno, R., Jayender, J.: 3D deep learning based classification of pulmonary ground glass opacity nodules with automatic segmentation. Comput. Med. Imaging Graph. **88**, 101814 (2021)
84. Pham, V.-T., Tran, T.-T., Wang, P.-C., Chen, P.-Y., Lo, M.-T.: EAR-UNet: a deep learning-based approach for segmentation of tympanic membranes from otoscopic images. Artif. Intell. Med. **115**, 102065 (2021)
85. Zhang, Q.: Automatic epicardial fat segmentation and quantification of CT scans using dual U-Nets with a morphological processing layer. IEEE Access **8**, 128032–128041 (2020)
86. Zhang, Q., Zhou, J., Zhang, B., Jia, W., Wu, E.: Automatic epicardial fat segmentation and quantification of CT scans using dual U-nets with a morphological processing layer. IEEE Access **8**, 128032–128041 (2020). https://doi.org/10.1109/ACCESS.2020.3008190
87. Marzola, F., van Alfen, N., Doorduin, J., Meiburger, K.M.: Deep learning segmentation of transverse musculoskeletal ultrasound images for neuromuscular disease assessment. Comput. Biol. Med. **135**, 104623 (2021)
88. Ding, L., Zhao, K., Zhang, X., Wang, X., Zhang, J.: A lightweight U-Net architecture multi-scale convolutional network for pediatric hand bone segmentation in X-ray image. IEEE Access **7**, 68436–68445 (2019). https://doi.org/10.1109/ACCESS.2019.2918205
89. Ding, Y.: A stacked multi-connection simple reducing net for brain tumor segmentation. IEEE Access **7**, 104011–104024 (2019)
90. Civit-Masot, J., Luna-Perejón, F., Vicente-Díaz, S., Rodríguez Corral, J.M., Civit, A.: TPU cloud-based generalized U-Net for eye fundus image segmentation. IEEE Access **7**,142379–142387 (2019). https://doi.org/10.1109/ACCESS.2019.2944692
91. Rahman, T., et al.: Reliable tuberculosis detection using chest X-ray with deep learning, segmentation and visualization. IEEE Access **8**, 191586–191601 (2020). https://doi.org/10.1109/ACCESS.2020.3031384

92. Zeng, G., et al.: MRI-based 3D models of the hip joint enables radiation-free computer-assisted planning of periacetabular osteotomy for treatment of hip dysplasia using deep learning for automatic segmentation. Eur. J. Radiol. Open **8,** 100303 (2020). https://doi.org/10.1016/j.ejro.2020.100303

93. Al-Kofahi, Y.: A deep learning-based algorithm for 2-D cell segmentation in microscopy images . BMC Inform. **19**, 1–11 (2018)

94. Milletari, F.: Hough-CNN: deep learning for segmentation of deep brain regions in MRI and ultra-sound. Comput. Vis. Image Underst. **164**, 92–102 (2017)

95. Milletari, F., et al.: Hough-CNN: deep learning for segmentation of deep brain regions in MRI and ultra-sound Comput. Vis. Image Underst. **164**, 92–102 (2017)

96. Gibson, E.: Automatic multi-organ segmentation on abdominal CT with dense V-networks. IEEE Trans. Medi. Imaging. IEEE Trans. Med. Imaging, **37**(8), 1822–1834 (2018)

97. Zeng, Y., Tsui, P.-H., Wu, W., Zhou, Z., Wu, S.: Fetal ultrasound image segmentation for automatic head circumference biometry using deeply supervised attention-gated V-Net. J. Digit. Imaging **34**(1), 134–148 (2021). https://doi.org/10.1007/s10278-020-00410-5

Real Time Captioning and Notes Making of Online Classes

A. Vasantha Raman, V. Sanjay Thiruvengadam, J. Santhosh, and Thenmozhi Durairaj[✉]

Sri Sivasubramaniya Nadar College of Engineering, Chennai, India
theni_d@ssn.edu.in

Abstract. Due to the COVID-19 pandemic, all activities have turned online. The people who are hard of hearing are facing high difficulty to continue their education. So, the presented system supports them in attending the online classes by providing the real time captions. Additionally, it provides summarized notes for all the students so that they can refer to them before the next class. Google Speech to Text API is used to convert the speech to text, for providing real time captions. Three text summarization models were explored, namely BART, Seq2Seq model and the TextRank algorithm. The BART and the Seq2Seq models require a labelled dataset for training, whereas the TextRank algorithm is an unsupervised learning algorithm. For BART, the dataset is built using semi supervised methods. We evaluated all these models with rouge score evaluation metrics, among these BART proves to be best for our dataset with the following scores of 0.47, 0.30, 0.48 for rouge-1, rouge-2 and rouge-l respectively.

Keywords: Bidirectional Auto Regressive Transformer (BART) · Sequence to Sequence Model (Seq2Seq Model) · Transformers · Summarization · Speech-to- text

1 Introduction

In wake of the COVID-19 pandemic, all the activities have turned online. It came to be called the virus that changes the way we use the internet. With increasing concerns over social distancing, people are seeking ways to connect socially online. All the classes are being conducted online and going by the present trends, the online mode of study is here to stay. Global corporate outlook of the scenario is no different. Companies that once had whiteboard pitch meetings, have turned to screen sharing and presentations on online meeting platforms. College lectures is one such scenario, wherein both the students and teachers are involved in the process of imparting and enriching their knowledge throughout the course of lecture. Online meeting platforms are playing a key role in making sure that the activities stay afloat during the course of pandemic.

To further aid the same, we introduce a real time caption and lecture summarizer system. This research work helps reduce the stress on students by doing the

© IFIP International Federation for Information Processing 2022
Published by Springer Nature Switzerland AG 2022
L. Kalinathan et al. (Eds.): ICCIDS 2022, IFIP AICT 654, pp. 207–220, 2022.
https://doi.org/10.1007/978-3-031-16364-7_16

summarization on the contents of the class and helps in preparation of exams. This system can be considered as an extension of the idea brought by IBM Viascribe [9] module wherein the speech is converted to text real time. This idea used by Viascribe is expanded to new depths by adding a summarization feature to the module lecture endings. People who miss classes can refer to the summarized notes for reference in a quick glance. This proposed system can be used by Online Meeting Platforms such as Zoom, Google Meets and Microsoft Teams. This system can also be extended to offline classes to real time scenarios as well. People with hearing impairments feel very difficult to understand the online classes. The proposed system supports them in attending the classes by providing the real time captions along with summaries of the classes.

The main objective of the proposed system is to create an application that provides services of Real Time Captioning and Notes Making for a Lecture.

The focus of this research work is on speech to text conversion and on text summarization. To implement the functionality of speech to text, Google Speech to Text API [3] is used. For providing the utility of summarization, BART model [5] is used. The model was chosen, after consideration of other established models like Sequence to Sequence and Text Rank [7]. The chosen BART model was attuned to suit that of a lecture on Computer Science domain by training it over a course of NPTEL video lectures and its summaries, making up for a data set of a thousand eight hundred rows.

2   Related Work

IBM proposed a system named ViaScribe and Hosted Transcription Service [9] in 2013 which is used for real time captioning system using speech recognition. It is captioning software for digital audio and video that creates a written transcript and audio recording, as well as automatically captions live audio content as it occurs. It used ViaVoice engine in the process of doing so. It was observed to produce 80% accuracy. They were working on using cloud to extend the service. But, recently it was announced that IBM discontinued the Viascribe service

Microsoft proposed a system named 'Advances in online audio visual meeting transcription' [6] in 2019 which is used to identify the speakers while transcripting when people are talking simultaneously in a meeting. In this speech separation system, camera signal is not used. They used continuous speech separation method. The transcription system provides each word utterance with the attendee name. The challenge was to provide the real time transcription.

IBC proposed a system named 'Just in time prepared captioning for live transmission' [8] in 2017 which is an in-time transcription of audio in a live news broadcast. They done this using automatic speech recognition system. This work is done specific to news live broadcast. They are trying to increase accuracy. Their focus was primarily on suiting to accents of english like UK, Australia and US. The percentage accuracy of pieces marked ranged from 40.91% to 100%, while the percentage of words missing ranged from 0% to 68.52%

Facebook proposed a system named 'Wav2vec 2.0' [1] in 2020, Framework for self-supervised learning of representations of raw audio data. This uses multi layer convolutional neural network. There is no extension to other regional languages.

Review of speech to text (STR) recognition technology for enhancing learning [10] studies and overviews new speech to text technologies in the period of 1999 to 2014 which was proposed on 2014. This is built on speech recognition technology. Their focus was to analyze the results of study and to understand how Speech-to-Text Recognition technology can enhance learning. The review revealed that the STR-generated texts enable students understand the lectures much better and perform well in the examinations.

The invention of a new simple architecture, Transformer [2] was published on 2017, solely based on attention mechanisms. Self-attention is an attention mechanism relating different positions of a single sequence in order to compute a representation of the sequence. This aims on introducing transformers model to revolutionize Natural language Processing by overcoming the problems experienced by then most preferred Recurrent Neural Networks (RNN). RNNs generate a sequence of hidden states h_t, as a function of the previous hidden state h_{t1} and the input for position t. This becomes critical at longer sequence lengths, as memory constraints limit batching across examples. Extension of the Transformer to problems involving input and output modalities other than text.

Future n-gram prediction for CNN/DailyMail [12] was proposed in 2020. It is a pretrained Seq2Seq model called ProphetNet. Instead of optimizing one-step-ahead prediction in the traditional sequence-to-sequence model, the ProphetNet is optimized by n-step ahead prediction that predicts the next n tokens simultaneously based on previous context tokens at each time step.

Summarize text with pre trained encoders [11] was proposed in 2019. In this they used Bidirectional Encoder Representations from Transformers. Quadratic dependency (mainly in terms of memory) on the sequence length due to their full attention mechanism. BERT uses Transformers to perform the language modeling task. The paper's results show that a language model which is bidirectionally trained can have a deeper sense of language context and flow than single-direction language models. In the paper, the authors explored an innovative technique named Masked Language Modeling which allows bidirectional training in models.

NLP tasks like question answering and generation [4] was proposed on 2020 with improved performance and Using sparse attention mechanisms to reduce dependency on sequence length memory. In this approximated learning is achieved because of sparse attention mechanism. BigBird is a universal approximator of sequence functions and is Turing complete, thereby preserving these properties of the quadratic, full attention model. But, it Requires long context for the purposes of NLP operations including question generation, summarization etc.

A denoising autoencoder for pretraining sequence to sequence models [5] was proposed on 2019. Its trained by corrupting the text with an arbitrary noising function and learning a model to reconstruct it. This involves Exploring new methods to corrupt documents for pre-training and tailoring them to specific tasks including Question generation, Text generation, Summarization etc. It uses a standard Transformer-based neural machine translation architecture which, despite its simplicity, can be seen as generalizing BERT (due to the bidirectional encoder), GPT (with the left-to-right decoder), and many other more recent pretraining schemes. It achieves new state-of-the-art results on a range of abstractive dialogue, question answering, and summarization tasks, with gains of up to 6 ROUGE.

TextRank [7] system was proposed in 2004. It is based on bringing Order into Texts. Graph based ranking algorithm for text processing and Exploring new ways to link sentences/keywords. Graph-based ranking algorithms like HITS algorithm, Google's PageRank have been successfully used in citation analysis, social networks, and the analysis of the link-structure of the World Wide Web.

Google's Automatic Speech to Text Recognition Systems provide state of the art results [3]. They are built using attention based encoder-decoder architectures. They present results with a unidirectional LSTM encoder for streaming recognition.

3   Proposed System

The proposed system consists of the below mentioned modules.

1. Speech to Text Recognition
2. Data Collection
3. Data Annotation
4. Data Preprocessing
5. Choosing the best model
6. Fine Tuning
7. Summary Generation

The application gets the speech as input from the microphone and the Google Speech to Text API converts the speech into text. This is provided as real time captions to the clients. The captions are appended into a transcript. Once the meeting gets over, the transcript is sent to the text summarizer for the summarization task which is given as notes to the clients.

3.1 Application Workflow

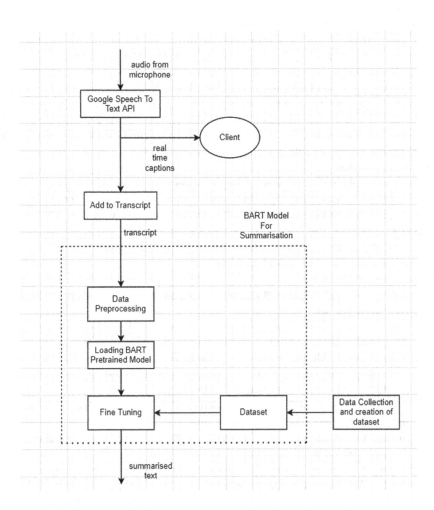

Fig. 1. Application workflow

Once the application starts, a new session can be started and the teacher can start talking. The speech to text transcription happens using the Google Speech to Text API. This real time caption is presented to the client in the application. The captions are appended to a transcript which is used for creating notes at the end of the lecture. Once the meeting ends, the transcript is sent to the summarizer model which generates the notes. This flow is depicted in Fig. 1.

Module Description

1. Google Speech to Text API
   The speech from the speaker's microphone is detected and sent to the Google Speech to Text API. The output is the real time captions which is sent to the clients. Along with this, the captions are added to the transcript.
2. Data Collection
   The audio from YouTube using YouTube Data API are collected. The audio files of 3 playlists were taken, namely Design and Analysis of Algorithms, Software Testing and Programming in C. A total of 182 audio files were collected. These audio files were given as input to the Google Speech to Text API and the corresponding transcripts were collected.
3. Data Annotation
   The transcripts collected from Google Speech to Text API is summarized manually. For each transcript, the corresponding summary is written and a csv file is made which constitutes (transcript, summary) pairs. This is the dataset which we use for choosing the summarization model. This dataset consists of 600 rows. Among these 600 rows, 300 rows are training set and the remaining 300 rows are testing set.
4. Data Preprocessing
   The dataset is loaded into a dataframe using pandas. The various steps done for preprocessing are
   - Cleaning The null values, escape characters, and symbols are removed. All the alphabets are converted into their lowercase letters.
   - Tokenization All the words are tokenized using a tokenizer and special tokens are added at the beginning and the end of each sentence. _START_is the start token and _END_token.
   - Stop words removal The words which don't add much value to the meaning are removed. Words like the, a, are, etc. are removed.
   - Stemming Its the process of converting the words into its root form. For example, eaten word is stemmed as eat.
5. Choosing the best summarization model The 3 models we are considering are BART, Sequence to Sequence model, and the TextRank algorithm. These 3 models are trained with the above annotated dataset of 300 rows. The 3 trained models are evaluated with the testing set of 300 rows using ROUGE score. The best performing model is chosen as our summarizer.
6. Fine Tuning and Building Dataset
   - Getting Audio Files: Audio files of the lectures/online classes are obtained from NPTEL HRD Swayam Youtube page. The playlists included Design and Analysis of Algorithms, Introduction to Programming with C, Software testing.
   - Google Speech To Text API: The input audio are files are given to Google Speech to Text which is pretrained model for speech to text, gives the transcripts for the input audio files.
   - Transcripts: The Google Speech to Text API gives the transcripts in form of text files.

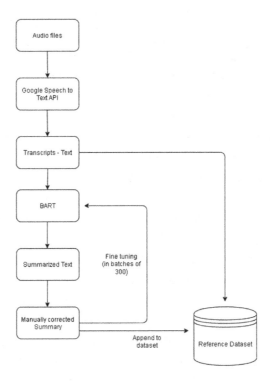

Fig. 2. Data processing flow

– BART Model: The BART model is initially downloaded from the huggingface transformers collection. It is pretrained on large Corpus of 16GB
  of data. The transcript is fed to this version of BART. And the output is
  obtained as summarized text.
– Manual Correction: This summary is manually edited to add on in the
  required words and remove the unwanted ones. This way, the BART
  model is fine tuned on the corrected version of the data and learns them
  over a period of time and over many iterations in batches of 300.
– Reference Dataset: This is built over many iterations in a semi supervised
  manner. This collected dataset can be used for future purposes. The same
  flow is depicted in Fig. 2
7. Summary Generation: After fine tuning the BART model, the model is
   exported as pickle file and its used in the application.

3.2 Architecture Diagram

For the task of summarization, as shown in Fig. 3, 3 models were considered,
namely BART, Sequence to Sequence model, and TextRank Algorithm. The
transcript from the lecture is sent to these summarizers and their outputs are
evaluated. According to the results obtained, the best model is chosen.

The train dataset consists of 1800 instances of (transcript, summary). The BART and the Sequence to Sequence models are fine tuned using this dataset. The test dataset comprises of 300 instances of (transcript, summary). The Rouge scores of each model is found by comparing the predicted summary of each model with the target summary.

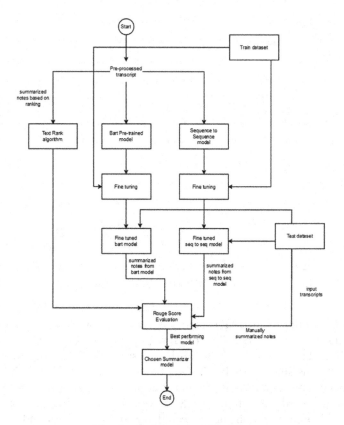

Fig. 3. Architecture diagram

4   Implementation

4.1   Dataset

Dataset consists of output text files given by Google Speech to text API, when read through Lecture content of 82 videos. These videos were got from YouTube. The playlists that we used are Design and Analysis of Algorithms, Programming in C, Software Testing.

The input is fed into pretrained BART model from which intermediate output is obtained. This undergoes manual supervision, and the revisioned form is used as target summary for fine tuning the data in the next iteration.

4.2   Process of Collecting Dataset

Since the ground truth for the summaries of the video transcripts were not available, the dataset was created using semi-supervised learning method. The transcripts were downloaded and the summaries were created manually for 300 rows. Then the model was fine-tuned using this as the dataset. Then, another 300 rows were given to the model to predict the summaries. Then, the predicted summaries were manually validated and again the model was trained with the updated dataset. Likewise, this process was repeated for 2100 rows.

4.3   Experimental Setup

The environment and the various libraries and API used for the project are explained. Primarily Google Speech to Text API is used for speech to text conversion and Huggingface transformers are used for summarization by BART model.

Python v3.7. Python 3.7, the latest version of the language, as a requirement for BART finetuning.

Fastai v2. fastai is a deep learning library which is used along with BART for finetuning.

TorchText. TorchText is internally used by transformer BART during the process of finetuning.

Huggingface Transformers. The BART pretrained model was downloaded from the huggingface transformers collection.

Google Colab. Google colab was used to fine tune the BART model over our dataset consisting of 2×2234 rows. The content consists of 120 videos of data related to computer science field, including software testing, Design And Analysis of Algorithms, Introduction to Programming with C.

Google Speech to Text. An application (Videoder) used to download the audios in any given format from youtube servers for the required playlist. This was then given to the Google Speech to Text API available in Google Cloud Platforms, from which the transcripts for the whole audios were returned. This transcripts were in turn used to fine tune the BART model for the purpose of summarization.

4.4   Model Building

The explanation about the implementation is mentioned below. The link to access the notebook is also given at the last section of implementation.

Importing the Packages. The first step involved importing the ohmeow-blurr, datasets, bert-score packages.

Loading the Dataset. The dataset is stored into a dataframe from a csv file.

Data Preprocessing. The null values are removed using dropna() function in pandas. The tokenizer object of BART is then created which takes care of the tokenization process.

The dataset is broken into batches called datablocks for processing. The default argument values for the batch transform were given using the function default_text_gen_kwargs(). The arguments involve max_length, min_length, number of return sequences, padding token id, beginning of sentence token, end of sentence token, cachable and many more. To view full list, refer to the implementation link. Once all the arguments are assigned values, the BatchTransform constructor is called with the architecture object, configuration object, tokenizer as parameters. The data is split into batches called as datablocks.

Loading BART Pretrained Model. The bart pretrained model, architecture, configuration objects are loaded using the ohmeow-blurr library from huggingface.

Fine Tuning. The BART Model is prepared for training by wrapping it in blurr's HF_BaseModelWrapper object and using the callback, HF_BaseModelCallback. A new HF_Seq2SeqMetricsCallback object can be used for specifying the Seq2Seq metrics, i.e. Rouge.

A Learner object is created which is used for fine tuning. The constructor takes the datablocks, model loss function, callback object and splitter as parameters. Here, CrossEntropyLoss is used as loss function.

Then, we freeze our model so that only the last layer group's parameter is trainable. Then the fine tuning is done using a function named fit_one_cycle() with the number of epochs, learning rate and the callback object as the parameters.

Obtaining Results. The results are obtained using the show_results() function.

Exporting the Model. The model was be exported as a pickle model using load_learner() function.

Summarize Using the Model. The blurr_generate (text) function can be used for summarizing the text.

5 Results and Performance Analysis

5.1 Rouge (Metric)

ROUGE is a set of quantitative metrics for evaluating the summary. The metrics compare an automatically produced summary against a reference summary. The commonly used Rouge metrics are Rouge-1, Rouge-2 and Rouge-L. The Rouge-1 measures the overlap of the unigrams, the Rouge-2 measures the overlap of the bigrams and the Rouge-L measures the overlap of Longest Common Subsequence (LCS) in the produced summary and the reference summary.

5.2 BART Model

The BART model is trained using semi-supervised learning. The train dataset consists of 1800 rows with transcripts, but not summary. So, the train dataset was divided into 6 blocks, each with 300 rows each. At each iteration, the predicted summary of the previous iteration model is manually corrected. This (transcript, summary) is used for fine tuning the next iteration BART model. The ROUGE score of the model got after each iteration is mentioned below. The results of the BART model at each iteration are shown below in Table 1. There is a good improvement in the Rouge scores after each iteration.

Table 1. Rouge scores of BART at each iteration

Iteration	Rouge-1	Rouge-2	Rouge-L
1	0.471191912847482	0.302830036989221	0.486972871206460
2	0.473210500215044	0.304601467425532	0.496998216586448
3	0.470217640597870	0.304199531679189	0.493619067435048
4	0.481398041623864	0.321774241800320	0.501677609785666
5	0.542424041559528	0.408314456403223	0.553228532702422
6	0.536735590282537	0.389777546282537	0.545027953284034

The graph of the Table 1 is presented below for each type of the Rouge scores. The graph for Rouge-1, Rouge-2 and Rouge-L scores are shown in Figs. 4a, 4b and 4c respectively.

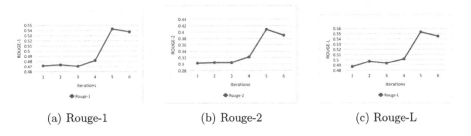

(a) Rouge-1   (b) Rouge-2   (c) Rouge-L

Fig. 4. Rouge score

Table 2. Rouge score comparison of models

Models	Rouge-1	Rouge-2	Rouge-L
BART	0.471191	0.302830	0.486972
Seq2Seq	0.191351	0.014464	0.203535
TextRank	0.463548	0.31577	0.521487

5.3 Inference

From the above results, for each iteration of the summarization, there is a gradual increase in the rouge score. The dataset is built progressively using semi-supervised learning. This shows that for a particular domain, our model performs well when there is more dataset. The ROUGE scores are calculated with the TextRank Algorithm, the Sequence to Sequence Model and with the BART model. The results are shown in Table 2. From the above results, the BART model outperforms and proves to be the best. Thus it can be shown that the semi supervised learning can be applied to domains that lack a good quantity of dataset to begin with, gradually building over time whilst also improving the ML model.

5.4 Screenshots of GUI

The Application Home Page is shown in Fig. 5a. The new session can be started by clicking the "Start new session recording" button in the home page. Then a new screen is displayed where the real time captions are printed as shown in Fig. 5b.

(a) Application Home Page

(b) After clicking start new session recording

(c) Displaying the transcript after the session ends

(d) Summarized notes of the transcript

Fig. 5. .

Once the meeting ends, the session can be stopped by clicking at the text. Then the home page is returned. The transcript can be viewed by pressing the button "View selected session transcript" as shown in Fig. 5c The transcript downloaded by clicking the button "Download text file". Then the transcript can be summarized by clicking "Summarize" button as shown in Fig. 5d

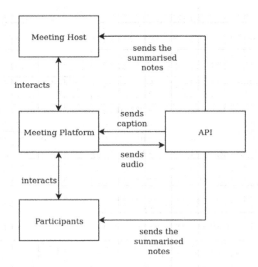

Fig. 6. Summarized notes of the transcript

6   Conclusion and Future Work

The design of the proposed system is discussed in the project. The application will perform real time captioning of lectures and provides summarized notes at the end. Google Speech to Text API is one of the tools available to perform speech to text conversion which is availed and used in the project. Similarly, the BART model for text summarization is used to provide the notes of the lecture at the end. The BART model for text summarization is used because it performs well for the dataset when compared to sequence-to-sequence model and text rank algorithm using rougescore evaluation metrics. Bart model performs with scores of 0.47, 0.30, 0.48 for rouge-1, rouge-2 and rouge-l whereas Sequence to Sequence performs with scores of 0.19, 0.01, 0.20 for rouge-1, rouge-2 and rouge-l respectively. Many different adaptations, tests, and experiments have been left for the future due to lack of time. As a part of the future work, this project could be extended by integrating this application with the existing online meeting platforms such as Zoom, Google Meet and Microsoft Teams. This involves exposing the application's service as an API across the web and performing seamless integration with online meeting platforms as depicted in Fig. 6.

References

1. Baevski, A., Zhou, H., Mohamed, A., Auli, M.: wav2vec 2.0: a framework for self-supervised learning of speech representations. Adv. Neural. Inf. Process. Syst. **33**, 2541–2551 (2020)
2. Vaswani, A., et al.: Attention is all you need. In: Proceedings of the 2020 Conference on Empirical Methods in Natural Language Processing (EMNLP), pp. 3862–3872 (2017)
3. Chiu, C.-C., et al.: State-of-the-art speech recognition with sequence-to-sequence models. IEEE Trans. Learn. Technol. **5**(3), 1206–1214
4. Zaheer, M.: Big bird: transformers for longer sequences. In: Proceedings of the 2019 Conference of the North American Chapter of the Association for Computational Linguistics: Human Language Technologies, pp. 505–516 (2020)
5. Lewis, M., et al.: Facebook AI 'BART: Denoising Sequence-to-Sequence Pre-training for Natural Language Generation, Translation, and Comprehension', pp. 7871–7880. Association for Computational Linguistics (ACL) (2019)
6. Yoshioka, T., et al.: Advances in online audio-visual meeting transcription. IEEE Trans. Learn. Technol. **4**(2), 1181–1192 (2019)
7. Mihalcea, R., Tarau, P: TextRank: Bringing Order into Texts, pp. 404–411. Association for Computational Linguistics (2004)
8. Renals, S., Simpson, M.N., Bell, P.J., Barrett, J.: Just-in-time prepared captioning for live transmissions, pp 27–35. IET Publication (2016)
9. Ranchal, R., et al.: Using speech recognition for real-time captioning and lecture transcription in the classroom. IEEE Trans. Learn. Technol. **6**(4), 299–311 (2013)
10. Shadiev, R., Hwang, W.-Y., Chen, N.-S., Huang, Y.-M.: Review of speech-to-text recognition technology for enhancing learning. Educ. Technol. Soc. **17**, 65–84 (2014)
11. Liu, Y., Lapata, M.: Text Summarization with Pretrained Encoders, pp. 3730–3740. Association for Computational Linguistics (ACL) (2019)
12. Yan, Y., et al.: ProphetNet: predicting future n-gram for sequence-to-sequence pre-training. In: Findings of the Association for Computational Linguistics, EMNLP 2020, pp. 2401–2410 (2020)

Disease Identification in Tomato Leaf Using Pre-trained ResNet and Deformable Inception

Arnav Ahuja[1], Aditya Tulsyan[1], and J. Jennifer Ranjani[2]([⊠]) [iD]

[1] Department of Computer Science and Information Systems and Department of Mathematics, Birla Institute of Technology and Science, Pilani 333031, India
[2] Virufy, San Francisco, USA
j.jenniferranjani@yahoo.co.in

Abstract. Disease in crops is a growing concern for the food and farming industry. In the last couple of decades, lack of immunity in plants and extreme climate conditions due to drastic climate changes have caused a significant increase in the growth of diseases among crops. On a large scale, these diseases cause eventual financial loss to the farmers due to a decrease in farming. Fast and early detection of the disease remains a challenge in most parts of the world because of the lack of robust research infrastructure. Automated techniques for disease detection are high in demand with a worldwide increase in image capturing and video recording devices, together with the continuous evolution in computer vision and machine learning. In this paper, a neural network model, trained on publicly available data which comprise both healthy and leaves with the disease, is proposed. The model adopts a new approach combining the ResNet and the InceptionNet architectures. Skip connections and the 1×1 convolutions in both the architectures are put to good use here. We achieve an accuracy of 99.08% in the PlantVillage dataset [17] and reasonable accuracy of 66.06% on the PlantDoc dataset [16] which is an increment of more than 25% from the approaches in the previous works. This paper also suggests a method to improve the detection of diseases in crops in the real world by augmenting the number of data points. We have discussed the use of deformable convolution, which is capable of learning various geometric transformations and can improve the performance of the architecture.

Keywords: Crop disease detection and classification · Tomato leaf · Deep learning · Convolutional neural network · Residual network · ResNet · Deformable convolution

1 Introduction

Agriculture is a vital contributor to the economy of any country as it is the major source of nourishment, raw material, and fuel. More than two-thirds of the

© IFIP International Federation for Information Processing 2022
Published by Springer Nature Switzerland AG 2022
L. Kalinathan et al. (Eds.): ICCIDS 2022, IFIP AICT 654, pp. 221–234, 2022.
https://doi.org/10.1007/978-3-031-16364-7_17

population all over the globe depend on agriculture directly or indirectly. Plant disease weakens the yield, and it affects the quality of food. Plant diseases and pests have become a vital challenge for the agriculture sector. Also, situations like less irrigation, global warming, dwindling pollinators, and plant diseases threaten the food sector. Tomato is one of the most vital and economically available crops, not only in India but across the globe, and its production has significantly expanded through these years [1]. Worldwide, tomato cultivation exposes it to many infections, instilling new diseases in plants. This crop is essentially defenseless and highly susceptible to many pathogens. Moreover, new viral diseases keep emerging [2] which drives a need to have a robust crop disease identification system. Through this work, we focus on recognizing diseases that attack the tomato plant.

2   Related Works

In the last few decades, multiple works were done by researchers in crop disease identification to identify the presence of pathogens accurately in different crops. Most models were able to attain quite a high accuracy but, a key concern is the ability of the models to work in non-laboratory i.e., in an environment like actual farmland. Among the previous work done by various researchers in this field, most publishers have used the PlantVillage dataset containing 54000 images from 14 different crops having over 25 plant diseases and taken in laboratory setup and in a particular time interval. In [3], transfer learning with the standard architectures was used, with a focus on the InceptionNet and MobileNet. Many approaches like [4] and [13] use the PlantVillage dataset in detecting the disease and have achieved good results. In [5], five different CNN architectures were used, namely, GoogLeNet, AlexNet, AlexNetOWTbn, VGG, and Overfeat architectures, in which VGG showcased the best results among all the architectures. Similarly in [18] the authors have compared architectures like ResNet, MobileNet, DenseNet and Inception (v3) for the PlantVillage dataset. They have done three types of classifications- binary classification (diseased or not diseased leaf), six class classification and 10 class classification. We use some of these models later for bench marking the performance of our architecture in comparison with the other architectures. For these papers, apart from the above-stated drawback of using images of plants clicked in laboratory conditions, the authors also used only one single architecture in the model and mainly did a hit and trial method to find the best architecture for various plants without giving out proper rationale. Each architecture comes with a set of disadvantages giving rise to the need for developing a model that works the best in most scenarios.

In [6], three classifiers such as support vector machines (SVM), extreme learning machine, and K-Nearest Neighbor, along with the latest Deep Learning models (DL) like ResNet (layers = 50 and 101), Inception (v3), GoogLeNet, SqueezeNet, and InceptionResNetv2, were used to classify and recognize seven distinct plant diseases. Analysis reveals that ResNet-50 with SVM delivers the most promising performance in terms of F1-score, specificity, sensitivity. In [7], plant diseases in cucumber were classified by basic CNN models and got the highest accuracy. According to [8], a new Deep Learning (DL) model, Inception (v3), was used for the detection of cassava disease. The conventional plant disease recognition and classification methods were replaced by Super-Resolution Convolutional Neural Network (SRCNN) in [9]. For tomato disease classification, SqueezeNet v1.1, & AlexNet models, were used in which AlexNet performed better among the two models in terms of the metric of accuracy in the work referenced in [10]. An analysis comparing the architectures was presented in [11] to select the best architecture for the detection of diseases in crops. And in [12], six different diseases in tomato crop, classified using VGG-16 and AlexNet architectures, were compared in terms of accuracy.

In the work mentioned above, researchers have used either PlantVillage or any other dataset containing images taken under laboratory conditions. Also, most publishers restrict their model to a particular architecture along with the conventional convolution technique that uses the kernel of fixed size as a filter. Other approaches like the deformable convolutional networks can improve the results of the models as shown in [14]. We have summarized the network used, dataset and metrics from various related works in Table 1.

Table 1. Summary on related works

Paper	Network used	Dataset	Metrics
Crop disease detection using deep learning [3]	MobileNet	PlantVillage	Accuracy = 99.6%
	InceptionNet	PlantVillage	Accuracy = 99.7%
ResNet based approach for classification and detection of plant leaf diseases [4]	ResNet	PlantVillage	Accuracy = 99.4% Precision = 0.965
Tomato crop disease classification using pre-trained deep learning algorithm [12]	AlexNet	PlantVillage	Accuracy = 97.49%
	VGG-16	PlantVillage	Accuracy = 97.23%
Tomato leaf diseases detection using deep learning technique [18]	ResNet	PlantVillage	Accuracy = 96.75% Precision = 0.968
	MobileNet	PlantVillage	Accuracy = 97.2% Precision = 0.972
	DenseNet	PlantVillage	Accuracy = 98.05% Precision = 0.980
	Inception (v3)	PlantVillage	Accuracy = 97.35% Precision = 0.973
Deep learning for image based cassava disease detection [8]	Inception (v3)	Custom Cassave Leaves Dataset	Accuracy = 92.9%
Disease detection on the leaves of tomato plants by using deep learning [10]	AlexNet	PlantVillage	Accuracy = 95.65%
	SqueezeNet	PlantVillage	Accuracy = 94.3%
A comparative study of fine tuning learning models for plant disease identification [11]	VGGNet	PlantVillage	Accuracy = 81.83%
	ResNet	PlantVillage	Accuracy = 99.5%
	Inception	PlantVillage	Accuracy = 98.08%
	DenseNet	PlantVillage	Accuracy = 99.6%

The proposed model aims to address the issues and drawbacks mentioned above. Our model is trained on plant images from real farmland setup with varying angles, resolution, and plants in different stages of growth. Secondly, we tried merging various network architectures according to their strength and developed a model to classify tomato leaves accurately. The paper is organized as follows: Sect. 3 describes the proposed architecture, Sect. 4 summarizes the two popular datasets along with the new mast dataset used in this work. Results from the various experiments conducted are summarized in Sect. 5. Section 5 and 6 are on conclusion and future work, respectively.

3   Proposed Model

In this paper, we propose an architecture based on the Residual network, ResNet-50, and Inception network architecture along with deformable convolution [14] for accurate prediction of disease in any tomato plant. ResNet architecture comprises core building blocks, known as residual blocks. Skip connections in the residual block address the decreasing gradient problem in deep neural networks. In the process of training our tomato plant images over ResNet-50, the residual shortcuts guarantee us the network integrity in our architecture if the coefficients of the regular connections converges to zero. These alternate connections improve the architecture by giving two options namely the normal path of these shortcuts. Also, skip connection ensures that our model will assimilate the function $f(x) = x$ which will ensure that the deeper layers perform as well as the shallow layers by taking care of the problem of vanishing and exploding gradients.

By referring to Inception, we mean Inception (v1), also known as GoogLeNet. Since, salient feature in a plant image is critical in identifying the type of diseases like bacterial spot, early or late blight will vary largely in size, shape, and position for every plant leaf image. It might be possible that the entire leaf inside the image has the spatial feature required to identify and classify the disease, or there can be a very tiny spot of infection in the whole leaf. Hence, we need filters of varying sizes to accurately extract the desired feature from the tomato plant and classify it into the type of disease. Multiple filters for a single input map demand a wide network rather than a deep network. Inception (v1) networks offer three different filter sizes such as $(1 \times 1, 3 \times 3, 5 \times 5)$, and hence it fits the best as per our needs. The output from every module is further concatenated and sent to the next Inception module. There are in total four such Inception modules in our architecture.

The major drawback of a typical Convolution Neural Network (CNN) is the sampling of the feature map by using a fixed size kernel. The CNN uses a fixed size window that is rectangular shaped to read the input feature map at set locations. The pooling layer of the Inception module uses the same set-size rectangular kernel to diminish the spatial resolution by a set ratio. This, in turn, introduces a variety of problems, namely the same receptive field of all the activation units in a given layer of the architecture even though there might be differently scaled objects available at various positions spatially. Vision-based tasks like object detection and segmentation require fine-localization for adapting to the object's scale. In such tasks, it is a desirable quality to have distinct receptive field sizes for every object. Hence, along with the Inception network having kernels of various sizes, we have used deformable convolution over our architecture for a more accurate extraction of features. Deformable convolutions offer a useful tool, as they add offsets that are 2-dimensional whereas standard convolutions offer a regular sampling of the grid locations. These offsets, often learned with the help of an additional convolutional layer, provide the sampling grid with a free-form deformation. Thus, the input features condition the deformation in a dense, local, and adaptive manner.

The ResNet-50 network comprises five stages, each with convolution and an identity block. Each convolution block has three convolution layers, and each identity block also has three convolution layers. The complete ResNet-50 has over 23 million trainable parameters. For our purpose, we have used a pre-trained ResNet-50 trained on the ImageNet dataset [15] and have replaced the last stage of ResNet-50 with three Inception (v1) blocks to extract salient features from the image accurately. We trained our network using a deformable convolutional technique for complete stage five i.e., Inception block with a varying number of convolution layers for the fourth stage of the network. We have done an ablation study in Sect. 4 to decide the number of trainable parameters, which give the most accurate and optimal result. A block diagram of the architecture is shown in Fig. 1. The first two layers of the architecture are the same as the first two layers of the ResNet-50 model. The third layer of the architecture consists of 4 bottleneck layers of the ResNet-50 architecture as well as 2 Inception modules. The fourth layer consists of 2 Inception modules and one bottleneck layer of the ResNet-50 architecture. Finally there are avg pooling and fully connected layers.

4 Dataset

4.1 PlantVillage Dataset

The most widely used dataset for the identification of diseases in plants is the PlantVillage Dataset [17], which consists of a large number of high-quality images taken under laboratory conditions. The dataset creators created a system for disease detection using GoogleNet and AlexNet architectures and obtained up to 99.35% accuracy. The dataset consists of 16,011 images in the tomato class, and these are the images we will use for classification in our model. PlantVillage dataset contains leaf images with a uniform, white background as they are curated in the laboratory setup. These images are not captured under natural conditions like farms, due to which their effectiveness in real-time situations is substandard.

4.2 PlantDoc Dataset

In contrast, we have used the PlantDoc Dataset [16] which contains data of diseased and healthy plants in real farmland scenarios. PlantDoc is comparatively

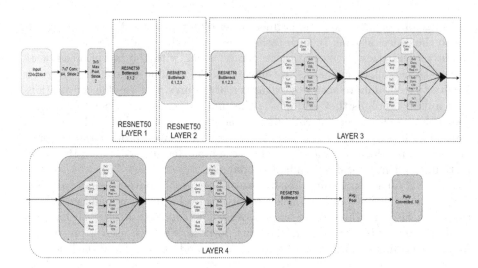

Fig. 1. Proposed InceptionResNet architecture for crop disease identification

Table 2. Number of training and testing images

Serial no.	Class name	PlantVillage		PlantDoc	
		Train data	Test data	Train data	Test data
1	Leaf Mold	809	143	85	6
2	Bacterial Spot	1808	319	101	9
3	Septoria Spot	1505	266	140	11
4	Early Blight	850	150	78	9
5	Late Blight	1623	286	101	10
6	Yellow Virus	2728	481	69	6
7	Mosaic Virus	317	56	44	10
8	Healthy	1352	239	54	8

close to the application that can help us use our model and identify diseases in crops and solve degradation of quality in plants. PlantDoc dataset originally contains 2,598 data points which consists of 13 plant categories and 17 types of diseases. For this paper, we restricted our study to one plant species i.e., tomato leaf, with eight classes containing one class for healthy leaves while the other seven with the diseased leaves. Out of 741 data points for tomato leaves, we segregated 672 data points for training our model and the remaining for testing the model i.e., 90%–10% split. Table 2 contains the exact distribution of tomato leaf for testing and training across different diseases.

Figure 2 contains sample images from different classes in both the PlantVillage and PlantDoc dataset.

Fig. 2. Samples from various classes to show gap between lab-controlled and real-life images

4.3 New Mask Dataset Using Data Augmentation

The PlantDoc dataset is a small dataset with about 700 images, hence, training a large model becomes difficult. Training only the PlantDoc dataset has led to lower accuracy in our models, as is evident in Tables 3 and 4, because of over-fitting in the images due to the small dataset size. However, the PlantDoc dataset mimics the images taken in real conditions, and hence it is necessary to get good accuracy in this dataset. To solve these two approaches were used-

Approach 1 - Basic augmentation like a random rotation of the image, flipping of the image, etc., was used. The torch.transform library was used in performing the random rotation and image flipping. Then, these images were used in random batches to train the models. However, this did not drive the accuracy significantly up, as the number of images originally was comparatively less for the parameters to learn any general trend. The need for a larger dataset arose.

Approach 2 - To solve this problem the PlantVillage dataset, which had an abundance of images, was put to use. Images from the PlantVillage dataset are superimposed on the PlantDoc dataset. Images of the same class are superimposed on top of each other, i.e., images of the yellow virus class of the PlantVillage dataset are superimposed on images with the yellow virus in the PlantDoc dataset. This led to the formation of the augmented dataset consisting of 14,000 images. These images were now used to train the ResNet and the InceptionResNet models. Samples from the augmented dataset are given in Fig. 3

Fig. 3. Samples from various classes in the New Mask Dataset

5   Experimental Results

In this section, we verify the performance of our model in identifying diseases in tomato leaf and summarize the experiments conducted together with the results. We have provided the code for further experiment and research[1]. The experimental analysis is summarized in several sections. In the first few sections, we have done an ablation study on the number of layers needed to be trained and the hyperparameters that give the most optimal result. Next, we have showcased the final result obtained from the conclusion of the ablation study.

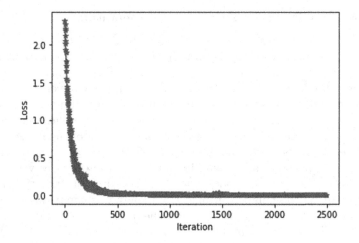

Fig. 4. Loss vs Iteration for the InceptionResNet model

5.1   Training

We have trained the model in random batches of a specific size for all three datasets. This training process takes place for the number of epochs passed as a parameter in the train function. The loss vs. iterations graph in Fig. 4 gives a snapshot of the neural network after every ten iterations, and it gives an idea of the performance of the network during training. Since the loss vs. iterations graph smoothly decreases during the iterations, it suggests the correctness of the model.

---

[1] github.com/arnavahuja/CropDiseaseIdentification.

Table 3. Results for ResNet-50 architecture

Architecture name	Description	Dataset	Accuracy
ResNet	Fully connected layer unfreezed	PlantVillage	0.726741
ResNet	Fully connected and one convolutional layer unfreezed	PlantVillage	0.925436
ResNet	Fully connected layer and two conv layer Unfreezed	PlantVillage	0.975007
ResNet	Fully connected layer unfreezed	PlantDoc	0.261029
ResNet	Fully connected and one convolutional layer unfreezed	PlantDoc	0.295189
ResNet	Fully connected layer and two conv layer Unfreezed	PlantDoc	0.272518
ResNet	Fully connected layer unfreezed	New Mask Dataset	0.74499 (NewData) 0.330968 (PD)
ResNet	Fully connected and one convolutional layer unfreezed	New Mask Dataset	0.83155 (NewData) 0.440379 (PD)
ResNet	Fully connected layer and two conv layer Unfreezed	New Mask Dataset	0.96711 (NewData) 0.624149 (PD)
ResNet	All layers unfreezed	PlantVillage	0.979115
ResNet	All layers unfreezed	PlantDoc	0.265165
ResNet	All layers unfreezed	New Mask Dataset	0.99179(NewData) 0.652945(PD)

5.2 Performance Evaluation

The calculate accuracy function takes the batch size as a parameter and gets a random batch from the validation data. The model then predicts the label of the validation data and then the number of correct labels is calculated. This accuracy is averaged over the entire validation dataset.

Table 4. Results for the InceptionResNet architecture

Architecture name	Description	Dataset	Accuracy
InceptionResNet	Inception blocks in layer 4	PlantVillage	0.981685
InceptionResNet	Inception blocks in layer 4	PlantDoc	0.424019
InceptionResNet	Inception blocks in layer 4	New Mask Dataset	0.95889 (NewData) 0.533121 (PD)
InceptionResNet	All layers unfreezed, along with bottleneck 0, 1, 2 of layer 4 changed to inception block	PlantVillage	0.990011
InceptionResNet	All layers unfreezed, along with bottleneck 0, 1, 2 of layer 4 changed to inception block	PlantDoc	0.589996
InceptionResNet	All layers unfreezed, along with bottleneck 4, 5 of layer 3 and bottleneck 0, 1 of layer 4 changed to inception block	New Mask Dataset	0.99086 (NewData) 0.660636 (PD)

5.3 Ablation Study on the Neural Network Architecture

The impact of various hyper-parameters on the architectures was experimented with, and we have tabulated their results in Table 3 and Table 4.

In Table 4, we have summarized the results for the InceptionResNet architecture, which has produced promising results. It has given an accuracy of 66.036%, which is greater than the previously achieved accuracy. Thus, it testifies the success of the InceptionResNet architecture and the new augmented dataset.

The success of this architecture is due to the unique 1×1 convolutions in the Inception block along with the skip connections in the ResNet. The 1×1 convolutions ensure that the number of trainable parameters does not significantly increase while increasing the number of kernels.

5.4 Benchmark Results

Benchmark results are a test of how our model performs in comparison with others.

The benchmark used here is the accuracy for the standard architectures in the same situations. The standard architectures like the VGG, ResNet, Alexnet, XceptionNet all give lower accuracies than our proposed architecture InceptionResNet (Table 5).

Table 5. Benchmark results of standard architectures

Architecture name	Description	Dataset	Accuracy
AlexNet	Trained entire network due to its small size	PlantVillage	0.983786
AlexNet	Trained entire network due to its small size	PlantDoc	0.508655
AlexNet	Trained entire network due to its small size	New Mask Dataset	0.96529 (NewData) 0.504191 (PD)
VGG	Trained the entire classifier of the VGG which consists of all fully connected layers	PlantVillage	0.800348
VGG	Trained the entire classifier of the VGG which consists of all fully connected layers	PlantDoc	0.413526
VGG	Trained the entire classifier of the VGG which consists of all fully connected layers	New Mask Dataset	0.73766 (NewData) 0.281480 (PD)
Xception	Trained last block (no. 12 of XceptionNet) $-68,01,706$ params	PlantVillage	0.980904
Xception	Trained last block (no. 12 of XceptionNet) $-68,01,706$ params	PlantDoc	0.477558
Xception	Trained last block (no. 12 of XceptionNet) $-68,01,706$ params	New Mask Dataset	0.84175 (NewData) 0.460062 (PD)

5.5 Hyperparameter Study

Table 6 illustrates the ablation study done on different hyper-parameters like learning and batch size for both the PlantVillage and the new mask dataset. We have experimented with these hyperparameters for both the ResNet and the InceptionResNet architectures. The number of epochs used was 1500 for all the experiments. Only the results of the PlantVillage dataset and the new mask dataset are relevant here as the PlantDoc dataset has low accuracy in itself and does not require a hyperparameter study.

Table 6. Hyperparameter Study

Model	Dataset	Learning rate	Batch size	Accuracy
ResNet	PlantVillage	0.0001	64	0.958135
		0.0001	128	0.964829
		0.00001	64	0.971923
		0.00001	128	0.979115
	New Mask Dataset	0.0001	64	0.964921
		0.0001	128	0.991799
		0.00001	64	0.984289
		0.00001	128	0.974923
Inception ResNet	PlantVillage	0.0001	64	0.990011
		0.0001	128	0.989294
		0.00001	64	0.969211
		0.00001	128	0.975479
	New Mask Dataset	0.0001	64	0.980857
		0.0001	128	0.968935
		0.00001	64	0.971294
		0.00001	128	0.990868

6   Conclusion

In this paper, we propose a novel neural network model based on standard convolution and deformable convolution techniques to identify disease in the Tomato plant in a real-time manner. The model we developed was trained and tested over images taken in actual farmland conditions and its augmented form to make our model more robust.

We look forward to our proposed system making an important contribution to the area of research in agriculture and crucially help farmers across the globe to solve the problem of infections in plants, especially fruits and vegetable bearing plants. As informed to our readers, at the beginning of this paper, more than 30% of the farm produce becomes trash because of various diseases. Hence, through this paper, we are hopeful that our work will help solve the problem of food scarcity in several underdeveloped nations that eventually leads to malnutrition among children.

7   Future Work

In future work, these models can be trained entirely using deformable convolutions. Deformable convolutions offer an advantage over traditional convolutions as they use learnable parameters, and the shape of the convolution kernels is not fixed. However, due to our limitations in terms of computational capacity, we were able to train only the last layers in the deformable models. Thus in the future, we can train the entire network with deformable convolutions from scratch, as the initial results look promising. Also, our work can be extended to various other groups of plant species apart from tomatoes. Moreover, numerous diseases exist, which we could not cover due to dataset limitations. Hence, our model can be made more robust and much more exhaustive.

Acknowledgement. This work was carried out when Jennifer Ranjani J was affiliated to Birla Institute of Technology and Science, Pilani Campus, Pilani, Rajasthan - 333031.

References

1. Food and Agriculture Organization of the United Nations: Value of Agricultural Production-Tomatoes. Food and Agriculture Data (2015). www.fao.org/faostat/en/#data/QV/visualize
2. Hanssen, I., Lapidot, M., Thomma, B.: Emerging viral diseases of tomato crops. Mol. Plant Microbe Interact. **23**, 539–548 (2010)
3. Kulkarni, O.: Crop disease detection using deep learning. In: 2018 4th International Conference on Computing Communication Control and Automation (ICCUBEA), Pune, India, pp. 1–4 (2018). https://doi.org/10.1109/ICCUBEA.2018.8697390
4. Kumar, V., Arora, H., Harsh, Sisodia, J.: ResNet-based approach for detection and classification of plant leaf diseases. In: 2020 International Conference on Electronics and Sustainable Communication Systems (ICESC), Coimbatore, India, pp. 495–502 (2020). https://doi.org/10.1109/ICESC48915.2020.9155585
5. Ferentinos, K.: Deep learning models for plant disease detection and diagnosis. Comput. Electron. Agric. **145**, 311–318 (2018). https://doi.org/10.1016/j.compag.2018.01.009
6. türkoğglu, M., Hanbay, D.: Plant disease and pest detection using deep learning-based features. Turk. J. Electr. Eng. Comput. Sci. **27**, 1636–1651 (2019). https://doi.org/10.3906/elk-1809-181
7. Fujita, E., Kawasaki, Y., Uga, H., Kagiwada, S., Iyatomi, H.: Basic investigation on a robust and practical plant diagnostic system. In: Proceedings of the 2016 15th IEEE International Conference on Machine Learning and Applications (ICMLA), Anaheim, CA, USA, 18–20 December 2016, pp. 989–992 (2016)
8. Ramcharan, A., Baranowski, K., McCloskey, P., Ahmed, B., Legg, J., Hughes, D.P.: Deep learning for image-based cassava disease detection. Front. Plant Sci. **2017**, 8 (1852)
9. Yamamoto, K., Togami, T., Yamaguchi, N.: Super-resolution of plant disease images for the acceleration of image-based phenotyping and vigor diagnosis in agriculture. Sensors **17**, 2557 (2017)

10. Durmuş, H., Güneş, E.O., Kırcı, M.: Disease detection on the leaves of the tomato plants by using deep learning. In: Proceedings of the 2017 6th International Conference on Agro-Geoinformatics, Fairfax, VA, USA, 7–10 August 2017, pp. 1–5 (2017)
11. Too, E.C., Yujian, L., Njuki, S., Yingchun, L.: A comparative study of fine-tuning deep learning models for plant disease identification. Comput. Electron. Agric. **161**, 272–279 (2019)
12. Rangarajan, A.K., Purushothaman, R., Ramesh, A.: Tomato crop disease classification using pre-trained deep learning algorithm. Procedia Comput. Sci. **133**, 1040–1047 (2018). ISSN 1877-0509. https://doi.org/10.1016/j.procs.2018.07.070
13. Saleem, M.H., Potgieter, J., Arif, K.M.: Plant disease detection and classification by deep learning. Plants **8**(11), 468 (2019). https://doi.org/10.3390/plants8110468
14. Dai, J., et al.: Deformable convolutional networks. In: 2017 IEEE International Conference on Computer Vision (ICCV), Venice, Italy, pp. 764–773 (2017). https://doi.org/10.1109/ICCV.2017.89
15. Deng, J., Dong, W., Socher, R., Li, L.-J., Li, K., Fei-Fei, L: ImageNet: a large-scale hierarchical image database. In: 2009 IEEE Conference on Computer Vision and Pattern Recognition, pp. 248–255 (2009). https://doi.org/10.1109/CVPR.2009.5206848
16. Singh, D., Jain, N., Jain, P., Kayal, P., Kumawat, S., Batra, N.: PlantDoc. In: Proceedings of the 7th ACM IKDD CoDS and 25th COMAD (2020). https://doi.org/10.1145/3371158.3371196
17. Hughes, D.P., Salath'e, M.: An open access repository of images on plant health to enable the development of mobile disease diagnostics through machine learning and crowdsourcing. CoRR abs/1511.08060 (2015). http://arxiv.org/abs/1511.08060
18. Chowdhury, M.E., et al.: Tomato leaf diseases detection using deep learning technique. In: Technology in Agriculture, London, United Kingdom. IntechOpen (2021). https://doi.org/10.5772/intechopen.97319. https://www.intechopen.com/chapters/76494

Allowance of Driving Based on Drowsiness Detection Using Audio and Video Processing

S. Sathesh$^{(\boxtimes)}$ ⓘ, S. Maheswaran ⓘ, P. Mohanavenkatesan,
M. Mohammed Azarudeen, K. Sowmitha, and S. Subash

Department of Electronics and Communication Engineering, Kongu Engineering College,
Perundurai, Erode 638060, Tamil Nadu, India
sathesh808@gmail.com, mohammedazarudeenm.18ece@kongu.edu

Abstract. In the last few years, more accidents are happened mainly due to the drowsiness of the driver. Various accident prevention technologies are developed but still accidents are happening. This is due to that, the technologies which are available in present are all detecting the drowsiness of the driver at the time accidents. So, there is a possibility for happening of accidents. If the drowsiness of the driver, can be predicted before driving, it will be very useful to prevent the accidents. In this method, there is a solution to detect the drowsiness of the driver, before the driver starts to drive. There is system in this method from which the driver got approval for driving. The system has two level verification process for the drivers to detect the drowsiness. The First level of verification process is captcha process using python libraries or audio listening process using gTTS library. The second level verification process is based on detecting the facial expression of the drivers using the haar cascade classifier with OpenCV library in python. This process has the accuracy level of 95% in pre-driving drowsiness detection. The above levels are the two levels which are used detect the drowsiness of the drivers before start to drive.

Keywords: Drowsiness · Accident · gTTS · Haar cascade classifier and OpenCV

1 Introduction

Drowsiness is the state where individual want to rest. Rest is a neurobiological need with unsurprising examples of languor and attentiveness. Security and insurance are assuming significant part in driving. Sluggishness results from the rest part of the circadian pattern of rest and alertness, limitation of rest, or potentially interference or fracture of rest. The deficiency of one night's rest can prompt outrageous momentary tiredness, while constantly confining rest by 1 or 2 h a night can prompted persistent languor. Dozing is the best way of lessening drowsiness. It is the state where individual wants to rest [1–3]. It has two unequivocal implications, alluding both to the state prior nodding off and ongoing condition alluding to being in that state free of the day-by-day mood. Individual can encounter tiredness if they had an adequate weakness and this can prompt street mishaps. The following Fig. 1 represents the percentage of accidents happened due to the drowsiness of the drivers.

© IFIP International Federation for Information Processing 2022
Published by Springer Nature Switzerland AG 2022
L. Kalinathan et al. (Eds.): ICCIDS 2022, IFIP AICT 654, pp. 235–250, 2022.
https://doi.org/10.1007/978-3-031-16364-7_18

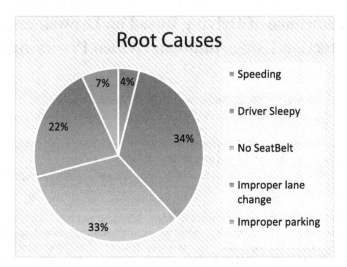

Fig. 1. Accidents percentage due to driver sleepy

The purpose of this project is to develop the simulation of drowsiness detection system. The focus of the project is to design a system that will approve the transports key for the drivers. By detecting the drowsiness of the driver before driving the accidents can be prevented. So, by two monitoring the state of the eyes, it is believed can detect the early symptom of driver's getting drowsiness, and to avoid the road accidents. The process of detecting the drowsiness between drivers is to analyze the opening and closet of the eyes. This process is based on the computer vision or OpenCV to detect image or video processing [4, 5]. In Facial recognition, this project is focused on the localization of the eyes, which involves both eyes and face by applying the existed image processing algorithm. The detection of the drowsiness will be determined once the position of the eyes located.

1.1 Facts About Drowsy Driving

Drowsy driving is responsible for 33% of all fatal accidents. In the previous year, over 168 million people drove their vehicle while drowsy [6]. According to National Sleep Foundation's (NSF) 2015 poll, about thirty seven percent people or more than hundred million people have fallen asleep at the wheel. There are around eleven million drivers who admit to dozing off or being too fatigued to drive, and they admit to having an accident or near-accident.

Mostly the accidents happened due to the drowsiness driving is caused by people who have their age between 18 and 29. Each year 100,000 accidents are the direct result of driver's fatigue which is estimated by National Highway Traffic Safety Administration. According to monitory losses in each year is 12.5 billion dollars due to vehicular crashes results in an estimated counts of 1,550 deaths, 71,000 injuries.

According to other European nations, Australia, Finland and England, drowsiness of driving represents 10 to 30 percent of all accidents. According to the National Sleep Foundation's sleep in America poll, more than 40% of drivers become irritated, impatient, and 12% of drivers tend to drive faster when they are drowsy. Most of the vehicle crashes occurred between at time of 4 to 6 a.m., midnight 2 a.m. and 2 to 4 p.m.

Nearly 23 percent of people agree that crashes of their vehicle is occurred between the above timings. Most of the people in the corporations saw the statistics of the accidents due to drowsiness in the daily newspapers. In United States, driving with drowsiness is the serious problem. The National Highway Safety Traffic Administration has found that determining a precise number of drowsy driving accidents and hospital reports to determine the prevalence of drowsy-driving crashes.

1.2 Causes of Drowsy Driving

There are many causes for drowsiness while driving [6]. The main causes are as follows,

- Lack of adequate sleep.
- The presence of untreated or unrecognized sleep disorders.
- Driving long distances alone.
- Driving at times of the day when you would normally be sleeping.
- Taking medicines, sometimes have drowsiness as a side effects.

The following Fig. 2 represents the accident caused by the drowsiness of the driver.

Fig. 2. Accident due to the drowsiness of driver

2 Existing Method

The drowsiness is detected using the facial expressions like tiredness face, yawning, eye blinking measurements, etc. In machine learning, various algorithms are used for the detecting the facial expressions. In [1] the driver's fatigue is monitored by his workshift and climatic conditions the studies shows that the driver fatigue during driving is more in Heavy traffic route (HTR) compared to monotonous route(MR). In [2] they proposed a method of multi model of detecting the drowsiness and distraction of the driver by using various types of sensors and it is separated into different modules which it is independent to each other. It can be enabled or disabled at any time to analyze the driver's fatigue and distraction. In [3] the analyzing of the drowsiness detection system is tested with the four variables namely kernel size, threshold value, lighting condition (morning, noon, afternoon, and night), and eye's characteristic (eyeglasses or not).

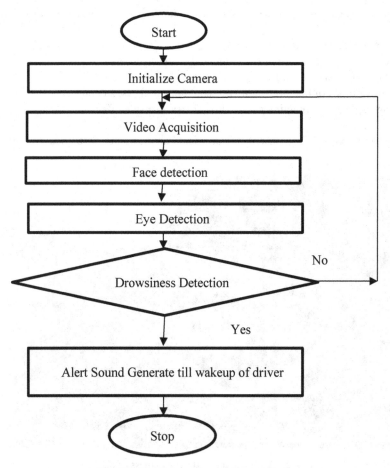

Fig. 3. Workflow of the existing method.

In [4, 5] the approach of detecting the drowsiness of the driver is detected by using Haar cascade and Open CV and the alertness is detected by how much time the eyes are in closed state.

The above process is explained in following Fig. 3 which represents the workflow of the existing method for the detection of drowsiness of the driver.

In [6–9] the windows of the transport are automatically opened or broke to let the air in and help the jammed one to breathe the air. To intercept the sleepiness of the driver in real time, a behavioral based approach is used. In real-time, the drowsiness detection is carried out by way of taking pictures on video. And the studies proposed the surveillance system to monitor the drowsiness which is developed by mobile application on the android operating system.

In [10–15] the studies shows that the drowsiness is detected by using smartphones by capturing live photos and videos and analyzing the drowsiness and make some alarm sounds to alert the driver.In [16, 17] the face spotting is executed with the aid of the local binary sample set of pattern algorithm which is in cascade classifier of OpenCV library. For object detection, Histogram of Oriented Gradients (HOG) that is used in laptop imaginative and prescient. In [18, 19] the picture is split into cells for the manner of encoding of capabilities. Then, instead of requiring a Raspberry Pi, this method is employed as a output which is based on software. This method has three parts of section as follows, Face, Eye and Drowsiness detection.

In [20], there is a slider under the seat position which is away from the crash area and doors are unlocked automatically.In [17, 21] this method's workflow is that the system monitors continuously by camera and it records the inattention of the diver by his fatigue. This system captures the video with the framework at 720p and 30 frames per second.To detect the driver's tiredness of a brief audible alert by the time rest. Until the vehicle is driven, the sluggishness identification cycle will continue. The sound will continue to buzz until the motorist is completely alert. This is affirmed by the framework which continues to take contribution which is completely alert processes which are all available for detecting the drowsiness of the drivers only when the driver have the correct intensity of the lightening for the face and it is also found the drowsiness while in the time of driving.

In [22] the methods used to find the drowsiness are facial expressions, local binary pattern, and EEG low channel peripheral signals. It cannot find the drowsiness of the driver before driving. In [23, 24] the unsupervised machine learning technique is used to find the accident-prone areas like pothole, bumps, sharp turns and normal. Arduino UNO and sensors are used for the detections. GPS is used to coordinates marking of automobiles. If the automobile is at distance of 400m from a danger spot, an alert is passed to the driver to slow down.

2.1 Limitations of Existing Method

The main drawback of existing method is that, all the systems are detect the drowsiness only at the time of driving. So, there may be possibilities for the occurring of accidents. This system doesn't work, if there is not a correct lightening effect in the face.

In public transports, if this system detects the drowsiness of the driver, the alarm starts sounds. So, which also disturbs the passengers in those transports.

Sometimes the sound of the alarm may not be heard by the driver. It may also lead to happening of the accidents.

3   Proposed Method

In this proposed system, the drowsiness of the drivers should be identified before the driver starts to drive. So, the major limitation will be overcome by this system. The main objective of this system is to give approval for taking the keys of the transports. If the system finds the drowsiness in face, then the driver should not get approval to take the keys of the transport.

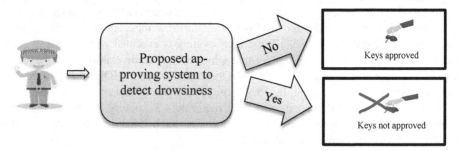

Fig. 4. Outline method of proposed system

The above Fig. 4 represents the outline method of this proposed method for allowance of driving for drivers. In this proposed method, there are two levels of verification process should be done.

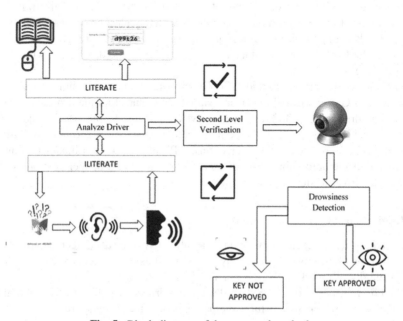

Fig. 5. Block diagram of the proposed method.

The above Fig. 5 represents the block of this proposed system in a step-by-step manner.

3.1 First Level Verification Process

The first level verification process is that, captcha entering process for literate people or audio listening process for illiterate people. The captcha process is nothing but, a captcha is displayed on the screen. The driver should enter the captcha correctly which is displayed on the screen. If the driver enters the captcha correctly, then the driver will move on to the second level checking process. The audio listening process is nothing but, an audio with the captcha will be played. The driver should listen the captcha audio correctly and then, the driver should repeat the audio correctly, which is listened by the driver.

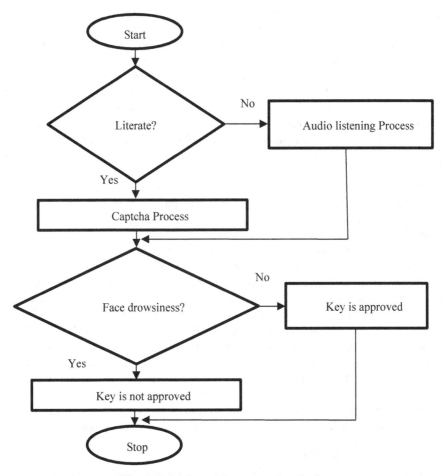

Fig. 6. Workflow of the proposed method

The above Fig. 6 represents the workflow of the proposed method. If the repeated captcha audio of the driver is correctly matched with system played captcha audio, then the driver will move on to the second level verification process. The packages used for the first level verification process are captcha, gTTS, Speech recognition. The captcha package is used to generate captcha randomly. The gTTS package is used for the creation of captcha audio file [25, 26]. It is also used for the conversion of text to speech process. The speech recognition package is used to recognize the speech by the drivers.

3.2 Second Level Verification Process

The second level verification process is that, the camera should capture video of the face of driver for some time before start to drive. After the capturing of the video, the haar cascade classifier in OpenCV python library will detect the face and eyes of the driver in the recorded video. The video should divide into frames. Then each frame is given as input. Haar cascade classifier is used, because of its computation speed of the features of the haar. This is because of the use of integral image which will also be called as summed area tables. Due to the presence of the adaboosting algorithm, this will be very efficient for the feature selection process.

To detect the face in the image, convert the image into grayscale. Because the OpenCV algorithm for detection of objects takes the input as gray images. Then perform detection using detect Multiscale function, which returns an array of detection with x, y co-ordinates, height, and width of the boundary box of the object. For the detection of eyes, the setting of cascade classifier in left eye and right eye. Then, detect using detect multiscale function. This can be done by extracting the boundary box of the eye. After that, the eye image is taken from frame. The left eye and right eye image data is given to the CNN classifier, which will predict the status of both the eyes. First the colour image is converted into grayscale image and resize into 24 × 24-pixel images. Then, normalization is done for better convergence. After that loading of model is done which will predict the status of both the eyes.

The Haar cascade classifier process involves four steps. Such as Haar features, Integral image, adaboosting, cascading. The following Fig. 6 represents the process of haar cascade classifier for both eyes and face. The second block haar features extraction involves three types of features. Such as, edge features, line features, four rectangle features. The feature Extraction process extracts 1,60,000 + features in the 24 × 24 window. So right now, it is a problem to calculate the huge set of features, i.e., 1,60,000 + features for every 24 × 24 window. This thing looks practically difficult. So, to overcome these difficulties, select only the features with useful information. This can be done by the adaboosting algorithm i.e., it eliminates the redundant features and features which has no useful information. So, using this algorithm, many of the features can be reduced. In the extracted feature's part, it is difficult to sum of all the black region and sum of all white region in each feature. So, there is a trick to solve this problem called integral image. The Fig. 7 represents the processes takes place in haar cascade classifiers. Using this classifier object should be detected or not.

The Fig. 8 represents the process of integral image which shows that how to calculate the integral image value of each pixel.

Sum of the all the pixels in Z = 1 + 4 − (2 + 3)

$$Z = W + (W + X + Y + Z) = (W + X + W + Y) => Z \qquad (1)$$

As shown in Eq. (1) is the calculation of the integral image of each pixel of the image. The adaboosting technique select certain number of features from 1,60,000 features and give weight to the features. A linear combination of all these features is used to decide it is a face or not. It combines all the weak classifiers and forms a strong classifier. Generally, 2500 features are used to form a strong classifier. The cascading process is nothing but, combining a small group of features i.e., out of 2500 features ten features are kept in one classifier, like that all features are cascaded.

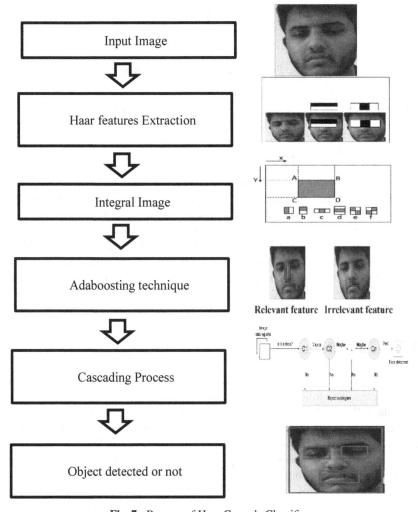

Fig. 7. Process of Haar Cascade Classifier

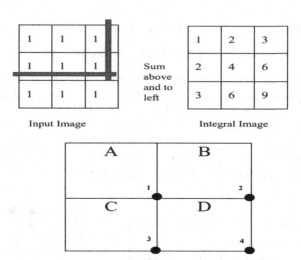

Fig. 8. Process of Integral image calculation

The following Eq. (2) represents the adaboost cascading process of the haar cascade classifier.

$$F(x) = \alpha 1 f1(x) + \alpha 2 f2(x) + \alpha 3 f3(x) + \ldots\ldots\ldots \tag{2}$$

F(x) represents strong classifier and f(x) represents weak classifiers. Now it is easy to detect the face at each classifier. Then, the score is essentially a worth we will use to decide how long the individual has shut his eyes. Thus, if the two eyes are shut, we will continue to expand score and when eyes are open, we decline the score. The maximum score value at which the drowsiness starts is called threshold value. A threshold value is fixed i.e., the closed score of the eyes become more than 15, which indicates that the person's eyes are closed for a long period of time. When the score crosses the threshold value, an alarm sounds and the driver should not allow to drive. Otherwise, drivers will be provided with key of the vehicle to drive.

4   Results and Discussion

The proposed method should detect drowsiness before driving using two levels of verification process. The following figures represent the first level verification process as follows. The following Fig. 9 represents the sample output one with correct captcha for the literate drivers i.e., the captcha verification process for the literate people. As shown in the Fig. 9, if the driver enters the captcha correctly, he will allow for the next level checking process. In this process, the captcha package in python is used for the creation of the captcha and displaying the captcha. There are two functions for the creation of captcha and the creation of the captcha audio file. Also, two functions for the random text captcha for image captcha and random text captcha for the audio captcha. Some of the drivers of public transports are illiterate. For those people, who may not know how to use the computer, the following audio listening process will be done. An audio with random text value is created and played by the gTTS (Google Text-to-Speech) library in python.

```
if literate enter 'y' else enter 'N' : y
```

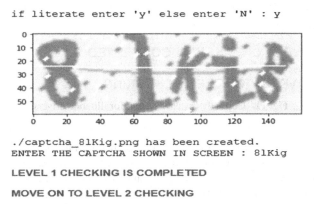

```
./captcha_81Kig.png has been created.
ENTER THE CAPTCHA SHOWN IN SCREEN : 81Kig
```

LEVEL 1 CHECKING IS COMPLETED

MOVE ON TO LEVEL 2 CHECKING

Fig. 9. Sample output with correct captcha for the literate drivers.

After hearing that audio, the driver should repeat the same audio correctly, which was played by the system. The Fig. 10 represents the sample output with incorrect captcha for the literate drivers i.e., the captcha verification process for the literate people.

As shown in the Fig. 10, if the driver enters the captcha incorrectly, he will not allow for the next level checking process. As said as above in this process, the captcha package in python is used for the creation of the captcha and displaying the captcha.

There are two functions for the creation of captcha and the creation of the captcha audio file. Also, two functions for the random text captcha for image captcha and random text captcha for the audio captcha. Some of the drivers of public transports are illiterate. For those people, who may not know how to use the computer, the following audio listening process will be done.

An audio with random text value is created and played by the gTTS library in python.

```
if literate enter 'y' else enter 'N' : y
```

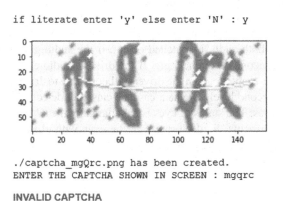

```
./captcha_mgQrc.png has been created.
ENTER THE CAPTCHA SHOWN IN SCREEN : mgqrc
```

INVALID CAPTCHA

Fig. 10. Sample output with incorrect captcha for the literate drivers.

```
if literate enter 'y' else enter 'N' : n
7 3 5 4 8
73548
Recognizing...
rec_text--> 73548
```

LEVEL 1 CHECKING IS COMPLETED

MOVE ON TO LEVEL 2 CHECKING

Fig. 11. Correctly matched output for illiterate drivers.

The above Fig. 11 represents the correctly matched output for the illiterate drivers. As shown in the above picture the audio captcha which is repeated by the driver is recognizing with the shown message recognizing. An audio with random text value is created and played by the gTTS library in python. The driver should repeat the same voice generated by the system. If the recognized voice of the driver is not matched with the system generated voice of the captcha, then, its through the message as invalid captcha and the driver of the vehicle should not be allowed to drive the vehicle.

```
if literate enter 'y' else enter 'N' : n
8 6 6 0 8
86608
Recognizing...
rec_text--> 16608
```

INVALID CAPTCHA

Fig. 12. Incorrectly matched output for illiterate drivers.

The above Fig. 12 represents the incorrectly matched output for the illiterate drivers. As shown in the above picture the audio captcha which is repeated by the driver is recognizing with the shown message recognizing. Then the recognized text is displayed and check whether both audio is matched or not. If the audio got matched, the driver allows to start the second level checking. If the driver's audio captcha is not matched with the system audio captcha, then the driver will not allow to drive the vehicle without checking with the second level. The haar cascade classifier detects the face of the driver by extract the features. The status of eyes is calculated by the CNN loaded model.

The Fig. 13 represents the drowsiness of the eyes with closed score status given by the CNN model and haar cascade classifier. The haar cascade classifier detects the face of the driver by extract the features. The status of eyes is calculated by the CNN loaded model. The algorithmic representation of the second level verification is given as follows.

i)   A video is captured using the cv2.videocapture() method in openCV.

ii) The captured video is converted to frames.
iii) Convert the frames into grayscale using cv2.cvtcolor(frame, cv2.color_BGR2GRAY).
iv) Prediction of both right eye and left eye, which is to be cleary open. In which the resizing and the reshaping process takes place.
v) After the prediction, breaking of the loop will be done.
vi) If both predicted values of the eyes are strictly zero, then closed score value is printed on the frame, else open score value is printed on the frame.
vii) The closed score is above the value of 15, then the alarm sounds. The driver is not allowed to drive the vehicle, otherwise driver will be allowed for the driving of the vehicle.

Fig. 13. Detection of drowsiness eyes with closed score.

In the Fig. 13 the drowsiness detected with the closed score of 25. So, the alarm sounds and the driver will not allow to drive i.e., the keys of the transport will not be given. Then the following Fig. 14 represents that, the extracted video does not have the drowsiness in the face of the diver, then second level verification is completed successfully i.e., the keys will be provided to the drivers.

0

LEVEL 2 CHECKING IS COMPLETED

KEY IS APPROVED

Fig. 14. Output of second level verification for video without drowsiness.

4.1 Efficiency of the System

The feedback of the fifty drivers is collected. The feedback of the driver before this system is that they should feel unsafe due to their drowsiness. But, after the arrival of this system, they feel very safe to drive at day and night times because this system finds the drowsiness of the driver, before the driving starts. So, it is very safe to drive the vehicles.

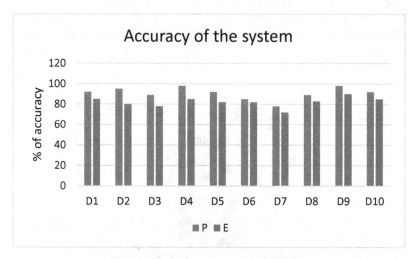

Fig. 15. Accuracy of the proposed system

Figure 15 Shows the percentage of accuracy of the system. Here, D1 to D10 refers drivers, P and E refers proposed and existing system.

5   Conclusion

The safety and protection are plays important role in driving. Sluggishness results from the rest part of the circadian pattern of rest and alertness, limitation of rest, or potentially interference or discontinuity of rest. Sleeping is the most effective way to reduce sleepiness. To reduce the accidents, drowsiness should detect before driving. So, this system prevents the accidents by not allowing the drowsy drivers. Using this system, there is some possibility of reduction in some percentage of accidents, which will be better. The above system successfully detected the drowsiness by using the two levels of verification process. First level verification process successfully verified and the second level of verification using facial features is also successfully verified. Drowsiness results from the rest part of the circadian pattern of rest and alertness, limitation of rest, as well as interference or discontinuity of rest.

References

1. Makowiec-Dąbrowska, T., et al.: Climate conditions and work-related fatigue among professional drivers. Int. J. Biometeorol. **63**(2), 121–128 (2018). https://doi.org/10.1007/s00484-018-1643-y
2. Craye, C., Rashwan, A., Kamel, M.S., Karray, F.: A multi-modal driver fatigue and distraction assessment system. Int. J. Intell. Transp. Syst. Res. **14**(3), 173–194 (2016)
3. Purnamasari, P.D., Kriswoyo, A., Ratna, A.A.P., Sudiana, D.: Eye Based Drowsiness Detection System for Driver. J. Elect. Eng. Technol. **17**(1), 697–705 (2021). https://doi.org/10.1007/s42835-021-00925-z
4. Ravi Teja, P., Anjana Gowri, G., Preethi Lalithya, G., Ajay, R., Anuradha, T., Pavan Kumar, C.S.: Driver drowsiness detection using convolution neural networks. In: Satapathy, S.C., Bhateja, V., Favorskaya, M.N., Adilakshmi, T. (eds.) Smart Computing Techniques and Applications. SIST, vol. 224, pp. 617–626. Springer, Singapore (2021). https://doi.org/10.1007/978-981-16-1502-3_61
5. Sathesh, S., Pradheep, V.A., Maheswaran, S., Premkumar, P., Gokul, N.S., Sriram, P.: Computer vision based real time tracking system to identify overtaking vehicles for safety precaution using single board computer. J. Adv. Res. Dynam. Control Syst. **12**(07-Special Issue), 1551–61 (2020)
6. Drowsy Driving. https://www.sleepfoundation.org/drowsy-driving. (Accessed 15 July 2021)
7. Weng, C.-H., Lai, Y.-H., Lai, S.-H.: Driver drowsiness detection via a hierarchical temporal deep belief network. In: Chen, C.-S., Lu, J., Ma, K.-K. (eds.) ACCV 2016. LNCS, vol. 10118, pp. 117–133. Springer, Cham (2017). https://doi.org/10.1007/978-3-319-54526-4_9
8. Adochiei, I.-R., et al.: Drivers' drowsiness detection and warning systems for critical infrastructures. In: 2020 International Conference on e-Health and Bioengineering (EHB). IEEE (2020)
9. Awasthi, A., Nand, P., Verma, M., Astya, R.: Drowsiness detection using behavioral-centered technique-A Review. In: 2021 11th International Conference on Cloud Computing, Data Science & Engineering (Confluence). IEEE (2021)
10. Fazeen, M., Gozick, B., Dantu, R., Bhukhiya, M., González, M.C.: Safe driving using mobile phones. IEEE Trans. Intell. Transp. Syst. **13**(3), 1462–1468 (2012)
11. Galarza, E.E., Egas, F.D., Silva, F.M., Velasco, P.M., Galarza, E.D.: Real time driver drowsiness detection based on driver's face image behavior using a system of human computer interaction implemented in a smartphone. In: Rocha, Á., Guarda, T. (eds.) ICITS 2018. AISC, vol. 721, pp. 563–572. Springer, Cham (2018). https://doi.org/10.1007/978-3-319-73450-7_53
12. Chaudhary, V., Dalwai, Z., Kulkarni, V.: Intelligent distraction and drowsiness detection system for automobiles. In: 2021 International Conference on Intelligent Technologies (CONIT). IEEE (2021)
13. Chhabra, R., Verma, S., Rama Krishna, C.: Detecting aggressive driving behavior using mobile smartphone. In: Krishna, C.R., Dutta, M., Kumar, R. (eds.) Proceedings of 2nd International Conference on Communication, Computing and Networking. LNNS, vol. 46, pp. 513–521. Springer, Singapore (2019). https://doi.org/10.1007/978-981-13-1217-5_49
14. Fino, E., Mazzetti, M.: Monitoring healthy and disturbed sleep through smartphone applications: a review of experimental evidence. Sleep and Breathing **23**(1), 13–24 (2018). https://doi.org/10.1007/s11325-018-1661-3
15. Lemola, S., Perkinson-Gloor, N., Brand, S., Dewald-Kaufmann, J.F., Grob, A.: Adolescents' electronic media use at night, sleep disturbance, and depressive symptoms in the smartphone age. J. Youth Adolesc. **44**(2), 405–418 (2015)
16. Khan, S., Akram, A., Usman, N.: Real time automatic attendance system for face recognition using face API and OpenCV. Wireless Pers. Commun. **113**(1), 469–480 (2020). https://doi.org/10.1007/s11277-020-07224-2

17. Bruce, B.R., Aitken, J.M., Petke, J.: Deep parameter optimisation for face detection using the Viola-Jones algorithm in OpenCV. In: Sarro, F., Deb, K. (eds.) SSBSE 2016. LNCS, vol. 9962, pp. 238–243. Springer, Cham (2016). https://doi.org/10.1007/978-3-319-47106-8_18

18. Lashkov, I., Kashevnik, A., Shilov, N., Parfenov, V., Shabaev, A.: Driver dangerous state detection based on OpenCV & dlib libraries using mobile video processing. In: 2019 IEEE International Conference on Computational Science and Engineering (CSE) and IEEE International Conference on Embedded and Ubiquitous Computing (EUC). IEEE (2019)

19. Parthasaradhy, P., Manjunathachari, K.: Accident avoidance and prediction system using adaptive probabilistic threshold monitoring technique. Microprocess. Microsyst. **71**, 102869 (2019)

20. Chaudhary, U., Patel, A., Patel, A., Soni, M.: Survey paper on automatic vehicle accident detection and rescue system. In: Kotecha, K., Piuri, V., Shah, H.N., Patel, R. (eds.) Data Science and Intelligent Applications. LNDECT, vol. 52, pp. 319–324. Springer, Singapore (2021). https://doi.org/10.1007/978-981-15-4474-3_35

21. Li, L., Chen, Y., Li, Z.: Yawning detection for monitoring driver fatigue based on two cameras. In: 2009 12th International IEEE Conference on Intelligent Transportation Systems. IEEE (2009)

22. Pimplaskar, D., Nagmode, M., Borkar, A.: Real time eye blinking detection and tracking using opencv. Technology **13**(14), 15 (2015)

23. Han, W., Yang, Y., Huang, G.-B., Sourina, O., Klanner, F., Denk, C.: Driver drowsiness detection based on novel eye openness recognition method and unsupervised feature learning. In: 2015 IEEE International Conference on Systems, Man, and Cybernetics. IEEE (2015)

24. Yang, C., Wang, X., Mao, S.: Unsupervised drowsy driving detection with RFID. IEEE Trans. Veh. Technol. **69**(8), 8151–8163 (2020)

25. Virtue, S., Vidal-Puig, A.: GTTs and ITTs in mice: simple tests, complex answers. Nature Metabolism. pp. 1–4 (2021)

26. Rodríguez-Fuentes, L.J., Varona, A., Penagarikano, M., Bordel, G., Diez, M.: GTTS Systems for the SWS Task at MediaEval 20MediaEval. Citeseer (2013)

Identification and Classification of Groundnut Leaf Disease Using Convolutional Neural Network

S. Maheswaran[1]([⊠]) [iD], N. Indhumathi[1], S. Dhanalakshmi[2], S. Nandita[1],
I. Mohammed Shafiq[1], and P. Rithka[1]

[1] Department of Electronics and Communication Engineering, Kongu Engineering College,
Perundurai, Erode 638060, Tamilnadu, India
mmaheswaraneie@gmail.com, {nandits.20ece,mohammedshafiqi.20ece,
rithikap.20ece}@kongu.edu

[2] Department of Electronics and Communication Engineering, SRM Institute of Science and
Technology, Kattankulathur, Chennai 603203, Tamilnadu, India
dhanalas@srmist.edu.in

Abstract. Groundnut is a major oilseed crop and food crop. In groundnut plant major diseases are occuring on the leaf part. The diseases of the leaf will greatly impact the quality and the production of the groundnut is also reduced. In order to reduce the diseases occurring and the problem happens due to the diseases, this paper uses Artificial Intelligence to identify the disease of groundnut leaves. This proposed model identifies the leaf diseases such as leaf spot, rust, groundnut bud necrosis, root rot and web blotch. In this model image processing and CNN are used along with the Artificial Intelligence. This proposed model was trained with large number of leaf diseased data sets collected from farms. The collected datasets are tested using CNN and the results of the dataset were evaluated. In this experiment than the traditional method Artificial Intelligence had a higher efficiency and accuracy. The accuracy of this model was as high as 96.50%. This research study can come up with an instance for the leaf disease identification of groundnut. The provided solution is a anecdote, scalable and accessible tool for the identification of disease and the management of diverse agricultural plants.

Keywords: Groundnut leaf disease · Artificial intelligence · Convolutional neural network · Image processing

1 Introduction

In India, foundation for all the human needs is agriculture and the Indian economy is also raised from this agriculture by means of foreign trade. India is suitable for many cultivatable crops, and groundnut is one among them. Groundnut is a leguminous crop which belong to the Leguminosae family. Groundnut or peanut is commonly known as the poor man's nut. Today, it is an important oilseed crop. It is the sixth important oilseed in the world. The rank of India in the production of groundnut is second. In

© IFIP International Federation for Information Processing 2022
Published by Springer Nature Switzerland AG 2022
L. Kalinathan et al. (Eds.): ICCIDS 2022, IFIP AICT 654, pp. 251–270, 2022.
https://doi.org/10.1007/978-3-031-16364-7_19

India, the availability of groundnut is throughout the year because it has two crop cycles harvested in March & October. Yet the backbone of India is agriculture, there is no further development and advanced technologies in the area of disease detection in leaves and so far. Due to the non-development of advanced technologies the problems in the production and the quality of the food crops are high caused by different diseases. Nowadays because of the emerging techniques many researchers are emerging towards the field of leaf disease detection and providing a sophisticated result. The yield and the quality of the groundnut crop is decreased by damaging the chlorophyll and green tissues due to the disease occurred in the leaves.

Identification of groundnut-leaf diseases needs a competent knowledge, and it is too easy to camouflaged them only by the observation of artificial visual. Anyway using this method, groundnut diseases in leaves cannot be properly analysed and manage in time. The best solution to control groundnut disease is quickly and accurately diagnosing the type of the disease and taking corresponding measures within time [1–5]. In traditional method, experts will determine and identify the plantleaf diseases on bared eyes. But the bared eye disease detection will give a low accuracy on the identification of the disease. So this technique to identify the disease is a loss. Mostly 85% of the diseases occurs on the plants leaf rather than the other parts. It is noticed that in India efficacy of a fiscal destruction. It occurs due to plant syndromes and it is about 15%, if it not controlled it may lead to the loss of 40% to 60%. In current years, artificial intelligence methods is an optimistic method and made many innovation in the fields of object detection, speech recognition etc., By using the advantages of the CNN we can greatly reduce the parameters and the storage requirements and thereby increasing the network efficiency [5–9]. Artificial Intelligence had made more advancement in the plant leaf disease detection.

This research study is mainly fascinated on the identification of single and mixed diseases on each leaf of the groundnut [10]. The types of leaf identification include healthy leaves and other diseases like rust, root rot, leaf-spot, budnecrosis,webblotch and combination of these diseases at the same time. In tradition and existing method, there is no better combination of artificial intelligence and deep learning so in this study deep learning is used as a base model and prediction is based on the artificial intelligence.

1.1 Problem Statement

Cercospora leaf spot is a major disease caused by fungi on leaves of groundnut crop. Although it is a major problem traditional farmers in Sierra Leone consider these spots on the groundnut leaves as sign of maturity of the crop. The most issues of the farmers is that there is a difficulty in the differentiation of early leaf spot and late leaf spot. The major leaf disease occur in groundnut are listed below:

1. LEAFSPOT

    - Early leaf spot
    - Late leaf spot

2. RUST
3. ROOTROT

4. GROUNDNUTBUDNECROSIS
5. WEBBLOTCH

Leaf Spot

Early Leaf Spot

Fig. 1. The images of Early Leaf Spot Disease (Color figure online)

Caused by- Cercospora arachidicola (fungi).

Symptoms:
Early leaf spot symptoms on the upper leaf surface are dark circular brown spots which are surrounded by a yellow halo as shown in Fig. 1a. Lower leaf early leaf spots are smooth and brown in color as shown in Fig. 1c. In Oklahoma most commonly occurring foliar disease is Early leaf spot.
  The fungal pathogens attack any portion of the plant above the ground.

Late Leaf Spot
Caused by: Cercosporidiumper sonatum-(fungi).

Symptoms:
All the aerial part of the plant is affected. The symptoms can be identified in all the airborne parts of the plant. The symptoms will appear on the lower leaves which looks like small brown dusty pustules (uredosori) as shown in Fig. 2a. On the leaf's upper surface small brown spots will appear. The rust spots can be seen on the stem and petioles. The symptoms like brown teliosori as dark spots will appear on the late season. Small shrivelled seeds will be produced by the plant on the severe infection.

a

b

c

Fig. 2. The images of Late Leaf Spot Disease (Color figure online)

RUST
Caused by: Pucciniaarachidis-(fungi).

Symptoms:
All the aerial part of the plant is affected. The symptoms can be identified in all the aerial parts of the plant. The symptoms will appear on the lower leaves which looks like small brown dusty pustules (uredosori) as shown in Fig. 3a. On the leaf's upper surface small brown spots will appear. The rust spots can be seen on the stem and petioles. The symptoms like brown teliosori as dark spots will appear on the late season. Small shriveled seeds will be produced by the plant on the severe infection.

Fig. 3. The images of Rust Leaf Disease (Color figure online)

Root Rot

Caused by: Macrophomina phaseolina (fungi).

Symptoms:
Over head the soiled dish brown lesions are identified as the symptoms as shown in Fig. 4a. Drooping is seen in the leaves which cause the death of the plant as in the Fig. 4b.

a   b

c

Fig. 4. Images of Root Rot Disease

Groundnut Bud Necrosis

Caused by: Groundnut bud necrosis virus (GBNV-Tospovirus).

Symptoms:
The first seen symptom is rings pot in 2–6 weeks as shown in Fig. 5a and Fig. 5c. The newly appearing leaves is identified with small ring spots with leaf inched inwards assign Fig. 5d.

The other symptoms noticed are necrotic spots and the lesions are developed in irregular shapes on the leaves of the plant. Stem and petioles exhibits necrotic spots.

Damage: This disease leads the plant to grows taunted with short internodes and auxiliary shoots.

Fig. 5. The Images of Groundnut Bud Necrosis Leaf Disease

WEB Blotch

Caused by: Phoma arachidicola – (fungi).

Symptoms:
The web blotch is identified as large and it is irregularly shaped with dark brown blotches with irregularly light brown margins as shown in Fig. 6a and Fig. 6b. It appears 1st on the upper leaf surface. Affected leafs will dry and become brittle and fall from the plant.

The disease is most severe on Spanish and Virginia.

Identification of above these diseases through bare eyes is not possible and it is not that much accurate compared to modern techniques for identification. Even the expert cannot accurately say the results of the disease identification. If there is not proper detection of these diseases, then the percentage of wastage of groundnut crops will increase. If the disease is identified incorrectly and according to them is identification if the farmers uses pesticides, then that will affect the groundnut crop badly. If the groundnut crops are affected, then that leads to decreased production of groundnut and that further affects the economy of the farmer as well as nation. So in order for the accurate identification of the leaf diseases this model is proposed.

AI has been nowadays a successful application and it in corporate image processing which is a path for finding many solutions. Now AI has set foot in to the field of agriculture. Currently, several computers based vision applications in Artificial Intelligence such as DBM (deep Boltzmann Machine), CNN (convolutional neural network), DBN,

Fig. 6. The images of Web Blotch Leaf Disease (Color figure online)

RNN has higher performing accuracy. For this proposed method the most prominent application is CNN. These days, to detect and execute the analysis of different objects and its automatic drawings CNN is used [11]. In CNN K-fold cross-validation technique is recently suggested for the splitting of the datasets and generalization in boosting for the CNN model. Generally, this research is highly-developed as an abrade rather than modifying the existing model.

2 Literature Review

A crop prediction model [12] was developed using ANN by using smart phones by the researchers Ravichandran and Koteshwari. This model has three layers and efficiency is based on hidden layers. First, this model is trained by algorithms like as Rprop, Almeida's and Silva algorithms, Delta-bar- delta. By training this model by these algorithms a most favourable configurations are found. This model is based on the efficiency so a rather way is required to examine the no of hidden layers. To examine the correct hidden layer's trial and error method was executed. Finally, it was predicted that more the no of hidden layers in ANN the more the accuracy of the model.

The study [13] says that datasets are used in CNN model for the classification and identification of rice plant diseases. Here for training the model 500 different infected images were collected and given for the processing from the rice field.

In [14], the researches uses deep learning to the problem. They tried this method in mango plant. The researchers selected five different diseases of mango leaf and collected the samples from different varieties of mango plant. From these collections they have addressed about 1200 datasets. Here CNN structure was used to evaluate the disease affected leaves. CNN model is trained with 80% of images and left 20% is used for testing the datasets from the 600 images of 1200 datasets. The left other 600 images, which were used to find the mango leaf disease and the accuracy of the model which shown the practicability of it in real world application. If more images are collected, then by fine calibrating the CNN parameters, the accuracy of the classification can be increased.

A prediction model is developed by Robinson and Mort [18] where the primary data like precipitation, humidity, cloud cover, temperature is fed using neural networks. These data are then converted into binary data and divided into two groups such as input string and output string for the neural network. This model was trained with 10 trail sets. Finally, most efficient values were predicted from the above mentioned parameters.

According to [15] different graph formats are formed for the CNN for the process of data of uniform encephalography which is used to predict the four classes of motor imaginaries. This data is used to disclose with the electro encephalography electrode. The data is conveyed by them with 2D to 3D transformation. Through these dimensions the structure was processed.

In order to make use of the deep learning a short term voltage stability is proposed [19]. They maintained a clustering algorithm to increase there liability.

In study [16] a aim taken using image processing for the detection of cotton leaves. Here, for the segmentation of the datasets K-means algorithms were used.

In [17] Research is based on the identification of leaf disease in banana plant. In this study for training the images used is 3700. In each class the data set is not balanced. Different experiments were carried out, for example the model is trained by using the datasets obtained from the coloured and grayscale images. Best accuracy is obtained at 98.6% by them in coloured images they acquired the accuracy of 98.6% in colour images.

The researcher Li et al. proposed a neural network with the mixture of back-propagation for the identification of cotton leaf disease. Initially the areas like stem and leaves disease are identified using the images with phone window application.

The researcher Wan et al. proposed a methodology by the recommendation of famous convolutional neural network (CNN) which is a highly robust technique to extract the features manually. Here the CNN is transformed into low space dimensional and the network structured information is protected. This proposed methodology is very effective for the detection of leaves disease.

3   Methodology of the Research

This study is used to create different innovations and ideas to design a product using qualitative or quantitative data. In this model DSRM model is used which is an ideal datasets representation. The below figure represents the processing model for this research. Here "problem-centered initiation" is used for this model. For solving the problem only, the problem-centered initiation can be used. Here the problem is observed by the research experts so it is applicable for this model and it deals with the domain of groundnut leaf disease identification. The below Fig. 7. Depicts the DSRM method proposed by the researchers which is accommodated to this research [20].

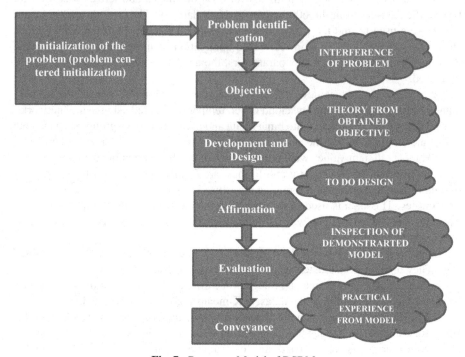

Fig. 7. Processes Model of DSRM

4   Proposed Methodology

4.1   Dataset Collection and Sampling Technique

The data used for this model are both primary and secondary data. Here primary is nothing but the images which are taken freshly with infection. Secondary data is nothing but analyzing the data collected by another researcher for the first time for another purpose (i.e. referring from other authors, institutions or other persons). In this model the researches has used judgmental model for that they have selecting 5 different infected sample and one healthy sample from a non probabilistic population. For this model, about 3600 images were collected and distributed into leaf spot, rust, root rot, groundnut necrosis, web blotch and healthy. These 6 classes were used provided with the balanced data set to train the model.

4.2   Groundnut Images Sample Digitization

In this model data acquisition system is used to create a clear impartial and clarified digital images for the database of groundnut plant leaf. The aim of this is to provide the computerized system with balanced or uniform illumination and lightning. The captured images from smart phones and digital camera are then transferred to the computer and stored in hard disk in PNG format [21].

4.3   Image Pre-processing

The first and the fore most basic task is to insert the pre-processed images into a network. The commonly used image processing technique in all image processing application are image augmentation, vectorization, image resizing, normalization. In this proposed method all these processes were carried out before going into deep learning using Raspberry pi.

4.4   Feature Extraction

Different short comes of Artificial Intelligence is solved by Deep learning, extracting the features manually by using CNN model. In CNN model there are different layers. Those layers will grasp the knowledge. By using the mechanism of filtering the values are extracted by matching the Data [22].

4.5   Dataset Partitioning and Model Selection Methodology

K-fold cross validation dataset partitioning is used in this model. Here the data are partitioned into k values. K+1 value should be obtained for future divisions. The k value is assigned as 10 as it is there commended value for Artificial Intelligence. Now the values of K are 10, therefore it is 10-fold cross validation where the datasets are divided by 10. D = 3600/10 = 360. So for each fold 360 data were used. 80% of images is used for the training from this continuous activity (2880 images) which provides the most significant performance and 20% remaining (720) images are used for the testing; therefore, the validation of the experimental model is done.

4.6   Tool Selection

For the collection of the Groundnut leaf diseased images to train this model, two devices are taken in use to capture the images, one is smart phone and another one is digital camera. This proposed evolution model was trained on the Raspberry pi and the training and the testing was carried out on CPU.

4.7   Technique for Evaluation

Routine structure of the models is evaluated by researchers using different techniques at different time periods like in the starting (developmental) stage and at end. The researches have 1st evaluated the procurement of the prototype by using theconfusion matrix and four evaluation metrics for the confusion matrix on the testing dataset. The four evaluation metrices are F1-score, Recall, Precision and Accuracy. For the subjective evaluation a questionnaire is used by the experts to estimate the performance of the prototype. The result of this evaluation finally says the practical applicability of the model.

5 Designation of Groundnut Leaf Disease Detection Model

Acquisition of Images using Cameras

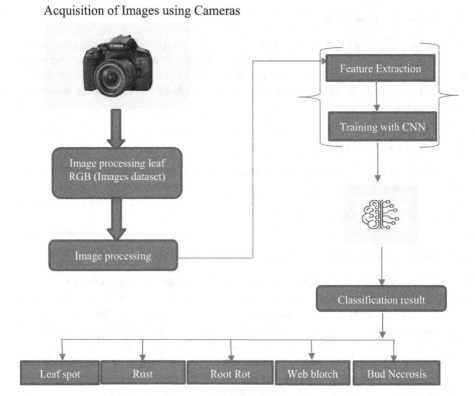

Fig. 8. Groundnut leaf diseases detection model.

For the designation of the model the 1st step made is the image acquisition of the images collected through the digital camera and smart phone from the fields. After this step for the further analysis, techniques of image pre-processing were applied to get the acquired images. Next to this step, all the pre-processed images along with the neural network were fed in the CNN algorithm. By using image analysis an image is extracted for the representation of best-suited extraction. Depends on the best-suited eradicated features, the testing and the training data used for the identification are extracted. At-last, a new image is classified by the trained knowledge into its class of syndrome. The steps and methods taken for training the model and result of training the model are shown in the below flow diagram. By understanding the flow diagram, the whole process undergone through this model can be clearly known. The trained model here is depicted the result by classifying the different leaf disease [23]. The Fig. 8 shows the groudnut leaf disease detection model.

6   The Architecture Proposed for the CNN of the Model

The architecture for this proposed CNN model has two sections namely classification and feature learning section. Normally, its starts from images feeding through input layer and stops with output layer. Between the input layer and the output layer there are many hidden layers consisting of different layers. Here, the output layer name will both epoch name of a groundnut images which is also the label of groundnut leaves disease. Generally, for in this proposed CNN architecture, every groundnut leaf images are added with the neurons and after that those images where augmented with equivalent weights. The processed augmentation output for the fore coming layers are prepared and it is then replicated to the upcoming layers. The layers of output show the predictions task for calculating the neurons for this model.

6.1   Convolutional Layer

The technique with convolutional layer of impartial community is the maximum famous and effective concept of Lin et al. Therefore, the convolution layer is small; with inside the case of multilayer perceptron incorporates absolutely linked in a couple of layers. Hiring the activation characteristic is nonlinear, and modified the nonlinear neural community with linear clear out. The convolution system for the image category is presented, along with its attributes and layers mentioned in the next section:

- The function maps reduce the digital illustration of parameters during system of weight sharing.
- Location correlation learns the connections around a nearby pixel.
- Objects may indicate spatial instability.

6.2   Pooling Layer

Consequently, the derived data with its spatial measurement is decreased via method of pooling. Common pooling, max pooling, and L2-norm pooling are the distinct varieties of pooling manner; especially multi-integration is a powerful and effective method as it has a very beneficial assembling speed [24].

6.3   ReLU Layer

The Rectified Linear Unit Layer (ReLU) with sensors exposed to loss of function or non-line filling space is provided below.

$$F(x) = 0; \quad x < 0$$
$$F(x) = x; \quad x > 0$$

The full community now no longer has an impact on the convolution layer of the reception area, and offers a decision-making feature with the indirect structure. There is a low level of full linear feature, so we load the tanh feature and use of the given below eqn. At ReLU, the training method has changed over and over again, yet no distinct variations can be important in making accuracy easier.

$$F(x) = \tanh(x)$$

$$F(x) = \frac{1}{\left(1 + e^{-x}\right)}$$

6.4 Fully Connected Layer

The final layer of the convolution is fully connected layer. In this layer the input is fed and the inputs are the number of vector. This layer is named as fully connected as every input is coupled with every output. Finally from the last layer N- dimensional input and output are collected and the length of predefined vector with a feed-forward network [25].

6.5 Loss Layer

The output of the version and the performance of the goal feature are all calculated by replacing the layer loss. Input for forward and backward policy symbols are used. Based on input and target values, losses are calculated (going forward). In the background, the element of gradient loss associated with the elements is calculated. Therefore, the weight in DCNN and its gradient loss factor is determined by the use of a back distribution algorithm. Therefore, a decrease in stochastic gradient is used in weight loss backpropagation and is stated as follows [26] (Figs. 9, 10 and Table 1).

Start:
    Parameter of initial value and learning rate β should be selected
Repeat:
    Repeat the procedure till the minimum approximation is received.
    Generate randomly shuffle examples in a training set
for j=1,2,3...n do :
$$F = f - \beta \nabla Q_j(f)$$
    End For
End Repeat
    End

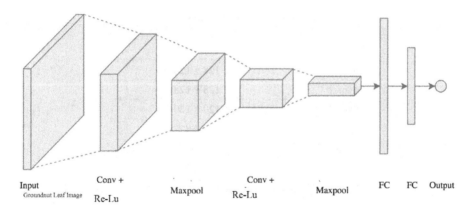

Input
Groundnut Leaf Image

Conv +
Re-Lu

Maxpool

Conv +
Re-Lu

Maxpool

FC   FC   Output

Fig. 9. The architecture of Basic Proposed Network

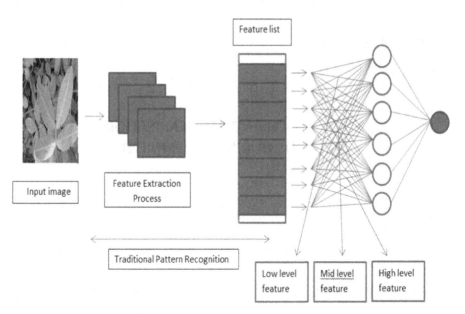

Feature list

Input image

Feature Extraction
Process

Traditional Pattern Recognition

Low level
feature

Mid level
feature

High level
feature

Fig. 10. Architecture of proposed method.

Table 1. Layer description

Layer	Description
Layer-1	128 * 128 * 3
Layer-2	128 * 128 * 3
Layer-2	120 * 120 * 3
Layer-2	88 * 88 * 5
Layer-3	35 * 35 * 25
Layer-3	30 * 30 * 25
Output layer-classification	• Rust • Webblotch • Leafspot • Root rot • Healthy • Budnecrosis

7 Experimental Results

See Fig. 11 and Tables 2 and 3.

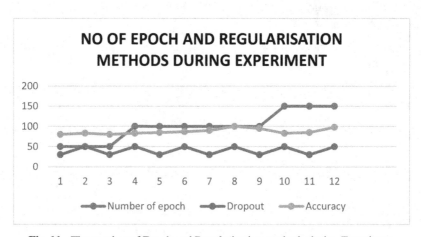

Fig. 11. The number of Epoch and Regularisation methods during Experiment

Table 2. The data that were collected during the experiment for the accuracy and the dropout for the model in different number of epochs.

	Number of epoch	Dropout	Accuracy
1	50	30	80
2	50	50	83
3	50	30	80
4	100	50	83
5	100	30	85
6	100	50	87
7	100	30	90
8	100	50	100
9	100	30	95
10	150	50	83
11	150	30	85
12	150	50	98

To get an efficient model we undergone different experiments by customizing various parameters. The parameters dataset color, number of epoch, augmentation, optimizer and dropout. In this project the model is trained by 3 different no of epochs 50, 100, 150. The model gave the best performance in in 100 epochs. A dropout is added to the CNN to give additional performance (2.75%). So during the experimental process 0.25 and 0.5 drop out percentage is used and the best performance is seen in the 0.5 dropout. In regularization method a most important experiment is carried out, that the usage of optimization algorithm will reduce the loss by updating the iterations according to the gradient. From the above graph the regularization method and the no of epoch are observed is identified. For this study, Adam and RMS Prop which are most recently used optimization algorithms. 2.5% of loss is reduced by Adam optimization algorithm as shown in graph. Highest validation accuracy is observed at the 100 the poch as 0.960 for the corresponding training accuracy by the researchers.

As per the Training at 100th epoch we have observed the highest accuracy as 0.98. The above graphs show all the success rate of validation and training data sets achieved during the experiment in the Fig. 12.

Table 3. The values used while training the model and the accuracy obtained at the values while validating it are listed below.

X-Values	Train	validation			
0	0.3	0.4	81	0.97	0.89
0	0.35	0.5	82	0.96	0.89
0	0.4	0.4	83	0.97	0.89
1	0.45	0.43	84	0.98	0.9
2	0.55	0.53	85	0.97	0.91
3	0.58	0.56	86	0.97	0.92
4	0.6	0.5	87	0.96	0.93
5	0.5	0.45	88	0.95	0.92
6	0.62	0.59	89	0.95	0.91
7	0.65	0.63	90	0.96	0.93
8	0.6	0.65	91	0.95	0.91
9	0.6	0.67	92	0.96	0.9
10	0.6	0.68	93	0.97	0.92
11	0.7	0.73	94	0.97	0.93
12	0.71	0.75	95	0.96	0.94
13	0.72	0.77	96	0.98	0.93
14	0.73	0.75	97	0.97	0.94
15	0.7	0.73	98	0.97	0.92
16	0.75	0.7	99	0.98	0.95
			100	0.98	0.96

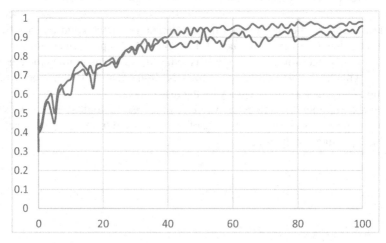

Fig. 12. The Training Accuracy Vs Validation Accuracy

8   Results and Discussion

The final analyze of this experimental model is done using the K-fold cross validation. The above method produced 20% of best performance.15% of performance is being provided by RGB image dataset with augmentation. As color is an important factor in the identification of leaf disease, training the model by coloured dataset will takes a longer time to train and to increase the performance even in the complex layer. The CNN model and grayscale dataset achieved the accuracy of 97.5%. The Adam optimization method and the no of epoch gave 15% and 5.2% model performance. At last different pre-processing techniques were used by the researchers tore move the noise. The overall performance of the model is 95.8% accuracy for the detection of the leaf diseases in groundnut crops plants.

9   Conclusion

This Artificial Intelligence based model was implemented in Raspberry pi. We have obtained an efficient model using different parameters and methods. The overall performance of the groundnut leaf disease detection was 70% satisfied by the evaluator and 15% extremely satisfied and remaining 10% is somehow satisfied. From this it shows that problem solving ability of this model is high and helps the farmers to identify the groundnut leaf diseases. Also this model reduces the time taken for identification plant leaf disease and the complexity in it.

References

1. Qi, H., Liang, Y., Ding, Q., Zou, J.: Automatic identification of peanut-leaf diseases based on stack ensemble. Appl. Sci. 11(4), 1950 (2021)
2. Prajapati, B.S., Dabhi, V.K., Prajapati, H.B.: A survey on detection and classification of cotton leaf disease. In: Proceedings of International Conference on Electrical, Electronics, and Optimization Techniques (ICEEOT), Chennai, India, March 2016, pp. 1–2 (2016)
3. Pawan, D.S.R.G., Warne, P.: Detection of diseases on cotton leaves using K-mean clustering method. Int. Res. J. Eng. Technol. 2(4), 426–428 (2015)
4. Wallelign, S., Polceanu, M., Buche, C.: Soybean plant disease identification using convolutional neural network. In: Proceedings of Artificial Intelligence Research Society Conference, Melbourne, May 2018 (2018)
5. Jha, K., Doshi, A., Patel, P., Shah, M.: A comprehensive review on automation in agriculture using artificial intelligence (2019)
6. Zhang, M., Li, J., Li, Y., Xu, R.: Deep learning for short-term voltage stability assessment of power systems. IEEE Access 9, 29711–29718 (2021)
7. Schwemmer, M.A., Skomrock, N.D., Sederberg, P.B., et al.: Meeting brain-computer interface user performance expectations using a deep neural network decoding framework. Nat. Med. 24(11), 1669–1676 (2018)
8. Ethiopian Institute of Agricultural Research. Cotton Research Strategy, EIAR. Addis Ababa, Ethiopia (2017)
9. Sonal, P., Patil, P., Rupali, M., Zambre, S.: Classification of cotton leaf spot disease using support vector machine. Int. J. Eng. Res. Appl. 4(5), 92–93 (2014)

10. Inga Hilbert, O.-I.: The cotton supply chain in Ethiopia. Freiburg **38** (2018)

11. Ravichandran, G., Koteshwari, R.S.: Agricultural crop predictor and advisor using ANN for smartphones, pp. 1–6 (2016)

12. Hevner, A.R., March, S.T., Park, J., Ram, S.: Design science in IS research. Manag. Inf. **28**, 75–105 (2004)

13. Arivazhagan, S., VinethLigi, S.: Mango leaf diseases identification using convolutional neural network. Int. J. Pure Appl. Math. **120**, 11067–11079 (2018)

14. Hou, Y., Jia, S., Zhang, S., et al.: Deep feature mining via attention-based BiLSTM-GCN for human motor imagery recognition. arXiv preprint arXiv:2005.00777 (2020)

15. Lun, X., Jia, S., Hu, Y., et al.: GCNs-net: a graph convolutional neural network approach for decoding time-resolved eeg motor imagery signals. arXiv preprint arXiv:2006.08924 (2020)

16. Jia, S., Hou, Y., Shi, Y., et al.: Attention-based graph ResNet for motor intent detection from raw EEG signals. arXiv preprint arXiv:2007.13484 (2020)

17. Sonal, P., Patil, P., Rupali, M., Zambre, S.: Classification of cotton leaf spot disease using support vector machine. Int. J. Eng. Res. Appl. **4**(5), 92–93 (2014)

18. Yang, L., Yi, S., Zeng, N., Liu, Y., Zhang, Y.: Identification of rice diseases using deep convolutional neural networks. Neurocomputing **276**(1), 378–384 (2017)

19. Amara, J., Bouaziz, B., Algergawy, A.: A deep learning-based approach for banana leaf diseases classification. In: Proceedings of Datenbanksysteme für Business, Technologie und Web (BTW 2017), Stuttgart, Germany, March 2017, pp. 80–89 (2017)

20. Lyu, Y., Chen, J., Song, Z.: Image-based process monitoring using deep learning framework. Chemom. Intell. Lab. Syst. **189**, 7–19 (2019)

21. Ni, Z., Cai, Y.-X., Wang, Y.-Y., Tian, Y.-T., Wang, X.-L., Badami, B.: Skin cancer diagnosis based on optimized convolutional neural network. Artif. Intell. Med. **102**, 1–7 (2020)

22. Chollet, F.: Deep Learning with Python. Manning Publications Co., Shelter Island (2018)

23. Guo, T., Dong, J., Li, H., Gao, Y.: Simple convolutional neural network on image classification. In: Proceedings of IEEE International Conference in Big Data Analysis, Boston, December 2017, pp. 721–730 (2017)

24. Singh, K.K.: An artificial intelligence and cloud based collaborative platform for plant disease identification, tracking and forecasting for farmers. In: 2018 IEEE International Conference on Cloud Computing in Emerging Markets (CCEM), pp. 49–56 (2018). https://doi.org/10.1109/CCEM.2018.00016

25. Kumar, U., Singh, P., Boote, K.J.: Effect of climate change factors on processes of crop growth and development and yield of groundnut. Adv. Agron. **116**, 41–69 (2012)

26. Jha, K., Doshi, A., Patel, P., Shah, M.: A comprehensive review on automation in agriculture using artificial intelligence. Artif. Intell. Agricult. **2**, 1–12 (2019). ISSN:2589-7217

Enhanced Residual Connections Method for Low Resolution Images in Rice Plant Disease Classification

K. Sathya$^{(\boxtimes)}$ ⓘ and M. Rajalakshmi ⓘ

Coimbatore Institute of Technology, Coimbatore 641014, India
sathya.k@cit.edu.in

Abstract. Recent advancements in both raw computing powers as well as the capabilities of cameras in recent times, it has now become possible to capture images of very high quality. This improvement however does come at the cost of the overall space required to store such high-quality images. One possible solution to this problem would be the storage of the images in low resolution and then upsampling the images to obtain the original resolutions. Despite advances in computer vision, deep learning models for super resolution have been introduced to address the challenges and thus provide promising improved performance results. This paper will explore a novel self-supervised deep learning architecture entitled Enhanced Residual Connections for Image super resolution (ERCSR) that is capable of upsampling extremely low-quality images to their higher quality. Experiments on the different data sets such as DIV2k and rice plant images are made to evaluate this model and experimental results shows that our method outperforms image enhancement. Furthermore, rice plant images are subsequently passed through disease classification layers and achieves desired accuracy for super resolution images.

Keywords: Super resolution · Residual learning · Deep learning · Interpolation · Rice disease classification

1 Introduction

The term super resolution is a general indicative of the processes of converting a low resolution image (I_{LR}) to a high resolution (I_{HR}) or super resolution image (I_{SR}) image. This type of upsampling a I_{LR} image has a few use cases such as the need for conserving storage, surveillance where the images captured are usually very low in quality or even in the medical field where capturing high quality scans may be impossible. The term super resolution is also an umbrella term that encompasses both single image upscaling and video feed upscaling. While the fundamental ideas between the two kinds of upscaling remains the same, the main difference between them is the fact that to upscale a video feed at a reasonable frame rate, the overall architecture must perform much faster than what is needed from a single image.

© IFIP International Federation for Information Processing 2022
Published by Springer Nature Switzerland AG 2022
L. Kalinathan et al. (Eds.): ICCIDS 2022, IFIP AICT 654, pp. 271–284, 2022.
https://doi.org/10.1007/978-3-031-16364-7_20

Image super resolution of single images can be processed quite easily using very basic techniques such as the use of bicubic or bilinear interpolation [1, 2]. These interpolation techniques however, guess the correct pixel value in the I_{SR} image by means of relatively simple arithmetic on the local pixel information in the I_{LR} image. While this approach is somewhat feasible when considering a small upscaling factor such as x1.5 or x2, interpolation techniques quickly become unusable when considering upscaling factors of x3 or above. This is due to the fact that the local pixel information in the I_{LR} image will be directly responsible for several pixels in the SR image causing extreme blurring and loss of details.

In more recent times, most of the attempts at image super resolution involve some form of a complex deep learning model that tend to do a better job than simple interpolation for finding out I_{SR} pixel values [3, 4]. The deep learning models can be trained in a supervised manner if one has both the I_{LR} and I_{HR} image dataset. This however is quite difficult as finding I_{LR} images that have consistency in the amount downscaling throughout the entire dataset is rare.

The caveat of having large amounts of data for training a deep learning model is easily solved by using a standard and known downgrade functions on I_{HR} images like the bicubic interpolation. With this method, it is easily possible to obtain large amounts of data that are downgraded in the same method from I_{HR} images which can be found in abundance now. This fashion of pseudo self-supervised training has become the standard across the field with many competitions providing high quality freely to promote research in real world problems [5]. This paper uses Rice Plant Disease Image Dataset (RPDID) to validate the performance of the proposed method and achieved desired accuracy for super resolution images compared to low resolution images. By this exploration, the following key contributions of this paper can be summarized;

1. Remodeling the existing residual block defined in the architecture to make it more suitable for the specific case of super resolution.
2. Implementation of residual scaling to reduce the adverse effects of the residual networks to be able to maximize the architectures size.
3. Implementation of an architecture that is capable of producing super resolution images at various scales.
4. Comparison of the proposed architecture against various different models on various different datasets.

2 Related Works

The basic terminology used in the formulation of the image super sampling problem as well as go over the existing state of the art architectures in supervised image super sampling are discussed in this section.

The image super sampling problem can be approached in several different ways and in the state-of-the-art models there are several that are vastly different but achieve the same end goal. These different architectures are usually built on top of existing core architecture or training methodologies such as the CNN architecture or the GAN architecture. In this subsection we shall go over some of the more commonly used architecture types and recent examples in each category, looking at how each of the fundamental architecture are being used to achieve the overall goal.

2.1 SRCNN

SRCNN proposed by Dong, Loy, He, et al. [6] is a fairly straightforward architecture made up of CNN layers. SRCNN is divided into three separate parts: one layer for patch extraction, one for non-linear mapping and one for reconstruction. The SRCNN was made to be an end-to-end system that can easily upscaling the I_{LR} images into their I_{SR} images. And since the most difficult part of super resolution, the upscaling, is already done, the SRCNN architecture is used only to ensure that the I_{LR} upscaled image can be mapped to the I_{SR} image pixels properly. Mean Square Error (MSE) loss function is used for training the SRCNN architecture and Peak Signal-to-Noise Ratio (PSNR) is used as the evaluation metric. The FSRCNN architecture created by Dong, Loy, and Tang [7] works in very similar ways to the SRCNN but there are a few key differences. The main difference between the two architectures is that the FSRCNN architecture follows a post sampling architecture making the architecture to have a learnable deconvolutional upsampling function rather than the bicubic interpolation used in SRCNN. This change allows the FSRCNN to have no pre-processing stage to upsample the LR image before it can be passed through the model.

2.2 VDSR

Kim, Lee, and Lee [8] proposed an upgraded version of the SRCNN, VDSR which had the same overall architecture of SRCNN but was improved in certain areas. The VDSR's main area of improvement of SRCNN was the fact that it was a much deeper network utilizing 3×3 convolutional filters compared to the 1×1 in SRCNN. This methodology of deeper network with larger filters is similar to the VGG architecture [9] and allowed the VDSR to learn the residual between the I_{LR} and I_{SR} images. Gradient clipping is used to train the VDSR model to ensure that learning rates can be higher than if MSE is used.

2.3 Residual Networks

Residual blocks as used in the ResNet architecture, He, Zhang, Ren, et al. [10] have showcased that the overall performance of the model improves when compared to linear convolution networks such as the VGG networks. The same principle can be applied in the case of the super resolution problem, with the skip connections being able to pass on feature information from the I_{LR} image much smoother compared to linear models. The residual learning method was more efficient and further incorporated in several other deep learning based super resolution models [11–13].

2.4 CARN

The CARN architecture proposed by Ahn, Kang, and Sohn [14] develops on top of the traditional skip connections used in ResNet to incorporate a much smoother means for information travel. The novel" cascading" methodology, allows information gained from LR features on both a local and global level to be fed into multiple layers. The cascading nature of CARN cascading blocks allows each block to receive inputs from all the previous blocks which results in a more effective means of transfer compared to the original ResNet architecture. Furthermore, CARN uses a group-wise convolutions per block which allow the overall size of the model to decrease giving up 14x decrease in the number of computations required. The shared residual blocks in CARN also help in reducing the overall number of parameters used in the architecture.

2.5 Generative Model

Unlike the architecture, have seen thus far, Generative models built on top of the Generative-Adversarial-Network architectures [4] can also be used for LR to SR conversions. The main difference between these approaches is the fact that generative models do not try to minimize the loss between the LR and SR pixels, rather the try to optimize in a way that make the result more realistic.

SRGAN

The SRGAN architecture proposed by Ledig, Theis, Huszar, et al. [15] is a modification of the GAN architecture, that is designed to produce human eye pleasing 8 images from LR samples. The SRGAN architecture uses a novel ResNet based architecture SRRes-Net architecture as sort of a large pretrained model to fine tune upon and uses a multi task loss as the loss function. This loss is made up of three com-ponents, the adversarial loss that is passed on from the discriminator network, a MSE loss that works similar to non-generative models for capturing pixel similarity and finally a novel perceptual similarity loss that captures high level information. An interesting point about the SRGAN architecture is the fact that, while the model performs worse than other architecture when considering PSNR values, it still outputs a better-quality image to the naked eye (i.e., Mean Opinion Score was higher).

3 Problem Definition

The problem definition of image super resolution built around the ideology of recovering the I_{SR} image from the I_{LR} image. This definition remains true regardless of whether or not the I_{LR} images have been obtained through the means of a known downgrading function or not. Thus, the general modelling of the overall I_{LR} to I_{SR} architecture follows the trend shown in equations below.

D denotes function for degradation mapping that maps the low resolution image (I_{LR}) to the super resolution image (I_{SR}). δ here is the parameters of the degradation function such as the interpolation methods, scaling factors or noise function used.

$$I_{LR} = \mathcal{D}(I_{SR}; \delta) \tag{1}$$

$$\mathcal{D}(I_{SR}; \delta) = (I_{SR}) \downarrow s \subset \delta \tag{2}$$

where $\downarrow s$ is the degrading operation. Most of the datasets being used in research as of now follow this form of degradation. The most common form of degradation used is the bicubic interpolation method with anti-aliasing [16] but other forms of degradation are also used [17].

4 Proposed Architecture

The proposed architecture in terms of design and implementation is discussed in this section. From the previous section, several different methods can be employed to achieve the final result of super resolution. However, while several methods exist, most of the methods focus on the super resolution problem statement in a rigid manner where super resolution to different scales or from differing I_{LR} images will throw off the overall pipeline. These solutions treat the SR problem in a methodology that does not consider using the mutual relationship that exists between different scales of I_{LR} images being converted to different scales of the I_{SR} image. As such, the only possibility for these architectures to work with different scales is to be trained as such on, producing several scale-specific models.

The proposed model will build up on two different models that covered in the literature survey, namely the VDSR and the SRResNet. The VDSR model has been proven that it is capable of handling various scales for super resolution while utilizing a single network. VDSR has also shown that scale-specific networks underperform indicating that while the scales might differ, there exist a lot of information that is common between the different scales. The other network, SRResNet, has several useful architectural designs that can be used as well. Unlike VDSR, SRResNet does not require the input pipeline to begin with a computation and time consuming bicubic interpolated images.

However, the SRResNet presents a different issue as well; while the architecture is good at optimizing both time and computation needed in the preprocessing stage, the SRResNet architecture itself is exactly the same as the original ResNet architecture. The proposed architecture focuses on keeping the scale agnostic methodology of the VDSR while trimming down on the ResNet architecture so it is more suitable for the super resolution task as compared to a more complicated vision problem such as object detection.

4.1 Proposed Residual Block

The use of residual networks in the field of computer vision has drastically gone up in recent years and there are several works that showcase the power of residual connections in for the specific use case of super resolution [7, 8, 13, 19, 20]. The SRResNet architecture by Ledig, Theis, Huszar, et al. [15] implemented a residual block very much similar to the original ResNet architecture, it is possible to further fine-tune the block to improve on the time and computational complexity as shown in Fig. 1.

The standout difference between the three different types of residual block is the removal of the batch normalization layers that are present in the side branch. While the batch normalization layers help immensely when used in a more high-level computer vision task such as object detection, when using it in a task that requires a more fine-grained learning procedure it fails as shown by the work done by Nah, Hyun Kim, and Mu Lee [20] in their image deblurring task. Since the role of the batch normalization layer is to aggregate the information over a set of images, when using it for a task such as image deblurring or super resolution, the aggregation tends to result in over-normalize the features, thereby getting rid of the flexibility of the model.

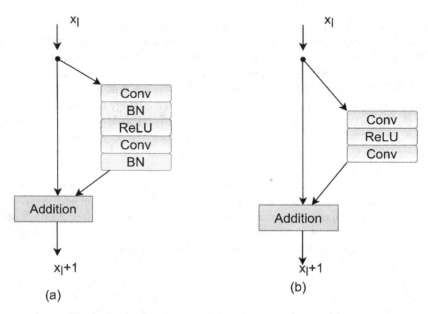

Fig. 1. Comparison between (a) SRResNet (b) Proposed block

An added advantage to dropping the batch normalization layers is also the fact that these layers are very heavy in terms of the computational requirements as they use the same number of computational re-sources as the CNN layers. Removal of the batch normalization layers give us a model which when having similar size and depth as the SRResNet architecture requires approximately 35 to 45% lesser memory during the training stage. Due to the reduced compute requirements when using the proposed residual block, it is possible to build up a larger model with more layers which results in better performance compare to SRResNet when trained on the same hardware.

4.2 Loss Function

Since the proposed model is primarily a very simple CNN only network, the loss function chosen for this implementation is the pixel loss methodology. The pixel loss is the simplest loss methodology when compared to loss functions used in other architectures such as the perceptual loss and the adversarial loss which are used in SRGAN. The pixel loss essentially is concerned with minimizing the difference be-tween the actual SR image and the generated SR image. The difference itself can be calculated using any distance metric and minimizing pixel loss generally leads to models which are optimized for generating images with high PSNR values.

The model utilizes L1 loss, Eq. 3, over the more common L2 loss, as this use case L2 loss was significantly better at convergence when compared to L1 loss. This choice in L1 loss comes more from an empirical standpoint after several experiments rather than a theoretical one, as generally minimizing L2 loss tends to provide higher PSNR.

$$\mathcal{L}(y - \hat{y}) = \sum\nolimits_{i=1}^{N} |y - \hat{y}_i| \qquad (3)$$

4.3 Proposed Model

Just like most computer vision task, the super resolution problem can also take advantage of the fact that increase in the number of parameters in the ERCSR model. There are two primary ways in which a model consisting of CNN layers can increase the number of parameters increasing the depth of the model (i.e., stacking CNN layers on top of each other) or by increasing the number of filters per CNN layer. In the former method, the relationship between the number of layers and the memory is linear. On the other hand, in the latter method, memory scales in a quadratic from with the increase in number of filters and so the overall model formation considers a balance between the two to achieve the maximum possible parameters given the limited hardware resources.

Figure 2 shows the high-level overview of the proposed model ERCSR with both the residual block structure and the upsampling block. While theoretically, increasing the number of filters used per CNN layer should accordingly scale the performance of the model, the work done by Szegedy, Ioffe, Vanhoucke, et al. [21] showed that there is actually an upper limit. Beyond this limit, any further increase in the number of features makes the training stage highly unstable and the performance of the model quickly decreases.

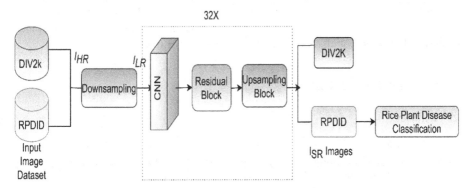

Fig. 2. Architecture of enhanced residual connections for image super resolution (ERCSR)

This issue occurs mainly due to the fact that once this threshold is crossed, by the time information reaches the average pooling the model primarily produces only zeros after training for a few thousand iterations. This phenomenon is unavoidable even when changing other hyperparameters such as the learning rate. In the proposed model," residual scaling" [21] is implemented in each residual block after the last CNN layer to stabilize the model. Residual Scaling involves scaling down the residuals produced before passing them on to the average pooling layer. Figure 3 shows the residual block structure broken down into its individual components.

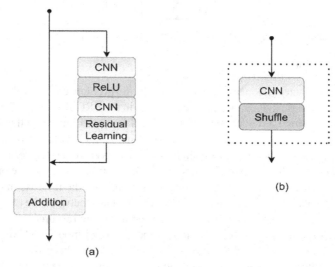

(a)

Fig. 3. (a) Scaled residual block (b) Upsampling block

The model itself consists of 32 CNN layers each with a filter size of 256 with a residual scaling factor equal to 0.1. The residual block itself is very similar to the one used in SRResNet but along with the removal of the batch normalization layers, the final ReLU activation is also removed. The presence of the ReLU activation interacted negatively once the scaling factor was introduced causing instability in the model and the removal of it did not affect the overall performance of the model.

5 Experimental Setup

5.1 Dataset

The DIV2k dataset [22] was used for the training and validation of this model. The dataset consists of 2k resolution images split into 800 training images, 100 validation images and 100 test images. The test images are not released as they are part of the challenge but the test image have a wide range of downgrades applied to them such as extreme blurring apart from the standard bicubic interpolation downgrades. The model is also tested on the Urban100 dataset [23], the B100 dataset [24] and the Set14 dataset [25].

5.2 Training

For training, the model takes in 48×48 RGB patches from the LR image and the corresponding HR image. Since there are only 800 images, a pre-processing stage of image augmentation is also done. In this augmentation stage, the trained data is arbitrarily flipped in both horizontal and vertical directions and this these flipped images have the mean RGB value of the DIV2k dataset subtracted from them. The model is trained with the ADAM optimizer with $\beta 1 = 0.9$, $\beta 2 = 0.998$ and $\epsilon = 10^{-9}$. The x2 scale model is trained first from scratch with the parameters and pre-processing techniques mentioned above. Once the x2 scale model converges, this model is then used as a pretrained model whilst training the x3 and x4 scale models.

5.3 Evaluation Metrics

To put it another way, PSNR refers to the ratio of a signal's maximum potential power to the distortion noise power that impacts the signal's quality of representation. Typically, the PSNR is expressed in terms of the logarithmic decibel scale since many signals have a broad dynamic range (ratio between the highest and smallest possible values of a variable quantity) In order to determine which image enhancement algorithm produces the best results, researchers may compare the outcomes of several algorithms on the same collection of test photographs. By measuring how closely an algorithm or combination of algorithms mimics the original image's PSNR, it is possible to more precisely determine that the method in question is superior.

$$MSE = \frac{1}{N^2} \sum_{w=1}^{N} \sum_{h=1}^{N} (I_{SR}(w, h) - I_{HR}(w, h))^2 \tag{4}$$

$$PSNR = 10 \log_{10} \frac{MAX^2}{MSE} \tag{5}$$

In Eq. 4 and 5, I_{SR} represents super resolution image, I_{HR} for high resolution image and N be the image size respectively. Here the PSNR is measured in terms of Decibel (dB).

6 Results and Discussion

The proposed model is tested against SRResNet which has been modified to showcase various different case scenarios. The SRResNet scores that are given in this section are from a individually retrained model that follows the exact architecture made by Ledig, Theis, Huszar, et al. [15]. One of the main reasons why SRResNet was re-trained was the fact that that original model was only trained using the L2 loss, and to establish a fair comparison with the proposed model SRResNet was retrained using L1 loss as well.

All the models (i.e., both variations of SRResNet and the proposed model) were trained solely on the DIV2k dataset and 300,000 steps in the train loop. Evaluation of each model is done at each step by using 10 randomly selected images from the evaluation dataset, the evaluation metrics was mainly the PSNR value. The data passed to all the models followed the same pre-processing technique. As a starting point for training the suggested model for the upsampling factors of 3 and 4, use the pre-trained network of 2. As seen in Fig. 4, this pre-training method expedites training and enhances performance.

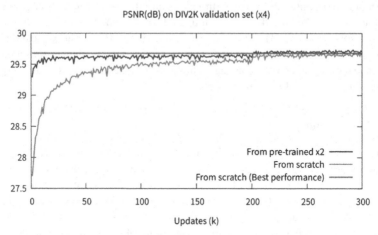

Fig. 4. Sample PSNR on DIV2K (Color figure online)

It is quicker to train an upscaling model using a pre-trained 2 model (blue line) rather than starting from scratch with random initialization (red line) (green line).

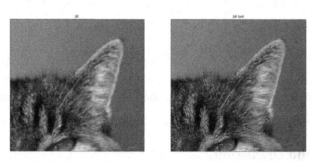

Fig. 5. Sample DIV2k dataset qualitative result

Figure 5 shows the qualitative result of the DIV2k dataset, which is shown in Table 1. In terms of all scale variables, SRResNet trained with L1 delivers somewhat improved outcomes than the original one trained with L2. The image augmentation methodology is in particular much more applicable for usage in the proposed model compared to SRResNet since the architecture is much lighter. The lack of batch normalization layers allows the proposed architecture to be further extended to produce better results than shown in Table 1 but the trade off in overall training time was not justifiable.

Beyond the DIV2k dataset, Table 2 shows the comparison of the proposed model to SRResNet on various another dataset.

Table 1. Quantitative results obtained from the DIV2k dataset

Scale	SRResNet (L1 loss)	SRResNet (L2 loss)	Proposed model
x2	34.30/0.9662	34.44/0.9665	35.12/0.9686
x3	30.82/0.9288	30.85/0.9292	31.05/0.9349
x4	28.92/0.8960	28.92/0.8961	29.17/0.9024

Table 2. Comparison of quantitative results (PSNR and SSIM) for DIV2k dataset

Dataset	Scale	SRCNN	VDSR	SRResNet (L2 Loss)	Proposed model
Set14	x2	32.42/0.9063	33.03/0.9124	–	34.02/0.9204
	x3	29.28/0.8209	29.77/0.8314	–	30.66/0.8481
	x4	27.49/0.7503	28.01/0.7674	28.53/0.7804	28.94/0.7901
Urban100	x2	29.50/0.8946	30.76/0.9140	–	33.10/0.9363
	x3	26.24/0.7989	27.14/0.8279	–	29.02/0.8685
	x4	24.52/0.7221	25.18/0.7524	27.57/0.7354	27.79/0.7437
B100	x2	31.36/0.8879	31.90/0.8960	–	32.37/0.9018
	x3	28.41/0.7863	28.82/0.7976	–	29.32/0.8104
	x4	26.90/0.7101	27.29/0.7251	26.07/0.7839	27.79/0.7437

6.1 Real World Performance

While the primary objective of this proposed study is to build an architecture which is capable of creating a SR image from LR images, it is important to test the generated images in a real-world setting. To this end, the proposed model uses the dataset that includes healthy and diseased leaves of rice plants for evaluating the performance of upsampling in terms of the classification accuracy. The RPDID dataset is a quite varied dataset, and to accurately test the model, the downsampling methodology is the same as the one use before in this work. These labelled images were then used as the testing data on various standard large pretrained image models.

The RPDID dataset is a quite varied dataset [27, 28], Table 3, and to accurately test the model, the downsampling methodology is the same as the one use before in this work. These labelled images were then used as the testing data on various standard large pretrained image models. One thing in common with all these models, however, was that for testing only the models trained on the ImageNet dataset were used to maintain consistency with labelling of the generated SR images. This setup allowed truly unbiased testing of the images generated by the architecture showcased in this work. As seen from Table 4 shows consistently got results close to the top accuracies reported by these individual models.

Table 3. Rice plant disease image dataset

Rice diseases	Total no of image samples
Leaf Blast	2219
Brown Spot	2163
Tungro	1308
Bacterial leaf blight	1624
Hispa	1488
Healthy	490
Leaf smut	565
Total	**9857**

Table 4. Comparison of average accuracy of RPDID classification

Model	Accuracy on LR images	Accuracy on generated SR images
VGG-16	54.4%	72.9%
Inception V3	58.8%	77.3%
EfficientNet B0	63.8%	74.2%
ResNet-50	56.76%	81.6%

Figure 5, also showcases both the accuracy and the loss for training and testing where we got the best results. From this, easily conclude the quality of the images generated by the architecture in this work can also extend the use case of such a model into a more real world setting without having to redesign the whole pipeline when it comes to image classification (Fig. 6).

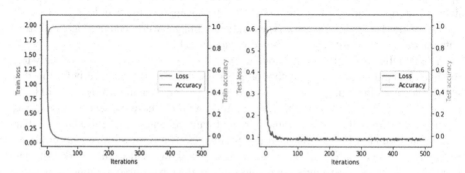

Fig. 6. Comparison of accuracy and loss over training and testing

7 Conclusion

To highlights the importance of image quality, a deep learning model for the use case of super resolution problem that works on top of existing architectures such as the SRResNet is used in this paper. The proposed model ERCSR improves upon the original architectural designs by remaking the core component of the SRResNet - the residual block. In addition to the newly improved residual block, the proposed model also employs residual scaling to combat against the issues that all residual networks have. This residual scaling allows the maximum possible expansion of the network without causing the model itself to get unstable to the point of unviability as well as allow the same model to be used for SR at different scales. The two factors, the improved residual block and the residual scaling allows the proposed model to achieve cutting-edge results while utilizing much lesser computation time and resources to other models. In future, in this way super resolution algorithm can also be applied for segmentation of blobs, object detection etc. for more real-world problems.

References

1. Freeman, W.T., Jones, T.R., Pasztor, E.C.: Example-based super-resolution. IEEE Comput. Graph. Appl. **22**(2), 56–65 (2002)
2. Freedman, G., Fattal, R.: Image and video upscaling from local self-examples. ACM Trans. Graph. (TOG) **30**(2), 1–11 (2011)
3. Wang, X., et al.: ESRGAN: enhanced super-resolution generative adversarial networks. In: Leal-Taixé, L., Roth, S. (eds.) ECCV 2018. LNCS, vol. 11133, pp. 63–79. Springer, Cham (2019). https://doi.org/10.1007/978-3-030-11021-5_5
4. Goodfellow, I., et al.: Generative adversarial nets. In: Advances in Neural Information Processing Systems, 27 (2014)
5. Lugmayr, A., Danelljan, M., Timofte, R.: Unsupervised learning for real-world super-resolution. In: 2019 IEEE/CVF International Conference on Computer Vision Workshop (ICCVW), pp. 3408–3416. IEEE (2019)
6. Dong, C., Loy, C.C., He, K., Tang, X.: Image super-resolution using deep convolutional networks. IEEE Trans. Pattern Anal. Mach. Intell. **38**(2), 295–307 (2015)
7. Dong, C., Loy, C.C., Tang, X.: Accelerating the super-resolution convolutional neural network. In: Leibe, B., Matas, J., Sebe, N., Welling, M. (eds.) ECCV 2016. LNCS, vol. 9906, pp. 391–407. Springer, Cham (2016). https://doi.org/10.1007/978-3-319-46475-6_25
8. Kim, J., Lee, J.K., Lee, K.M.: Accurate image super-resolution using very deep convolutional networks. In: Proceedings of the IEEE Conference on Computer Vision and Pattern Recognition, pp. 1646–1654 (2016)
9. Simonyan, K., Zisserman, A.: Very deep convolutional networks for large-scale image recognition. arXiv preprint arXiv:1409.1556 (2014)
10. He, K., Zhang, X., Ren, S., Sun, J.: Deep residual learning for image recognition. In: Proceedings of the IEEE Conference on Computer Vision and Pattern Recognition, pp. 770–778 (2016)
11. Zhang, Y., Tian, Y., Kong, Y., Zhong, B., Fu, Y.: Residual dense network for image super-resolution. In: Proceedings of the IEEE Conference on Computer Vision and Pattern Recognition, pp. 2472–2481 (2018)
12. Lim, B., Son, S., Kim, H., Nah, S., Mu Lee, K.: Enhanced deep residual networks for single image super-resolution. In: Proceedings of the IEEE Conference on Computer Vision and Pattern Recognition Workshops, pp. 136–144 (2017)

13. Tai, Y., Yang, J., Liu, X.: Image super-resolution via deep recursive residual network. In: Proceedings of the IEEE Conference on Computer Vision and Pattern Recognition, pp. 3147–3155 (2017)
14. Ahn, N., Kang, B., Sohn, K.-A.: Fast, accurate, and lightweight super-resolution with cascading residual network. In: Ferrari, V., Hebert, M., Sminchisescu, C., Weiss, Y. (eds.) ECCV 2018. LNCS, vol. 11214, pp. 256–272. Springer, Cham (2018). https://doi.org/10.1007/978-3-030-01249-6_16
15. Ledig, C., et al.: Photo realistic single image super-resolution using a generative adversarial network. In: Proceedings of the IEEE Conference on Computer Vision and Pattern Recognition, pp. 4681–4690 (2017)
16. Keys, R.: Cubic convolution interpolation for digital image processing. IEEE Trans. Acoust. Speech Signal Process. **29**(6), 1153–1160 (1981)
17. Zhang, K., Zuo, W., Zhang, L.: Learning a single convolutional super-resolution network for multiple degradations. In: Proceedings of the IEEE Conference on Computer Vision and Pattern Recognition, pp. 3262–3271 (2018)
18. Dong, C., Loy, C.C., He, K., Tang, X.: Learning a deep convolutional network for image super-resolution. In: Fleet, D., Pajdla, T., Schiele, B., Tuytelaars, T. (eds.) ECCV 2014. LNCS, vol. 8692, pp. 184–199. Springer, Cham (2014). https://doi.org/10.1007/978-3-319-10593-2_13
19. Kim, J., Lee, J.K., Lee, K.M.: Deeply-recursive convolutional network for image super-resolution. In: Proceedings of the IEEE Conference on Computer Vision and Pattern Recognition, pp. 1637–1645 (2016)
20. Nah, S., Hyun Kim, T., Mu Lee, K.: Deep multi-scale convolutional neural network for dynamic scene deblurring. In: Proceedings of the IEEE Conference on Computer Vision and Pattern Recognition, pp. 3883–3891 (2017)
21. Szegedy, C., Ioffe, S., Vanhoucke, V., Alemi, A.A.: Inception-v4, inception-resnet and the impact of residual connections on learning. In: Thirty-first AAAI conference on Artificial Intelligence (2017)
22. Timofte, R., Gu, S., Wu, J., Van Gool, L.: Ntire 2018 challenge on single image super-resolution: methods and results. In: Proceedings of the IEEE Conference on Computer Vision and Pattern Recognition Workshops, pp. 852–863 (2018)
23. Huang, J.B., Singh, A., Ahuja, N.: Single image super-resolution from transformed self-exemplars. In: Proceedings of the IEEE Conference on Computer Vision and Pattern Recognition, pp. 5197–5206 (2015)
24. Martin, D., Fowlkes, C., Tal, D., Malik, J.: A database of human segmented natural images and its application to evaluating segmentation algorithms and measuring ecological statistics. In: Proceedings Eighth IEEE International Conference on Computer Vision. ICCV 2001, vol. 2, pp. 416–423. IEEE (2001)
25. Yang, J., Wright, J., Huang, T.S., Ma, Y.: Image super-resolution via sparse representation. IEEE Trans. Image Process. **19**(11), 2861–2873 (2010)
26. Wang, Z., Chen, J., Hoi, S.C.: Deep learning for image super-resolution: a survey. IEEE Trans. Pattern Anal. Mach. Intell. **43**(10), 3365–3387 (2020)
27. Sethy, P.K., Barpanda, N.K., Rath, A.K., Behera, S.K.: Deep feature-based rice leaf disease identification using support vector machine. Comput. Electron. Agric. **175**, 105527 (2020)
28. Prajapati, H.B., Shah, J.P., Dabhi, V.K.: Detection and classification of rice plant diseases. Intell. Decis. Technol. **11**(3), 357–373 (2017)

COVI-PROC

K. Jayashree[✉], K. R. Akilesh[✉], D. Aristotle[✉], and S. Ashok Balaji[✉]

Department of Computer Science and Engineering, Rajalakshmi Engineering College,
Chennai, India
k.jayashri@gmail.com, {200701015,200701027,
200701031}@rajalakshmi.edu.in

Abstract. As confusing as the world is right now, there are not that many options but to live through it. With safety belts fastened and survival kits prepped, we're going to keep on rolling' into 2022. In the upcoming years it is going to be mandatory to wear masks to protect each and everyone. People who don't follow the safety protocols are a big danger to themselves and to the society. To address this issue we have come with a device called "CovidProc", which are software systems which can be implemented using various hardware products like cameras, speakers, etc. to check if the people who aren't following safety protocols. So, we plan to do our device with the help of AIML algorithms.

1 Introduction

An ecosystem is a community of living organisms in conjunction with the nonliving components of their environment, interacting as a system. Ecosystem means people and belongings in their surroundings. Initially reported to the World Health Organization (WHO) on December 31, 2019, the COVID was declared an outbreak followed by a pandemic. This was followed by the hands of black fungus, white fungus. The succeeding variants are the Delta and freshly brewing Omicron.

The most effective countermeasure at the play is to vaccinate along with booster shots. Wearing masks, maintaining a healthy diet and maintaining social distancing are the key operations that will reduce the probable risk factor involved. Maintaining social distancing at public places is a duty that is barely followed in most of the places. To maintain and improve the standard of living, the government is deploying the police force and medical personnel to make sure that the pandemic health care regulations are followed throughout at all places. But it is not possible for the small share of people to monitor and guide a huge population for following the guidelines at all the public spots. The purpose of the coding systems is to be a helping hand and monitor and guide people in public spots where the healthcare workers can not present by implementing them in various applications.

2 Proposed Architecture

The model consists of 3 interconnected machine learning models. Our model consists of 3 ML architectures: one for identification masks worn or not, next one whether the place

© IFIP International Federation for Information Processing 2022
Published by Springer Nature Switzerland AG 2022
L. Kalinathan et al. (Eds.): ICCIDS 2022, IFIP AICT 654, pp. 285–289, 2022.
https://doi.org/10.1007/978-3-031-16364-7_21

is crowded or not and the last one for the identification of the shirt color of the person who is not wearing a mask and not maintaining social distance. Basic idea is that in small shops in present Covid times no social distancing is followed and also many people have forgotten the habit of wearing masks. Though many policemen, society activists and commoners insist on the importance of social distancing and mask, many never follow. In order to monitor and create an awareness we have come up with a cost efficient as well as a productive model which identifies and) the people a who haven't worn mask and displays a message stating that "_____ SHIRT/D IS NOT FOLLOWING SOCIAL DISTANCING, PLS FOLLOW AS THE CASES ARE RISING UP AND IT'S ALL FOR YOUR BENEFIT " will come through the loud speaker and '___' here gets filled up by the color of the shirt/dress worn by the person who is not following social distancing and not wearing a mask. All the machine learning models can be trained by uploading a wide range of photos based on the condition.

3 Working Modules

3.1 AIML Component Design

The basic concepts used in our device is executed using the AIML models. The use of pattern detection and feature training is a great boost to our device. Using AIML pattern detection, our device can determine if the user or the person is wearing mask or not, the place is crowded or not and to determine the dress color of the person. The applied AI algorithms are fast to act and determines the results in seconds.

A. AI model for cough and mask detection to confront COVID-19

```
from tensorflow.keras.applications.mobilenet_v2 import preprocess_input
from tensorflow.keras.preprocessing.image import img_to_array
from tensorflow.keras.models import load_model
from imutils.video import VideoStream
import numpy as np
import argparse
import imutils
import time
import cv2
import os

def detect_and_predict_mask(frame, faceNet, maskNet):
```

Just an random output of what this model will do

The AI model we designed supports public safety by using deep learning mobilenet v2 model using the tensorflow and keras framework to detect safety equipment, such as masks and other personal protective gear, and signs of illness, such as coughing, in public places. Using PPE Detection our cough and mask detection model lower the risk of spreading COVID-19 which can help reduce costs and save lives. The customers who use our device enjoy the benefit of protecting their employees and customers from COVID-19 using a fast, accurate, and cost-effective method. Our model is trained to detect instances of coughing and mask-wearing. It processes video streams to identify instances of coughing and to detect mask-wearing.

B. CNN Based Methods in Crowd Counting

Using convolutional neural networks (CNN), we can observe and calculate the number of people that are present in a given image.

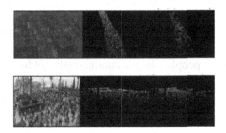

C. Color Detection Model

Here we have used a basic color detection model to identify the person who is not following social distancing and not wearing masks in crowded places. And our device will voice out the color of the dress, the person wearing which makes the person cautious.

Step 1: Input: Capture video through webcam.

Step 2: Read the video stream in image frames.

Step 3: Convert the imageFrame in BGR(RGB color space represented as three matrices of red, green and blue with integer values from 0 to 255) to HSV(hue-saturation-value) color space. *Hue* describes a color in terms of *saturation*, represents the amount of gray in that color and *value* describes the brightness or intensity of the color. This can be represented as three matrices in the range of 0–179, 0–255 and 0–255 respectively.

Step 4: Define the range of each color and create the corresponding mask.

Step 5: Morphological Transform: Dilation, to remove noises from the images.

Step 6: bitwise_and between the image frame and mask is performed to specifically detect that particular color and discard others.

Step 7: Create contour for the individual colors to display the detected colored region distinguishably.

Step 8: Output: Detection of the colors in real-time.

4   Output

4.1   Data Storage

A Biometric Data Storage Design: The data regarding shirt color, mask and face details will be stored in an organized manner for creating it easier for the engine to access it. Our model stores data in software designed storage such as to any server such as mysql or docker environment etc. We have designed our model to be strict as just pointing out people doesn't make any use to this society and that's why we have added a storage system so as to monitor the people who are continuously not following the social norms.

5   Related Work

Madurai City Police have introduced a new video analytic software for its closed-circuit television camera network that will detect people walking on road without wearing mask. With the new technology, the CCTV cameras installed under Thilagar Thidal and Vilakkuthoon police station limits will also detect those who have not worn the mask in the proper fashion. The software will automatically click the images of such violators and send an alert with their photographs to the mobile phones of the police personnel deployed in the bazaar through an android mobile application, "Cameras installed to measure the happiness of visitors at customer happiness centres of the Roads and Transport Authority (RTA) have been fitted with new AI technology to monitor the compliance of those visitors with the wearing of face masks. The step is part of the RTA's precautionary measures to curb the spread of COVID-19. We have successfully fitted the cameras with a new technology whereby they can also monitor whether clients are wearing face masks as part measures to curb the spread of COVID-19. The technology recorded an accuracy rate of 99.1% in this field." The RTA says the cameras generate analytical reports and daily indicators, which will help officials to take appropriate actions towards any failure of compliance with the precautionary measures in place. The cameras, which had been installed more than 18 months ago, can capture images at a rate of 30 frames per second, with a range of seven metres.

6 Conclusion

The objective of the proposed system is to implement social distancing with better ways to identify people not following the COVID protocols during this pandemic time.In the future we can also use the software embedded in security cameras which is available in most of the areas in today's world. In the upcoming periods many such times may come and we have to be equipped with proper technology and work together to overcome such difficulties. This work will be very useful in reducing the crowds and decreasing the chance to contact covid in crowded areas. This doesn't require much financial need to set it up. It is simple to install and use and it is cost efficient. This may reduce the chance to contact covid up to 60% and it may even become as low as 30% if everyone complies with the surveillance project. This is easy to use and doesn't require much technological knowledge to operate. This makes your job easier to implement COVID norms.

References

1. "Fundamentals of Machine Learning for Predictive Data Analytics: Algorithms, Worked Examples, and Case Studies" by John D. Kelleher, Brian Mac Namee, and Aoife D'Arcy
2. "Data Mining: Practical Machine Learning Tools and Techniques" by Ian H. Witten, Eibe Frank, and Mark A. Hall
3. "Machine Learning for Audio, Image and Video Analysis: Theory and Applications" by Francesco Camastra, Alessandro Vinciarelli
4. "The Architecture of an Embedded Smart Camera for Intelligent Inspection and Surveillance" by Michał Fularz, Marek Kraft, Adam Schmidt, Andrzej Kasiński
5. COVID-19 Government Response Tracker
6. Implementing Telehealth in rural Healthcare using IoT

GPS Tracking Traffic Management System Using Priority Based Algorithm

S. Manjula[1(✉)], M. Suganthy[2], and P. Anandan[3]

[1] Department of ECE, Rajalakshmi Institute of Technology, Chennai, India
manjulasankar@gmail.com
[2] Department of ECE, VelTech MultiTech Dr. Rangarajan Dr. Sakunthala Engineering Colege, Chennai, India
[3] Department of ECE, Saveetha School of Engineering, Saveetha Institute of Medical and Technical Sciences, Saveetha University, Chennai, India

Abstract. In recent years, there is an extreme increase in road vehicle usage which in turn a challenge to manage the traffic system. The current traffic system is not based on the vehicle density level and a pre-established time is distributed to the traffic lights for every lane crossing which had an issue like traffic congestion & wastage of time and this condition turns out to be worse in the peak hours, and it also increases the emission of CO2 in the environment. In this paper, real-time dynamic traffic management system based on density level of each junction with the help of approximate geo-density information is developed. The density level on each junction is then updated to the real-time database. Then, this real-time data is computed with an exclusive priority-based algorithm where the crossing with high density is prioritized over the other (emergency vehicles are given the maximum priority) while maintaining the congestion around the junction under control. The prototype of GPS based real-time traffic management system is implemented and tested.

Keywords: GPS · Priority based algorithm · Traffic system

1 Introduction

In our day-to-day life, people are living in an environment with full of pollution. The inhaled air is not free of dust or the drinking water is not free of impurities. Most of the reasons for this pollution is no other than the massive increase in our population. Due to this, the usage of vehicles is also increased which later made problems like traffic congestion, emission of CO_2 and also there is so much time is wasted while waiting in the queue for crossing the lane. The common working of a traffic light needs more than a little control and cooperation to assure that traffic and pedestrians move without difficulty and also safely. There is a series of different control systems used to conquer this problem, started at a normal time-based structure to computerized control and coordinating ways that self-adapt to reduce the time whoever using the junction.

© IFIP International Federation for Information Processing 2022
Published by Springer Nature Switzerland AG 2022
L. Kalinathan et al. (Eds.): ICCIDS 2022, IFIP AICT 654, pp. 290–300, 2022.
https://doi.org/10.1007/978-3-031-16364-7_22

2 Related Works

Various traffic management system was implemented for traffic congestion. A traffic light is usually monitored by a device installed inside the cabinet. Till now the Electromechanical controllers [1] are used in India. In this traffic system, for every lane in the junction, there's a time allotted constantly and it works in that case. Since the controller will monitor the signal timing in a pre-assigned manner. Even though there is no vehicle in that particular lane, the other lane in that junction has to wait and there is so much time is wasted. Some other ways are like using image processing technique [2], in that they find the density of vehicles on that particular junction and estimate the count. They used to monitor or control the traffic signal using that data. The major tasks of the [3] are detection of stationary objects and their location, by identifying those mobile vehicles, and the data is transmitted to the monitoring or controlling process. It is very useful in real-time traffic information by monitor the architecture by using wireless communications IoT.

In general, Electro-mechanical signal controllers use timers that have constant time intersection plans. The cycle of signalized timers is found by small motors that are attached to the timers. Cycle motors that are normally had a range from 35 s to 120 s. If a motor/gear in a timer fails, another cycle gear is used for replacing the failed one. A dial timer can control one intersection at a time because it has only one intersection time plot. Another way of using the load cell, which produces an electric signal based on the mass on the cell [4]. This helps us in calculating the number of vehicles approximately on the specific road and thus finds the amount of time for the particular lane to give the green signal to clear the vehicles on that lane.

One of the methods makes use of VANET to develop an Intelligent traffic system [5, 6] while another method makes use of the infrared signal to find vehicles and consider those rule breaches by any vehicle. Another method makes use of state-space equations to design a simulation of a domestic traffic network. In another system, to reduce the waiting time of the vehicle and to actively solve the traffic problem they use fuzzy logic [5, 7]. This method is proposed using Photo-electric sensors which are fixed on one side of the lane to control the vehicles crossing through that lane. In a smart toll system, the LED shows red color when the IR sensor detects the vehicle, by placing that sensor at a specific place. Every vehicle must have an active RFID tag [8, 9] installed which will store the vehicle registration no. as a digital code. Thus, the RFID readers can detect the vehicles. The number of vehicle counts will be stored in the database. This traffic light signaling is done by micro controllers. RFID has 16-bit information [10]. The information like vehicle number, driver's information, and the account details fed in RFID tag and the LCDs shows the data. The database matches the vehicle's details and the amount will deduct and then allows the vehicle to pass through. The method used in [11, 12] will be image processing-based signal controlling. The signal crossing time will change as per the vehicle load in the specific lane.

A system based on a PIC microcontroller stimulates the vehicle density using IR sensors and achieves an active timing period with different levels. Likewise, to solve the problem of emergency vehicles stuck in the crowded junction, the portable controller device is used. In [10, 13], Arduino UNO will circularly check each road of the junction and if it detects any traffic on the road hen it allows it to cross the junction by signaling a green light to it and vice versa, the road will be signaled as red light, if the junction detected has no traffic. Surveying the existing system related to traffic management system shows that they have used different technologies and sensor modules. Finally, they started with IoT [14] which is the emerging and promising one that requires for today's fast world. Each system gives its solutions for the problem of manual traffic management system where each technique and technology have its merits and demerits for process and system design. Every system has its drawbacks which can be overcome by other systems and hence the emerging technology for traffic management systems using IoT will give accurate and live information about the density of the traffic. Existing works are summarized in Table. 1.

Table 1. Summarize of literature survey

Ref.	Model	Description	Limitations
[1]	Electromechanical controllers	Using controllers, time slot is allotted for every lane	If no vehicle is in the lane, other vehicles have to wait to cross. Time consumption is a major problem
[2]	Image processing technique	Using this technique, find the count of the vehicles	Implementation is difficult
[3]	RFID + Sensors	Transmit data through wireless communication	Security of smart objects is a major problem and is limited to some situations
[4]	Load cell	Based on the number of vehicles on the specific road, amount of time for a particular lane is calculated	Slow process
[5]	IOT + Load Cell + RF Transmitter and Receiver	Through IOT and RFID, datas are stored in the cloud and analyze the data. Based on traffic data and using load cell, calculate the required time to clear the traffic. For emergency vehicle, RF transmitters and receivers are used to clear the traffic on particular lane	Security is a major problem
[6]	Fuzzy logic	Intelligent traffic system was implemented using Fuzzy control logic	Not effective solution
[7]	Image processing algorithms	Edge detection algorithms and object counting methods are used	It is very lengthy process. Implemented this system for cars only
[8]	RFID	Using RFID tag in the vehicle, RFID reader counts the number of vehicles in the database. According to number of vehicles, traffic signal can be changed	RFID range is very limited
This work	GPS based real time traffic system	Based on GPS, density of vehicles in particular location is identified	Effective and low cost solution

3 Proposed System

A GPS based real-time traffic management system is proposed to eliminate drawbacks from existing works. In this proposed system, all vehicles have GPS module.

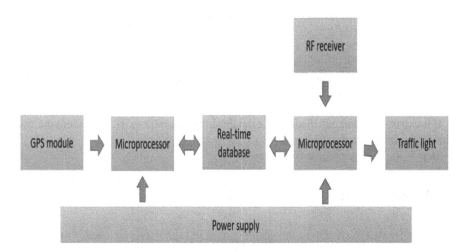

Fig. 1. Block diagram of proposed system

The block diagram of the proposed system is shown in Fig. 1. The proposed system consists of the sensing part and decision part. In this system, the traffic density is sensed by using the GPS in the geo-fenced location and fetch the values to the database and the values are processed, decision will depend on the priority-based algorithm. Ambulance's gps module is sensed in the geo-fenced location, instead of incrementing the density, it automatically flags the ambulance's entry to the database. With this algorithm, the emergency vehicles can be given separate priority.

3.1 Density Sensing

The power supply for the microprocessor is directly fed from the vehicle's power source (Battery) as shown in Fig. 2. The microprocessor tracks the geo-location regularly, but only updates its count only when it reaches the geo-fenced region *

The steps of density sensing part algorithm are as follows

Step 1: Acquire GPS coordinates from the GPS module.

Step 2: Process the collection of 4 geo-fence locations for each traffic junction of 4 crossings from JSON file.

Step 3: Check if the GPS coordinates fall into any of the geo-fence locations extracted from the JSON file.

Step 4: If true, Increment the user count in the real-time database and wait for 15 s, re-check for the existence of the vehicle in the same geo-fence location.

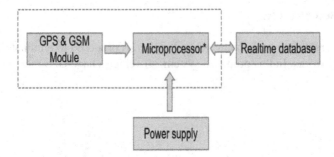

Fig. 2. Block diagram of sensing part

If it exists, no change is made to the database. If it doesn't exist, decrement the user count in the real-time database.

Step 5: Repeat the process from Step 1.

3.2 Decision Making

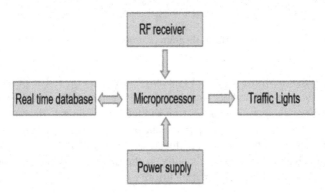

Fig. 3. Block diagram of decision part

The density values from the real-time database are computed through the priority-based algorithm in the microprocessor. This algorithm grants a time span of 15 s for each prioritized crossing, with a priority of the crossing done every 10 s. RF receiver functionality only works to prioritize emergency vehicles.

The Algorithm used in Decision making part are as follows

Step 1: Acquire the data from the database.

Step 2: Find the maximum density count from the available crossing data.

Step 3: Check for any priority signal for emergency vehicle. If a priority signal exists, assign the signal priority to that crossing. If it doesn't exist, assign the signal priority to the crossing with maximum density.

Step 4: Wait for 15 s, • If the signal priority doesn't change, increment the time counter, • If the signal priority changes set the time counter to zero.

Step 5: If the time counter reaches four (4), reset the time counter, reassign signal priority to the next maximum density crossing.

Step 6: Repeat the process from Step-1 (Fig. 3).

The proposed model is to sense the density of the traffic at each road crossing in real-time. For the sensing of the traffic density an embedded kit with a Global Positioning System (GPS) module is to be installed in the vehicle(s). This embedded kit is designed to track the movement of the vehicle locally and only updates it to the internet only when it is in the specified geo-fenced location. So, whenever the vehicle (GPS) enters the pre-mapped geo-fenced location at each side of the road crossing, it updates its count to the real time database (which in this case is Google firebase). The traffic density which is available on the internet (Real-time database) is then computed with a priority-based algorithm at each road crossing junction. With this algorithm (which is to be discussed in the next slide), the traffic is managed by regulating the congestion at the higher traffic density side. This algorithm is overridden when an emergency vehicle is to be detected by the use of RF communication between the emergency vehicle and traffic controller at the junction. The overridden algorithm automatically prioritizes the side at which the emergency vehicle is detected. Any glitch/abnormality in working of this methodology returns the traffic controller to a fail-safe mode where the controller works with a typical timer system.

4   Circuit Design

As declared, the GPS modules are set up like a u-blox NEO-6M GPS engine. The ROM version is ROM 7.0.3, the type number of NEO-6M is NEO-6M-0-001 and the PCN reference is UBX-TN-11047-1. The NEO-6M module, For serial communication, has one UART interface, but the common UART baud rate is 9,600. Because the GPS signal is RHCP and the model of the GPS antenna is varying from common whip antennas (used for polarized signals). The patch antenna is the most popular one compared to others. They are flat and are mounted on a metal plate, normally have a ceramic body. The patch antennas are usually cast in housing. Always know, the optimal execution of the GPS receiver depends on the position of the mounting of the antenna. The patch antenna always should be installed parallel to the geographic horizon. Make sure the antenna is not having a partial view, make it has a maximum view of the sky. Keep as same as for every antenna to have a narrow line with the many satellites (Fig. 4).

In the Raspberry Pi 3 range, the latest product is Raspberry Pi 3 Model B+, it's having a 64-bit quad-core processor running at 1.4 GHz. Also, it has a dual-band which is 2.4 GHz, Bluetooth 4.2/BLE. It also has faster Ethernet, 5 GHz wireless LAN, and PoE capability via a separate PoE HAT. The dual-band wireless LAN has a separate consent certification, permitting the board to build it into final products with exceptionally decreased wireless LAN consent analyzing, increasing both cost and time. When compared to Raspberry Pi 2 Model B and Raspberry Pi 3 Model B, the Raspberry Pi 3 Model B+ will have the same mechanical footprint. Circuit Diagrams of sensing part and decision part are shown in Fig. 4 and 5.

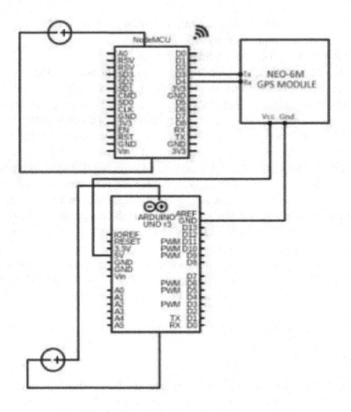

Fig. 4. Circuit diagram of sensing part

The Relay Module we use has 5 V Dual Channel and it is used for low-level triggers only. In between the input terminal and GND, signal voltage will be there which refers to the low-level trigger and the trigger gets short-circuited away from the Channel connection by the GND −ve trigger.

In Fig. 7, Node MCU – ESP8266 board, GPS Module takes 5 V 2 A dc power as an input. The development board is enabled by internetworking through WiFi.

The U-bloxNeo-6m GPS module (Fig. 7) senses its location data from the GPS satellite. When the sensed GPS location is within the Geo-fenced location, the microcontroller gets instructed to upload the density count to the database.

The data from the real-time database is passed to Raspberry pi 3 b+, then the data are compared and finds the highest priority lane, which gets a green signal then the other lane gets a red signal. In the sensing part, the location of the vehicles detected using a GPS module within the geo-fenced location and the count of the vehicle is transferred to the database.

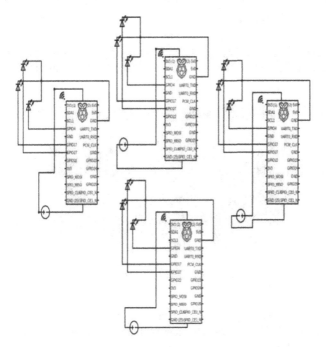

Fig. 5. Circuit diagram of decision part

5 Software Design

This section comprises a webpage companion which displays the current traffic density status (i.e.) it shows the vehicle count at each crossing of the junction. The information from the main traffic density sensing part is transmitted to an online server-less database from where it is redirected to the decision-making system. This data is displayed in the web console for any troubleshooting purpose.

The Firebase/Real-time Database (Fig. 9) lets you develop prosperous, cumulative applications by permitting secure access to the database directly from client-side code. Data is endured locally and even while offline, those events keep on fire, providing the end-user a responsive experience. The Real-time Database synchronizes the local data differs from the remote updates that appeared while the client was offline, combining any conflicts perpetually after the device regains its connection. This Database gives an adaptable, expression-based rule language called Firebase Security Rules, to describe how your data should be maintained when the data can be used or decoded. When combined with Firebase verification, developers can assign who has permission to which data and how they can design it. Our Database is a NoSQL database and as such has contrasting enhancement and performance compared to a mutual database. This Real-time Database API is designed to only permit actions that can be concluded rapidly. This qualifies you to develop a great real-time knowledge that can provide millions of users without gets uncomfortable on interest.

6   Experimental Results

In this, the description of the results obtained through the proposed model is discussed here. The hardware Decision Part Module is designed as shown in Fig. 6.

Fig. 6. Decision making

The data from the real-time database is passed to Raspberry pi 3 b+ (Fig. 6), then the data are compared and finds the highest priority lane, which gets a green signal then the other lane gets a red signal. The hardware module of Sensing Part as shown in Fig. 7.

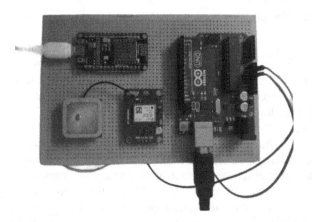

Fig. 7. Density sensing

In sensing part, the location of the vehicles detected using GPS module within the geo fenced location and the count of the vehicle is transferred to the database. The GPS tracking traffic system is successfully implemented as shown in Figs. 8 and 9.

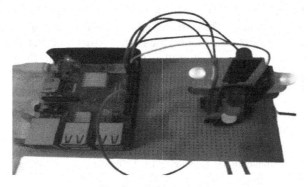

Fig. 8. Working model

Fig. 9. Real time database

7 Conclusion

The Real-time dynamic traffic management system based on density using GPS and Raspberry pi 3 b+ has been successfully implemented. The proposed model is easy to adopt and it is user-friendly, the main feature is to give a green signal to high priority based on the density level and giving way to the emergency vehicle. Hence this traffic system is dynamic waiting time, emission of CO_2, congestion can be reduced. The emergency vehicle has a higher priority and also gets a huge advantage in the implementation of this model.

The proposed traffic control system is the very basic step towards achieving automation in the field of the traffic control system. With various advancements taking place in today's world, people are in search of automated systems which not only saves their time but also a lot of energy in different forms. The saving of fuel (petrol, diesel, natural gas), reduction in time of operation of automobile engines, reduction in the emission of the harmful gases in the atmosphere. Thus, this system helps in reducing the number of accidents that take place just because of this inappropriate transport system and making way to a better traffic management system. The proposed system aims to save the number of manpower wasted at the signals and hence making effective utilization of time.

Additional work can be done in this area and made available to build a real-time dynamic traffic management system. Further research can be made to make this algorithm better optimized to deliver more precise results based on the past density data. As for the future implementation, the management controllers could be grouped to form a cluster based system, where the controllers would be able to communicate with others to provide an Artificial Intelligence-based management system.

References

1. Sangeetha, K., Kavibharathi, G., Kishorekumar, T.: Traffic controller using image processing. Mediterran. J. Basic Appl. Sci. **3**(1), 76–82 (2019)
2. Kastrinaki, V., Zervakis, M., Kalaitzakis, K.: A survey of video processing techniques for traffic applications. Image Vis. Comput. **21**(4), 359–381 (2013)
3. Al-Sakran, H.O.: Intelligent traffic information system based on integration of Internet of Things and agent technology. Int. J. Adv. Comput. Sci. Appl. **6**(2), 37–43 (2015)
4. Varun, K.S., Kumar, K.A., Chowdary, V.R., Raju, C.S.K.: A perceptive model of traffic flow: using Arduino boar. Adv. Phys. Theor. Appl. **72**, 1–7 (2018)
5. Chandana, K.K., Meenakshi Sundaram, S., D'sa, C.: A smart traffic management system for congestion control and warnings using Internet of Things. Saudi J. Eng. Technol. **2**(5), 192–196 (2017)
6. Chinyere, O., Francisca, O., Amano, O.: Design and simulation of an intelligent traffic control system. Int. J. Adv. Eng. Technol. **1**(5), 47–57 (2011)
7. Al Hussain, A.: Automatic traffic using image processing. J. Softw. Eng. Appl. **10**(9), 765–776 (2017)
8. Ghazal, B., El Khatib, K.: Smart traffic light control system. In: IEEE 3rd International Conference on Electrical, Electronics, Computer Engineering and their Applications (EECEA) (2016)
9. Roychowdhury, P., Das, S.: Automatic road traffic management system in a city. Trends Transp. Eng. Appl. Sci. **1**(2), 38–46 (2014)
10. Kavyashree, M., Mamatha, M., Manasa, N.M., Vidhyashree, H.E., Nagashree, R.N.: RFID based smart toll collection system. Int. J. Eng. Res. Technol. **8**(11), 177–180 (2020)
11. Jadhav, P., Kelkar, P., Patil, K., Thorat, S.: Smart traffic control system using image processing. Int. Res. J. Eng. Technol. **3**(3), 280–283 (2016)
12. Rahishet, A.S., Indore, A., Deshmukh, V., Pushpa, U.S.: Intelligent traffic light control using image processing. In: Proceedings of 21st IRF International Conference (2017)
13. Verma, G., Sonkar, R., Bowaria, L.: Smart traffic light system. Int. J. Sci. Technol. Eng. **4**(10), 96–101 (2018)
14. Smart Traffic Management System using Internet of Things (IoT).pdf - Smart Traffic Management System Using Internet of Things (IoT Final Year Project | Course Hero, 14 May 2019 (2019)

Accident Detection System Using Deep Learning

J. Amala Ruby Florence$^{(\boxtimes)}$ and G. Kirubasri$^{(\boxtimes)}$

Department of Computer Science and Engineering, Sona College of Technology, Salem,
Tamil Nadu, India
{amalarubyflorence.20dsm,kirubasri.cse}@sonatech.ac.in

Abstract. India has around one and a half lakh persons die due to road accidents per year. One of the main reasons for this number is no timely availability of help. Automatic accident detection can shorten the response time of rescue agencies and vehicles around accidents to improve rescue efficiency and traffic safety level. The ability to detect and track vehicle can be used in applications like monitoring road accidents. The proposed system uses YOLOv5 deep learning algorithm to detect the vehicles from the real time CCTV surveillance video. The primary focus of the system is to build a model that detects various class of vehicle by using custom dataset. The dataset consists of 1000 images with various condition such as rainfall, low visibility, luminosity, and weather conditions. The proposed framework uses YOLOv5 to detect vehicle with improved efficiency in real time object detection system. This model can further be extended to analyse and classify accidents using 3D Convolutional Neural Network based on the severity of the accident and to alert the nearest hospital.

Keywords: Object identification · YOLOv5 · Accident detection · Convolutional neural network · 3D-CNN algorithm · Webapp

1 Introduction

India loses 1.5 lakh lives due to road accidents every year according to recent survey. According to recent statistics provided by Ministry of Road Transport and Highways, India tops the world in road crash deaths and injuries. It has 1% of the world's vehicles but accounts for 11% of all road crash deaths, witnessing 53 road crashes every hour killing 1 person every 4 min. The country accounts for about 4.5 lakh road crashes per annum, in which 1.5 lakh people die.

Many proposed approaches for detecting these incidents include traffic flow modelling, vehicle activity analysis, and vehicle interaction modelling. Over-speeding is caught using surveillance cameras and radars. Despite all these precautions, the main cause of this proportion is a lack of timely assistance. Traffic flow modelling techniques are used in traditional detection approaches. The performance of traffic flow depends on data quality. In addition, these techniques are based on a single type of feature that is unable to match the accuracy and real-time performance requirements of accident detection. Image recognition technology provides high efficiency, flexibility in installation, and minimal maintenance costs for traffic situation awareness.

© IFIP International Federation for Information Processing 2022
Published by Springer Nature Switzerland AG 2022
L. Kalinathan et al. (Eds.): ICCIDS 2022, IFIP AICT 654, pp. 301–310, 2022.
https://doi.org/10.1007/978-3-031-16364-7_23

Automatic accident detection can increase rescue efficiency and traffic safety by reducing the time it takes for rescue agencies and vehicles to respond to an accident. Existing accident detection system suffer from tradeoff between computational overhead and detection accuracy. Some of the early detection techniques, such as YOLO CA, Gaussian mixture model (GMM), Single Shot Multi-Box Detector (SSD), and Convolutional Neural Network (CNN) have similar accuracy and precision compared to other models. Vision Based Detection requires a large storage capacity, and several parameters such as weather conditions and image resolution must be considered to attain accuracy. Vision-based crash detection systems have expanded significantly in recent years as technology progresses.

This system intends to detect accidents using a vehicle collision system in the initial phase. This methodology was proposed to develop an effective model for detecting different classes of vehicles such as cars, motorcycles, buses, and other vehicles. This research focuses on using CCTV surveillance cameras to detect vehicles in real time. The data was gathered from a variety of sources such as YouTube videos and search engines. The model is trained with a custom dataset using the YOLOv5 object detection algorithm that is based on computer vision. As a result, the proposed model is effective at detecting objects in real time under a variety of situations including rainfall, poor visibility, daytime, nighttime.

The motivation of this paper is to shorten the response time of rescue agencies and vehicles around accidents to improve rescue efficiency and traffic safety level. The major objective of this project is to create a model to detect road accidents, analyse and classify the severity of the accident and to build an alerting system. The model is evaluated using several measures such as precision, recall, mAP, F1 score etc.

2   Literature Analysis

A real-time automatic detection accident detection system. Boxy Vehicle dataset was used for training the model. The images from the dataset were annotated with 2D Bounding box. Custom annotation is made on images for training ResNet152 in YOLOv3.Mini YOLO is used in detection as its efficient and has high recall. Bounding Box and probabilities are obtained for each frame. Soft labeling was done to choose highest probable class. The processed images are tracked using SORT for the status of damage throughout the entire frames. The Velocity and position of each objects are accurately obtained. For each detected frame it is classified in a model as damaged or undamaged. The system is trained on SVM with radial basis kernel to identify the damage status. This model considered only the static motion of accidents. Sequence model was not addressed in this model and field testing was performed [1].

An unsupervised method for detecting traffic accidents in first-person footage. Main innovation is that they used three separate ways to discover anomalies by forecasting the future locations of traffic location and then evaluating the predictive accuracy, performance, and consistency metrics. One of the system's biggest flaws is that it assumes fixed cameras and unchanging backgrounds in videos, which is fine for surveillance but not for vehicle-mounted cameras. Second, it presents the issue as a one-class classification problem, which relies on time-consuming human annotation and only identifies anomaly categories that have been expressly trained. They provide a new dataset of different traffic incidents, An Accident Detection (A3D), as well as another publicly available dataset was used to evaluate this approach [2].

Identifying images of different vehicle crashes, the model uses deep learning methods such as Convolutional Neural Networks. Once training the model with a large amount of visual data is done the model can more accurately classify the severity of the crash by minimizing the classification error. The optimal strategy in generating the model is determined by comparing the accuracy of the model utilizing various Activation functions with the optimizers in CNN. By comparing activation functions with each optimizer, this work focuses on the comparison of various activation functions and optimizers and their effect on the prediction accuracy in classifying [10].

Recently, Zhenbo Lu proposed a framework using residual neural network (ResNet) combined with attention modules was proposed to extract crash-related appearance features from urban traffic videos which were further fed to a spatiotemporal feature fusion model, Conv-LSTM. The model achieved an accuracy of 87.78% on the validation dataset and an acceptable detection speed [12].

Sergio Robles-Serrano proposed DL-based method capable of detecting traffic accidents on video. The model considers that visual components occur in a temporal order to represent traffic collision events. As a result, the model architecture is composed of a visual extraction of features phase followed by a transient pattern identification. Convolution and recurrent layers are used in the training phase to learn visual and temporal features utilizing built-from-scratch and public accessible datasets. In public traffic accident datasets, an accuracy of 98% was attained in the detection of accidents, demonstrating a strong capacity in detection irrespective of the road structure [13].

Kyu Beom Lee proposed an Object Detection and Tracking System (ODTS) used in conjunction with the Faster Regional Convolution Neural Network (Faster R-CNN) in this paper for Object Detection and the Conventional Object Tracking algorithm was introduced and implemented for automatic monitoring and detection of unexpected events on CCTVs in tunnels. With a sample of event photos in tunnels, this model has been trained to achieve Average Precision value of 0.85, 0.72, and 0.91 for target items like cars, person, and fire. The ODTS-based Tunnel CCTV Accident Detection System was then tested using four accident recordings, one for each accident, based on a trained deep learning model. As an outcome, the technique can detect all incidents in under ten seconds [7].

In another deep learning-based detection real-time system with the use of a Computer Vision algorithm. The system uses YOLOv4 to perform quicker object detection in real time, and it was tested in a variety of situations, including rainfall, poor visibility, daytime, nighttime and snow using the proposed dataset. In the first phase, the dataset's preprocessing processes are completed. Different types of vehicle photos are recorded after the first phase. The Precision, Mean Average Precision (map) and Average Intersection Over Union (IoU) of the model with which it can recognize the various types of Vehicles were found in the final phase, based on the properties computed or computed in the earlier phases. When we compare the standard results to our results, we can see that the average precision of class Bus is far higher than any other model, that of class Motorcycle is significant compared to almost all other models, that of class Car is greater than

YOLOv2's performance, and only the average precision of class Truck is lower than the other models due to various types of trucks labelled in the dataset [14].

Shilpa Jahagirdar proposed an automatic road accident detection algorithm that use real-time surveillance videos to detect accidents. None of these systems are robust enough to address all accident instances since they do not consider variable lighting circumstances, changing weather conditions, and different traffic patterns. Many of these strategies have been detailed and compared in this study. For Motion detection, Background subtraction, optical flows were utilized. For feature extraction the parameters that were considered were velocity orientation, acceleration, vehicle area, Histogram, Traffic trajectory. For Accident Recognition Machine learning tools, statistical calculation and comparison to thresholds were made. The major challenge in this paper was to differentiate collision and occlusion factor that needs to be addressed particularly and it will help to reduce false accident detection rate [15].

3   Proposed Approach

A Convolution neural Network (CNN) is a type of artificial neural network used in scene classification, object detection, segmentation, and recognition and processing. CNN is the basic building block of almost all the object detection models. Most of the models like Mask-R CNN, SSD or YOLO are built using convolution neural networks. It is also called as ConvNet. These neural networks are best suited for image, video analysis and also in document recognition, speech synthesis and speech recognition.classification and location algorithms can be divided into two kinds namely Two stage model and One stage model.

3.1   Two Stage Model

R-CNN, Fast R-CNN, Faster R-CNN, and Faster R-CNN with FPN are examples of two-stage models. These techniques use selective research and the Region Proposal Network (RPN) to choose approximately 2000 proposal regions in the image, and then use CNN to detect objects based on the attributes retrieved from these regions. Although these region-based models effectively find objects, extracting solutions takes a long time. In R-CNN, they used a selective search to suggest regions for each item in the input image which was extensively time consuming. Fast R-CNN significantly enhanced R-CNN's efficiency but it still utilizes selective search as regional proposal network. Fast R-CNN share the convolutional layers for region proposal. Faster RCNN employed a fully convolutional neural network for region proposal.

3.2   One Stage Model

YOLO (You Only Look Once) and SSD (Single Shot Multi Box Detector) are the example One stage model. These algorithms implement location and classification by one CNN, which can provide end to end detection service. Because of eliminating the process of selecting the proposal regions, these algorithms are very fast and still have guaranteeing accuracy.

YOLOv3 is built on the Darknet Architecture and features 53 layers that have been trained on the ImageNet dataset. Upsampling and residual connections are used in YOLOv3. At three distinct scales, the detection is carried out. Although it is more effective at recognizing smaller things, it takes longer to process than earlier generations.

The CSPDarknet53 backbone, spatial pyramid pooling extra module, PANet path-aggregation neck, and YOLOv3 head make up the YOLOv4 architecture. The backbone of the YOLOv5 model is a Focus structure with a CSP network. The SPP block and PANet make up the neck. It uses GIoU-loss and has a YOLOv3 head (Fig. 1).

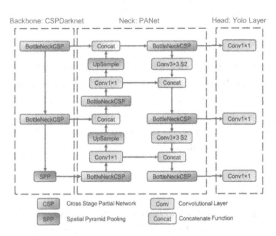

Fig. 1. Architecture of YOLOv5 framework

3.3 Architecture of YOLOv5

The YOLOv5 model consists of Focus structure and CSP network as the backbone. The neck is composed of SPP block and PANet. It has a YOLOv3 head using GIoU-loss. YOLOv5 is a convolutional neural network (CNN) that detects objects in real-time with great accuracy. This approach uses a single neural network to process the entire picture, then separates it into parts and predicts bounding boxes and probabilities for each component. These bounding boxes are weighted by the expected probability. It then delivers detected items after non-max suppression.

4 Experimental Design

All the experiments were conducted on Google Colabratory which used virtual GPU with 12.72 GB Main memory and 68.40 GB Disk space. Video was processed using cv2 and program was written in python 3.8 (Fig. 2).

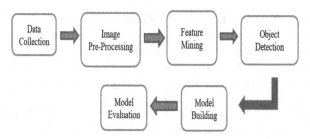

Fig. 2. Flow Diagram of proposed model

4.1 Data Collection

The dataset for this project is a collection of 100 mp4 films containing CCTV footage of various road incidents from around the world, which was gathered from YouTube videos and the CADP dataset. These videos have various types of vehicles such as Car, Bus, Motorcycle, Truck etc. The object was classified into four classes namely Car, Motorcycle, Bus, and others. These videos were shot in various weather and sunlight conditions, such as broad daylight, night, hail, snow, rain etc.

4.2 Image Pre-processing

In this Phase the images are annotated based on categories manually using online tool makesense. From the annotated image the labels of bounding box were collected and stored. Once label and respective images were collected the image and label dataset is split in the ratio of 80:20 for training and validation set. A dataset yaml was prepared to contain all the classification types and the path of training and validation set. We then download YOLOv5 darknet framework and the standard weight files for training and testing. Now we create a custom cfg file based on the needs of our training model, which includes four classes: car, bus, motorcycle, and others. So, in our standard YOLOv5 cfg file, we set batch = 64, subdivisions = 64, width = 416, height = 416, max batches = 8000 (number of classes * 2000), steps = 7200 (90% of max batches), and set filters = 27(number of classes + 5) *3 in every convolutional layer above yolo layer and set classes = 4 in every yolo layer below convolutional layer. We are now making changes to the YOLOv5 darknet framework's standard Makefile. Set GPU to 1, CUDNN to 1, OPENCV to 1, and CUDNN HALF to 0. Then we first break these videos of our dataset into frames. We do this with the help of cv2. Now for every second of each video we have 5 frames, now we have to annotate these frames and label the various kinds of vehicles traversing on the roads in various weather conditions.

The makesense tool then generates a.txt file for each image including the label of the class along with the coordinates of the x-centre, coordinates of the y-centre, width of the rectangle box, and height of the rectangle box for each of the vehicles indicated in the frame. Now, we divide these.txt and.jpg files in 80:20 ratios for training, testing and save them in the file path images and name them individually.

4.3 Feature Mining

By using the custom YOLOv5 darknet framework, we train our model for 8000 iterations with 30 epochs. At the end when our model training is completed, we get the 'best and last weights' file which can be trained and tested on our dataset and can be used to detect vehicle in videos.

4.4 Object Detection

After training our model we use the best weights file that we got from training our model and use our custom cfg file to test the model with the predefined classes of vehicle such as car, motorcycle, bus, others with the predicted accuracy percentage.

4.5 Model Building

Once the model is trained and tested, for better accuracy we alter the filters and custom cfg file to improve the performance.

4.6 Model Evaluation

The accuracy of object detection model is evaluated using performance measures such as Precision, Recall, F1 Score and mAP.

5 Proposed Architecture

The videos are collected from the CCTV surveillance camera and split into frames with dimension 1280 × 720. YOLO v5 expects annotations for each image in form of.txt file where each line of the text file describes a bounding boxto define the location of the target object (Fig. 3).

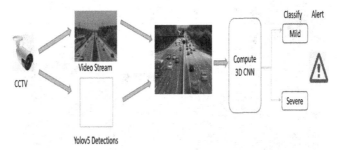

Fig. 3. Architectural design of proposed model

The images are annotated and labeled in the format of 'class x_centery_center width height'. The model is trained with the custom YOLOv5 darknet framework weights. we get the best and last weights file for our custom dataset that was to used to train the classes of vehicle. The model is tested with best weights file and evaluation of the model is done to validate and to improve the efficiency of vehicle identification.

This will be extended to identify the accident images and classify the accidents as mild and severe using 3D CNN Algorithm. Based on the classification, with the help of webapp we build a mail alerting system.

6   Results and Discussion

YOLO performs well when compared to Fast R-CNN. In comparison to v4 and v5, the time spent training in YOLOv3 is considerably lesser. When all factors are considered, v4 has some positive performance characteristics and requires less training time than v5 (Fig. 4, Table 1).

Table 1.   Analysis of various CNN Models

S. No.	Method	Average precision	Average IoU
1	Fast R-CNN	80.3	0.63
2	Faster R-CNN	86.79	0.69
3	SSD	89.65	0.71
4	YOLOv3	91.17	0.73
5	YOLO-CA	91.51	0.74
6	YOLOv5	91	0.915

Fig. 4.   Results of detected image frames

7 Accident Detection Module

The model detects the object and classifies according to its class. By detecting the classes this model can be further extended to detect and classify the accidents based on the severity by using 3D-CNN. Using the 3D volumetric representation of the image, 3D-CNN extracts spatial information from the inputs in three dimensions. It takes advantage of interslice context leading to improved performance. Once the model is trained to classify the accidents based on its severity it can further be employed in real time monitoring and alerting by building a webapp Api to alert nearest hospital with the location details from the mounted surveillance camera along with the severity information to quicken the response time and to improve the traffic safety level.

8 Conclusion

In this we have built a object detection system to identify the various types of vehicles using YOLOv5 from the real time CCTV surveillance video. After training the model we were able to classify the vehicle. We trained the model with 1000 images in future we can increase the dataset to improve the efficiency of the system. By Comparing the results, we can see the positive trend in the precision and accuracy level. YOLOv5 provides great accuracy in detection. As YOLOv5 was tested with larger batch size, it has higher interference speed than most of the detectors. In further step we can build end to end system to identify the accident images and to analyze the severity of it using 3D-CNN Algorithm.

References

1. Pillai, M.S., Khari, M.: Real-time image enhancement for an automatic automobile accident detection through CCTV using deep learning. Soft Comput. **25**, 11929–11940 (2021). https://doi.org/10.1007/s00500-021-05576-w
2. Nancy, P., Dilli Rao, D., Babuaravind, G., Bhanushree, S.: Highway accident detection and notification using machine learning. IJCSMC **9**, 168–176 (2020)
3. Yu, Y., Xu, M., Gu, J.: Vision-based traffic accident detection using sparse spatio-temporal features and weighted extreme learning machine. IET Intell. Transp. Syst. **13**, 1417–1428 (2019)
4. Tian, D., Zhang, C., Duan, X., Wang, X.: An automatic car accident detection method based on cooperative vehicle infrastructure systems. IEEE Access **7**, 127453–127463 (2019)
5. Diwan, A., Gupta, V., Chadha, C.: Accident detection using mask R-CNN. IJMTST **7**, 69–72 (2021)
6. Yadav, D.K., Renu, Ankita, Anjum, I.: Accident detection using deep learning. IJECCE (2020)
7. Lee, K.B., Shin, H.S.: An application of a deep learning algorithm for automatic detection of unexpected accidents under bad CCTV monitoring conditions tunnels. IEEE (2019)
8. Singh, C., Mohan, K.: Deep spatio-temporal representation for detection of road accidents. IEEE (2019)
9. Yao, Y., Xu, M., Wang, Y., Crandall, D.J., Atkins, E.M.: Unsupervised traffic accident detection in first-person videos. IEEE (2020)

10. Mothe, S., Teja, A.S., Kakumanu, B., Tata, R.K.: A model for assessing the nature of car crashes using convolutional neural networks. Int. J. Emerg. Trends Eng. Res. **8**, 859–863 (2020)
11. Rajesh, G., Benny, A.R : A deep learning based accident detection system. In: ICCSP (2020)
12. Lu, Z., Zhou, W., Zhang, S., Wang, C.: A new video-based crash detection method: balancing speed and accuracy using a feature fusion deep learning framework. IJAT (2020)
13. Robles-Serrano, S., Sanchez-Torres, G., Branch-Bedoya, J.: Automatic detection of traffic accidents from video using deep learning techniques. MDPI (2021)
14. Sindhu, V.S.: Vehicle identification from traffic video surveillance using YOLOv4. IEEE (2021)
15. Jahagirdar, S.: Automatic accident detection techniques using CCTV surveillance videos: methods, data sets and learning strategies. IJEAT **9**, 2249–8958 (2020)
16. Mothe, S., Teja, A.S., Kakumanu, B., Tata, R.K.: A model for assessing the nature of car crashes using convolutional neural networks. IEEE (2020)

Monitoring of PV Modules and Hotspot Detection Using Convolution Neural Network Based Approach

B. Sandeep, D. Saiteja Reddy, R. Aswin, and R. Mahalakshmi[✉]

Department of Electrical and Electronics Engineering, Amrita School of Engineering, Amrita Vishwa Vidyapeetham, Bengaluru, India
d_mahalakshmi@blr.amrita.edu

Abstract. The use of solar photovoltaic systems in green energy harvesting has increased greatly in the last few years. Fossil fuels reaching the end are also growing rapidly at the same rate. Despite the fact that solar energy is renewable and more efficient, it still needs regular Inspection and maintenance for maximizing solar modules' lifetime, reducing energy leakage, and protecting the environment. Our research proposes the use of infrared radiation (IR) cameras and convolution neural networks as an efficient way for detecting and categorizing anomaly solar modules. The IR cameras were able to detect the temperature distribution on the solar modules remotely, and the convolution neural networks correctly predicted the anomaly modules and classified the anomaly types based on those predictions. A convolution neural network, based on a VGG-based neural network approach, was proposed in this study to accurately predict and classify anomalous solar modules from IR images. The proposed approach was trained using IR images of solar modules with 5000 images of generated solar panel images. The experimental results indicated that the proposed model can correctly predict an anomaly module by 99% on average. Since it can be costly and time-consuming to collect real images containing hotspots, the model is trained with generated images rather than real images. A generated image can be used more efficiently and can also have custom features added to it. In the prediction process, the real image is processed and it is sent to the model to determine bounding boxes. It provides a more accurate prediction than direct use of the real image. Here we have used CNN custom model and TensorFlow libraries.

Keywords: Convolution neural network · PV modules · Detection of hotspots · CNN · TensorFlow · Solar panels

1 Introduction

The benefits of solar photovoltaic systems are numerous, and they are extremely useful. Their uniqueness lies in their lack of moving parts (in the sense of classical mechanics). There is no leakage of fluids or gasses (except in hybrid systems). They do not require fuel, which is one of the best features of these systems. They respond rapidly and achieve

© IFIP International Federation for Information Processing 2022
Published by Springer Nature Switzerland AG 2022
L. Kalinathan et al. (Eds.): ICCIDS 2022, IFIP AICT 654, pp. 311–323, 2022.
https://doi.org/10.1007/978-3-031-16364-7_24

full output immediately [1]. Although the cells run at moderate temperatures and do not emit pollution while producing electricity. Solar cells require little maintenance if properly manufactured and installed. Solar energy has gained a reputation as a clean, renewable energy source. Solar energy can help to reduce carbon dioxide levels in the atmosphere. Solar energy has no negative impact on the environment [2]. Solar power requires no additional resources than a source of light to function, in addition to not emitting greenhouse gasses. As a result, it is both safe and environmentally friendly.

PV cells and solar energy, like other renewable energy sources, have intermittent difficulties. It cannot be transformed into power constantly at night or in foggy or rainy conditions [3]. In other words, PV cells may not be able to generate enough electricity to fulfill the peak demand of a power grid. Solar energy panels are a less reliable power source because of their intermittency and instability. In addition to PV cells, inverters and storage batteries are required. An inverter is a device that turns direct electricity into varying electricity that may be used on your power grid. Storage batteries help on-grid connections because they provide a consistent supply of electric power. On the other side, this greater expenditure may be able to address the PV cells' intermittency issues [4]. PV plants' ability to produce energy may be affected by the failure of modules or components (wiring, bypass diodes, etc.), malfunctions, or accumulated dust and soiling on their surfaces; the last condition can decrease efficiency by more than 18% compared to normal conditions. For example, consider the PV system installed in Pavagada solar park (Karnataka) with a size of 53 km^2 (13000 acres). When working at its best, it can produce almost 2350 MWh per year. Assuming that the system operates normally with soiled modules due to dust deposits, debris, and bird droppings, the energy production could be decreased by 300 MWh/year, or 14.64% [5].

For PV systems to perform optimally, monitoring systems should be used to detect anomalies in normal operation and to check the cleanliness of modules using periodic cleaning. Using a convolution neural network to monitor solar modules with an infrared camera is performed. In addition to being able to monitor the condition of photovoltaic modules and other components, this technology can be used for pre-diagnosis and diagnosis of faulty working conditions by using UAVs (drones), electric equipment in PV systems [6, 7]. CNN can be used to detect any flaws or malfunctions in PV modules, e.g., conditions that result in local overheating, without interfering with the usual operation of the system [8]. Dirt, dust, and bird droppings on the surface can cause local overheating by changing the thermal exchange conditions between the module and the environment. In this work, machine learning is applied to construct a tool that automatically identifies abnormal operating conditions of PV modules [9]. The paper proposes to use machine learning as a tool in this work. In this project, we will develop a system for automatic classification of thermograms by using a convolutional neural network (CNN), distinguishing between overheating caused by external factors or by faults on the surface of the photovoltaic module [10]. Thermographic images can now show an anomaly state caused by a fault [11]. The importance of photovoltaic energy generation, the benefits of this energy generation, the various methods for identifying power losses in this generation method, and, most importantly, the most commonly raised problem of PV hotspots generation in PV modules and the causes of these hotspots are all discussed in the above literature surveys [12, 13].

Therefore, the solar panel must be monitored for failures, particularly hotspots. This paper proposes a method to identify hotspots using TensorFlow. Hotspot images are generated using the same processing method as solar panel images. Training models are validated with validating images, followed by testing of images in which the model determines the hotspot type and localization of the hotspot. In real solar panel images, the model can detect and classify the type of hotspot and its location based on the training with different types of hotspots containing images and subsequent validation for any particular generated image. In order to prevent panel damage, action needs to be taken immediately based on an analysis of the hotspot type and location.

2 System Model

2.1 Introduction

The input data is generated in python and after that, it is split into train data and test data. Now the model has sufficient data and it is split into train data and test data so that better accuracy can be obtained. In below Fig. 1, the process of the project is shown.

2.2 Block Diagram

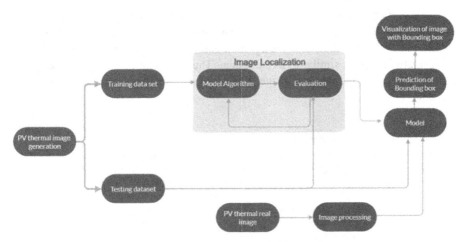

Fig. 1. Block diagram for hot spot localization.

Figure 1 shows a block diagram of the hotspot localization process, which includes all the key steps. It starts with the data generation, then it splits into training and testing data sets. Then comes the algorithm, and then after that, the evaluation, which is nothing but checking the model performance. The model is again trained and improved, and this is how Image Localization is achieved. Following that, real images are converted into processed images and fed into the model. Next, the Image localization model predicts the bounding box is followed by analysis and visualization of the results.

2.3 Data Generation

PV images with hot spots are generated through Python with an image ratio of 120 ×
240 pixels. During generation, 10,000 unique images were generated and split into 5000
training images and 5000 test images. Each image size is 120 × 240 pixels.

```
<?xml version="1.0" encoding="UTF-8"?>
<annotation>
    <folder>images</folder>
    <filename>images/training/images1.png</filename>
    <path>C:images/training/images1.png</path>
    <size>
        <width>120</width>
        <height>240</height>
        <depth>3</depth>
    </size>
    <segmented>0</segmented>
    <object>
        <name>spot</name>
        <pose>Unspecified</pose>
        <truncated>0</truncated>
        <difficult>0</difficult>
        <bndbox>
            <xmin>43</xmin>
            <ymin>142</ymin>
            <xmax>68</xmax>
            <ymax>167</ymax>
        </bndbox>
    </object>
</annotation>
```

Fig. 2. Generated image.   **Fig. 3.** XML file.

PV thermal image is shown in Fig. 2 along with the XML file for the bounding box
label in Fig. 3. With a dark blue background and a red hotspot at a random location,
the generated image is similar to a real image. The bounding box is a box containing
the hotspot location in the given image. XML files are used to view the image with the
bounding box.

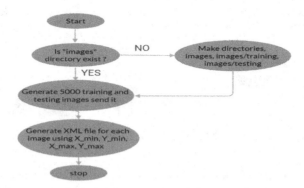

Fig. 4. Data generation flowchart.

Figure 4 shows a flowchart for Data Generation. if there is no images directory, it
creates directories such as images, images/training, images/testing to hold the images in
the relevant directory. Python was used to create the generated images. 10,000 unique

images were generated during the generation process and then divided into 5000 training and 5000 test datasets. These are converted into Tensorflow datasets. Once we have converted these into Tensorflow datasets, the model takes the test and training datasets separately. In addition to images, XML file coordinates are also generated, including X_min, Y_min, X_max, and Y_max. These coordinates can be used to draw bounding boxes. These XML files are also placed in directories along with images. For training and validation of models, these XML files serve as labels for images.

2.4 Training and Testing Data

In the second stage, when data is generated, the model must be trained and tested to improve accuracy and results. The generated data is separated into two subgroups, as seen below. During generation, 10,000 unique images were generated and split into 5000 training images and 5000 test images. With 5000 training images, we were able to improve the model's accuracy. Test images could be any number for validation purposes only. We chose the 5000 testing images. The generated images were created programmatically using Python.

2.4.1 Training Dataset

Fig. 5. Training dataset.

Figure 5 shows a flowchart for Training Dataset. The training images with XML files are converted into CSV as they need to be understandable by TensorFlow then the assigning of bounding box coordinates is done in CSV values and after that, Mapping of images with the file path and sending it to TensorFlow. Now as the data set is created now, the data is divided into batches each batch containing 64 images and fed to the model.

2.4.2 Testing Dataset

Fig. 6. Testing dataset.

Figure 6 shows a flowchart for Testing Dataset. Similar to the training data the test data also will be there as the data set used to offer an unbiased assessment of a final model's fit to the training dataset. The test dataset serves as the gold standard against which the model is judged. It's only used once a model has been fully trained (using the train and validation sets). The test set is typically used to compare models (For example, for many

Kaggle competitions, validation and training sets are released initially together, then the test set is released toward the end of the competition and the model's performance on the test set determines the winner). The validation set is frequently used as the test set, but this is not a good practice. The test set is well-curated in general. It includes properly sampled data from a variety of classes that the model might encounter in the actual world.

2.5 Image Processing

Fig. 7. Real PV thermal image (Source: Internet)

As the model has to detect hotspots in real PV thermal images like the one in Fig. 7, the model has been trained using the generated data and validated. It is necessary to process the real images (in a similar way to the generated image) so that the model can understand them. Hence, the real PV thermal images must be processed.

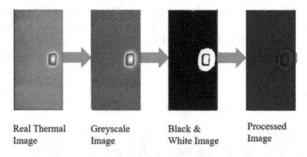

Real Thermal Image   Greyscale Image   Black & White Image   Processed Image

Fig. 8. Real thermal image to processed image.

Figure 8 shows how real images are converted into processed images by Image Processing.

- Real Thermal Image
- Grayscale Image
- Black & White Image
- Processed Image

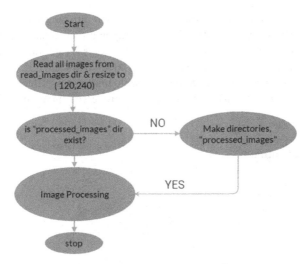

Fig. 9. Processed image generation.

Figure 9 shows the processed image Generation. First, all real images in the real_images directory are read and resized to 240 × 120. The processed_images directory is created if it does not exist and every image is sent to the image processing function, after which the processed images are sent to the processed_images directory.

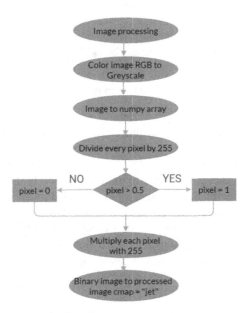

Fig. 10. Image processing.

Figure 10 shows how real color images are converted into grayscale images by using a filter. The image is converted into a NumPy array and then each pixel is divided by 255, and then the result is compared with 0.5. If the pixel is greater than 0.5, the set pixel value equals 1 then it is black, and else the pixel is 0 then it is white. 255 is multiplied by each pixel again. This logic transforms the grayscale image into a Black & White image (Binary image). A binary image is converted to a processed image by using c-map as "jet".

2.6 Image Localization

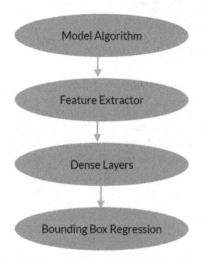

Fig. 11. Image localization model algorithm.

Figure 11 illustrates how the model algorithm works, which includes the feature extractor and dense layers followed by the bounding box regression. Tensorflow Training batches are sent to a feature extractor that has Conv2d filters and average pooling filters for extracting key features that are required to train the model. In the Dense layer, which follows the feature extractor, each node takes the flattened image output and multiplies with weights, adds bias, and sums all of these to the activation function, resulting in outputs of the function. The bounding box is predicted by adjusting the weights and biases in the dense layer. A Bounding Box regression has four nodes and outputs four values at the same time: X_min, Y_min, X_max, and Y_max.

2.7   Model Flowchart

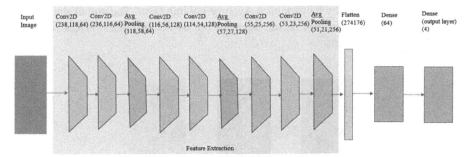

Fig. 12. Image localization model flowchart.

The model flow chart in Fig. 12 shows the model layers and the resulting architecture which is inspired by the VGG architecture, which is two convolutions followed by average pooling, and the cycle is repeated three times. Initially, there are 64 filters, followed by 128, and finally, 256 filters. It shows the flow of the model algorithm.

```
Model: "model_1"

Layer (type)                    Output Shape              Param #
=================================================================
input_2 (InputLayer)            [(None, 240, 120, 3)]     0

conv2d_7 (Conv2D)               (None, 238, 118, 64)      1792

conv2d_8 (Conv2D)               (None, 236, 116, 64)      36928

average_pooling2d_2 (Average    (None, 118, 58, 64)       0

conv2d_9 (Conv2D)               (None, 116, 56, 128)      73856

conv2d_10 (Conv2D)              (None, 114, 54, 128)      147584

average_pooling2d_3 (Average    (None, 57, 27, 128)       0

conv2d_11 (Conv2D)              (None, 55, 25, 256)       295168

conv2d_12 (Conv2D)              (None, 53, 23, 256)       590080

conv2d_13 (Conv2D)              (None, 51, 21, 256)       590080

flatten_1 (Flatten)             (None, 274176)            0
...
=================================================================
Total params: 19,283,076
Trainable params: 19,283,076
Non-trainable params: 0
```

Fig. 13. Model summary.

This Fig. 13 shows that the model summary displays all of the model's layers as well as their output shapes. It takes a 240 * 120-pixel input image. The size of the output image is reduced by 2 × 2 in each conv2d layer using the 3 × 3 Filter. The resulting image's size is reduced by half for each Average pooling layer. The output shape after flattening is 274176. This will be the input for the 64 neutron dense layers. The last dense layer has 4 neutrons.

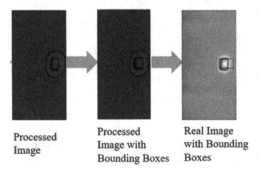

Processed Image   Processed Image with Bounding Boxes   Real Image with Bounding Boxes

Fig. 14. Predicting bounding box for processed image and real image

Figure 14 shows when the processed image is sent to the model it predicts output X_min, Y_min, X_max, Y_max for drawing the Bounding Box. By using these values, a bounding box can be drawn for the processed image as well as the real image.

3 Results

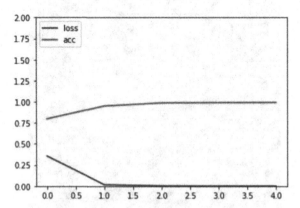

Fig. 15. Accuracy and bounding box loss results.

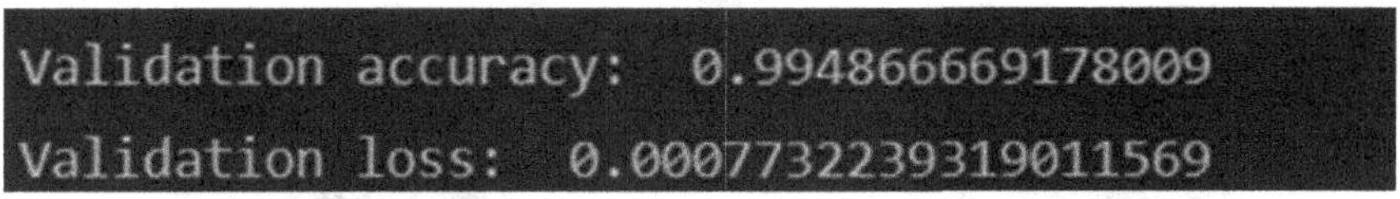

```
Validation accuracy:  0.9948666669178009
Validation loss:  0.00077322239319011569
```

Fig. 16. Validation accuracy and loss values

The results are shown in Fig. 15 as in graphs i.e. bounding box loss and accuracy. In Fig. 16 the validation accuracy and loss results are given. An epoch is one training iteration, so in one iteration all samples are iterated once. Figure 15 shows the graphs of Bounding Box Loss and accuracy. Figure 16 shows validation accuracy and loss. As it is observed the validation accuracy is 0.99486. The model is working perfectly.

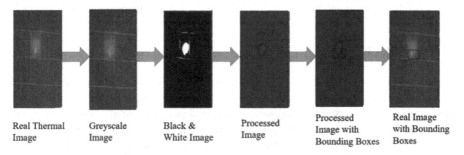

Real Thermal   Greyscale   Black &   Processed   Processed   Real Image
Image   Image   White Image   Image   Image with   with Bounding
                                                                Bounding Boxes   Boxes

Fig. 17. Step by step process through images

The Total process of this paper is shown in Fig. 17 as initially the PV thermal image is converted into a grayscale image using some filters and then based on the pixel values the grayscale image is converted into a black & white image. Now, this image is converted into a processed image using the model algorithm which is our model understandable image format. As the model is trained with this type of data set it can easily detect the hotspot and draw bounding boxes around it. And based on the bounding box coordinates the model can draw the same boxes in the real thermal image also. Thus, the hotspot is detected in the real PV thermal image and the model detection accuracy is good at 0.994 (Fig. 18a, 18b).

These are the bounding box results for other images.

Fig. 18a. Real image 1   **Fig. 18b.** Real image 2

4   Conclusion

There is an analysis of the accuracy results of both the training and testing data sets, and for the hotspot, localization is done by drawing bounding boxes, and the real PV image with hotspots is converted to the generated image format, and finally, according to the bounding box coordinates, the hotspots are detected in real thermal images.

With the generated images of PV modules, the model is trained. It is then used to process the real images of PV modules. Real thermal images are converted into grayscale and then into black and white. After that, the processed images are generated using a color map jet. As the model is trained with 5000 images the accuracy is very good. Thermal images can be converted to processed images, which is the format that is used to train the model, and the bounding boxes can be easily drawn around the hotspot. After getting the bounding box coordinates the same boxes can be drawn on a real image. So, the hotspot detection and localization for real thermal images is done.

References

1. Gautam, M., Raviteja, S., Mahalakshmi, R.: Household energy management model to maximize solar power utilization using machine learning. Procedia Comput. Sci. **165**, 90–96 (2019)
2. Sophie, N., Barbari, Z.: Study of defects in PV modules. Energies, 107 (2019)
3. Mani, M., Pillai, R.: Impact of dust on solar photovoltaic (PV) performance: research status, challenges, and recommendations. Renew. Sustain. Energy Rev. **14**, 3124–3131 (2010)
4. Kudelas, D., Taušová, M., Tauš, P., Gabániová, L'., Koščo, J.: Investigation of operating parameters and degradation of photovoltaic panels in a photovoltaic power plant. Energies **12**, 3631 (2019)
5. Cristaldi, L., et al.: Economical evaluation of PV system losses due to the dust and pollution. In: Proceedings of the 2012 IEEE International Instrumentation (2012)
6. Alsafasfeh, M., Abdel-Qader, I., Bazuin, B., Alsafasfeh, Q.H., Su, W.: Unsupervised fault detection and analysis for large photovoltaic systems using drones and machine vision. Energies **11**, 2252 (2018)

7. Li, X., Yang, Q., Lou, Z., Yan, W.: Deep learning-based module defect analysis for large-scale photovoltaic farms. IEEE Trans. Energy Convers. **34**(1), 520–529 (2019). https://doi.org/10. 1109/tec.2018.2873358

8. Ren, X., Guo, H., Li, S., Wang, S., Li, J.: A novel image classification method with CNN-XGBoost model. In: Kraetzer, C., Shi, Y.-Q., Dittmann, J., Kim, H.J. (eds.) IWDW 2017. LNCS, vol. 10431, pp. 378–390. Springer, Cham (2017). https://doi.org/10.1007/978-3-319-64185-0_28

9. Zhao, Q., Shao, S., Lu, L., Liu, X., Zhu, H.: A new PV array fault diagnosis method using fuzzy C-mean clustering and fuzzy membership algorithm. Energies **11**, 238 (2018)

10. Pei, T., Hao, X.: A fault detection method for photovoltaic systems based on voltage and current observation and evaluation. Energies **12**, 1712 (2019)

11. Jadin, M.S., Taib, S.: Recent progress in diagnosing the reliability of electrical equipment by using infrared thermography. Infrared Phys. Technol. **55**, 236–245 (2012)

12. Shivani, P.G., Harshit, S., Varma, C.V., Mahalakshmi, R.: Detection of broken strands on transmission lines through image processing. In: 2020 4th International Conference on Electronics, Communication, and Aerospace Technology (ICECA), Coimbatore, India, pp. 1016–1020 (2020)

13. Shivani, P.G., Harshit, S., Varma, C.V., Mahalakshmi, R.: Detection of icing and calculation of sag of transmission line through computer vision. In: 2020 Third International Conference on Smart Systems and Inventive Technology (ICSSIT), Tirunelveli, India, pp. 689–694 (2020)

Literature Review on Human Behavioural Analysis Using Deep Learning Algorithm

R. Poorni[1][✉] and P. Madhavan[2][✉]

[1] Department of Computer Science and Engineering, Sri Venkateswara College of Engineering, Pennalur, Sriperumbudur Tk., Tamil Nadu, India
poorniram21@gmail.com
[2] Department of Computing Technologies, SRM Institute of Science and Technology, Kattankulathur, Chennai, Tamil Nadu, India
madhavap@srmist.edu.in

Abstract. Human behaviour analysis is the active area of research in computer science and engineering which determines the behaviour of humans using various algorithms. The input can be taken from the real time environment to analyze the human predictions. Deep learning plays a vital role as the input data involves a lot of computational images and spatial and temporal information upon which the predictions can be made. In this paper, we discuss the various techniques, concepts and algorithms that are implemented on a various field of image analysis and on real world input data to visualize the behaviour of a human.

Keywords: Real time environment · Deep learning algorithms · Behaviour analysis

1 Introduction

In today's digital world, the factual information that we get from our day to day is getting massive. In the current trend, a statistics says that the data produced and used will reach around 175 ZB in the year 2025. This increasing amount of data leads to the evolution of various new fields such as data mining, data science, artificial intelligence and its subset areas, etc. The major impulse of artificial intelligence is the data. The data is fed into the AI system to learn, analyze and perform the tasks accordingly. The fact that the amount of data to be handled becomes tremendous and more the number of interpretations that can be done with the data which lets the system to learn by itself that leads to machine learning. By the use of data the algorithm learns by itself and improves the predictions or decision making from the experience gained. A model is built based on training data or sample data to make decisions without any external interventions. Machine learning is closely intertwined with statistical computing as the majority of the decisions made are based on the probability of their occurrence. It uses the neural network to duplicate the human brain to perform a task without being programmed by the user. The model can

© IFIP International Federation for Information Processing 2022
Published by Springer Nature Switzerland AG 2022
L. Kalinathan et al. (Eds.): ICCIDS 2022, IFIP AICT 654, pp. 324–331, 2022.
https://doi.org/10.1007/978-3-031-16364-7_25

be trained with the training set with the desired output it produces for the corresponding input which makes it supervised learning. Whereas unsupervised learning is the model created to work on the prediction without providing it with the structured data. The model learns by itself to identify the structure of the input fed into it and produce the result. When the model works over the dynamic environment with changing input and output reinforcement model is implemented. Various machine learning algorithms are used to develop applications based on the above models for decision making.

When the data used and processing gets huge, deep learning which is the subset of machine learning is used. The rising level of data in the current era will give more convenience to explore the models that can be implemented using deep learning. It helps to solve complex problems with more detailed representation of images with the help of neural networks. Neural network acts as a heart of deep learning algorithms as it contains more numbers of hidden layers into it. Each layer acts as a processing unit in a level of hierarchy. That is, the first layer will compute for a lower level of element from the image or an object. The result of this first layer is fed into the second layer as an input and in the second layer the next level of element is extracted and the output is fed into the next layer and so on until the desired output is achieved. The depth or the number of layers of the deep learning algorithms are considered based on Credit Assignment Path (CAP). It analyzes how to choose the data points so that the error or backtracking gets minimal. Each hidden layer is given with the weight to the input which will be adjusted based on the corresponding result obtained and the accuracy of the output. Since there is a huge amount of data and there are a wide range of parameters included, the deep learning model takes time to get trained. But testing the model will take less time. The output of every node is activated by the activation function which helps to remove the unwanted data from the useful information. The various research work done under the domain are discussed.

2   Related Work

With the help of deep learning, image processing, face identification and analysis and Neural Network algorithm, many earlier studies on human behaviour analysis have been done. In this section, some recent research work done by various authors are discussed.

[1] Nilay Tufek, Murat Yalcin, Mucahit Altintas, Fatma Kalaoglu, Yi Li, and Senem Kursun Bahadir conducted research on "Human Action Recognition Using Deep Learning Methods on Limited Sensory Data". Here, the accelerometer and the gyroscope data is used in a lesser amount to build an action recognition system. The Convolutional Neural Network, Long-Short Term Memory (LSTM) and various combinatorial algorithms were used and their performances were compared. The accuracy of the model was increased by augmenting the data. By using a 3 layer LSTM model, the accuracy of the model was increased by 97.4%. Over the collected data set the system produced 99% of accuracy. To analyse the performance precision, recall and f1-score metrics were

also used. For the purpose of evaluation on the classification an application is developed using a 3 layer LSTM network. To detect human activity a unique set of wearable devices were used. When the dataset is high, the 3 layer LSTM model provides high accuracy to sensory data when compared to KNN model. To improve the test accuracy, precision, recall and F1 Score radically the data augmentation is done on a small size dataset. The time series sensory data plays a vital role for the implementation of the system. To analyze and recognize more intricate behaviour, multiple sensor data can be used.

[2] Kai Zhang and Wenjie Ling proposed a research work on "Joint Motion Information Extraction and Human Behavior Recognition in Video Based on Deep Learning". Here it mainly focuses on human behaviour analysis and identification from videos. A two channel deep convolutional neural network model is designed for the structure and joint motion feature extraction. The network structure also simulates the brain visual cortex which processes the visual signal. The static information is processed by the spatial channel network and the dynamic information is processed by the temporal channel network. The spatial and the temporal information are extracted separately. The recognition of the two channel model is compared with the single channel model to verify the superiority of the dual channel structure. KTH behavioural dataset are used to perform the experiments and shows that the human behaviour analysis using deep learning has achieved more accuracy based on joint motion information. The joint motion feature extraction using deep convolutional neural networks removes the extra calculations and multifaceted computations that are involved in conventional feature extraction methods. This model can be applied to a wide variety of video data using which a more detailed understanding of action behaviour can be done.

[3] Dersu Giritlioglu, Burak Mandira, Selim Firat Yilmaz, Can Ufuk Ertenli, Berhan Faruk Akgur, Merve Kiniklioglu, Aslı Gül Kurt, Emre Mutlu, Seref Can Gurel, Hamdi Dibeklioglu proposed a paper on "Multimodal analysis of personality traits on videos of self-presentation and induced behavior". In this paper a multimodal deep architecture is presented to estimate the personality traits from audio-visual cues and transcribed speech. For the detailed analysis of personality traits audio visual dataset with self presentation along with recordings of induced behaviour is used. The reliability of various behaviours and their combinational use is assessed. The face normalisation is done by marking 68 boundaries of the face region from the video using OpenFace. The frontal view of the face is obtained by performing translation, rotation and scaling. The X and Y coordinates are identified and are shape normalized by using linear warping. Finally a facial boundary region with the resolution of 224 * 224 pixels are obtained. The normalized facial images are modelled with the spatio temporal pattern using ResNext model and CNN-GRU. The facial appearance and facial dynamics are captured from the input video. To access the temporal data ResNext model is used. The feature map is divided into smaller groups from the ResNet. The random temporal sampling of 1.5 s is used during training and the same size data is used during the validation phase. The second layer of the deep learning architecture is modelled with CNN-GRU for facial videos. For the modelling, the action unit and the head pose uses LSTM and to obtain the gaze feature the model uses RCNN. The body pose is obtained with the 2D coordinates of the traced points. The

audio of the video is extracted using pyAudio analysis framework. The speech language is automatically identified by the Google speech application programming interface. The testing performed over the SIAP and FID dataset. The openness and Neuroticism provides the social desirability effect. Based on the mean score, the participants give the rating of openness by 39.7% and neuroticism by 21.2%. For personality analysis, the face related models appear to be more reliable.

[4] Karnati Mohan, Ayan Seal, Ondrej Krejcar, and Anis Yazidi proposed a paper on "Facial Expression Recognition Using Local Gravitational Force Descriptor-Based Deep Convolution Neural Networks". The proposed work contains two methods. Identification of local features from the image of faces with the help of local gravitational force descriptor and embedding the descriptor into the traditional deep convolutional neural network model which involves exploring geometric features and holistic features. The final classification score is calculated using the score level fusion technique. The pixel value of an image is considered to be the mass of the body. The gravitational force of the pixel is considered to the centre pixel on its adjacent pixel. A pixel is selected along with its adjacent pixels as the GF of the image. The DCNN learns by itself with the help of back propagation with more accuracy. The methods such as VGG-16 and VGG-19 are developed based on the single branch convolutional layers connected sequentially. No edge information is provided as it focuses only on the receptive fields and so there is not enough information on spatial structure of the face. This problem is addressed using multi convolutional networks. The first part of the architecture extracts the local features of the object from the image and the second part extracts the holistic features. The former part contains three convolutional layers that are in order as max pooling, average pooling and zero padding. The latter part consists of five convolutional layers which are merged and sent to the layers for classifying facial expression. This can extract the features automatically. Each layer is embedded with the filter and the bias value which is then fed into the activation function. The overfitting problem is avoided by applying max pooling to the feature maps obtained. The gradients of the facial images are trained independently and the probability of each layer is calculated. The final prediction of the basic expression is done by score level fusion. The model is trained using keras framework. The data is augmented to improve the performance.

[5] Yeong-Hyeon Byeon, Dohyung Kim, Jaeyeon Lee and Keun-Chang Kwak proposed a paper on "Explaining the Unique Behavioral Characteristics of Elderly and Adults Based on Deep Learning". In this paper, it analyzes the behaviour of the elderly people depending on their physical condition. The silver robots provide customized services to those who are in need to improve the betterment of the elders. When there is a change in the human body it gets reflected as pose evolution images. The classification is done based on convolutional neural networks which is trained based on the elderly behaviour. The gradient weighted activation map is obtained for the result obtained after classification and a heatmap is generated for the purpose of analysis. The skeleton heatmap and RGB video matching are analyzed and the characteristics are derived from this. To efficiently store the movement of the human based on reconstruction of the

skeleton with coordinate points based on sensor data, the skeleton data type is used. The human body is modelled using kinect v2 with 25 joints. This also includes the head, arms, hips and legs. A sequence of skeletons is created like a video to get more information about the behaviour of a person. For the effective analysis of the skeleton sequence data both the temporal and the spatial information is used and to extract the appropriate information conversion methods are used. The skeleton sequence is converted into a color image called PEI method which is less readable for humans. The two dimensional convolutional neural network is used to extract the spatial and the temporal information from the skeletal sequenced image. The skeleton sequence is captured at the continuous time interval which has a 3D data type. Project the 3D coordinates into RGB space to convert the 3D image into 2D model. The product of number of joints, dimensionality of the coordinates and number of skeleton frames over time gives the skeleton sequence. The skeleton sequence can be converted into image by denoting the joint coordinate dimension with temporal dimension. The action analysis is conducted over the dataset of NTU RGB+D which has over 114480 data samples. The study was conducted over the average age of 77 years who performed 55 behaviours in their day to day life. After training the model identified the input data belongs to the elderly behaviour or young adults. For a more distinct behaviour the accuracy obtained was higher when compared to the normal behaviour.

[6] Angelina Lu, Marek Perkowski, proposed a research work on "Deep Learning Approach for Screening Autism Spectrum Disorder in Children with Facial Images and Analysis of Ethnoracial Factors in Model Development and Application". Here, a Autism Spectrum Disorder screening solution is developed with the help of facial images by using VGG-16 transfer learning based deep learning technique. The kaggle ASD facial image dataset is used to implement the model. The resulting model can identify the children with ASD and normally developed children. The deep learning model of Visual Geometry Group VGG 16 has produced the classification with high accuracy. The reusing of model is achieved with the help of transfer learning and gives the output of one task to another. The model majorly focuses on the quality of detecting ASD in children with the help of facial images and the race factor of the children involved in the experiments. The accuracy and the F1 score achieved are 95% and 0.95 respectively. Computer vision with facial images can be applied in screening the ASD children and further can be implemented to a user-friendly mobile application which helps the people in case of inaccessible medical emergencies.

Comparison

TITLE	INPUTS USED	ALGORITHMS/ METHODS USED	ACCURACY
Human Action Recognition Using Deep Learning Methods on Limited Sensory Data	Accelerometer and the gyroscope data	Convolutional Neural Network, Long-Short Term Memory (LSTM) and various combinatorial algorithms	3 layer LSTM model - 97.4% OVERALL PERFORMANCE - 99% of accuracy
Joint Motion Information Extraction and Human Behaviour Recognition in Video Based on Deep Learning	Human behaviour analysis and identification from videos	Two channel deep convolutional neural network model	Joint motion information —97%
Multimodal analysis of personality traits on videos of self-presentation and induced behaviour	Audio visual dataset with self presentation along with recordings of induced behaviour is used	Multimodal deep architecture	Openness - 39.7% neuroticism - 21.2% Face related models —98%
Facial Expression Recognition Using Local Gr avitational Force Descriptor- Based Deep Convolution Neural Networks	Image of faces - pixel value of an image is considered to be the mass of the body	Facial Expression Recognition Using Local Gravitational Force Descriptor- Based Deep Convolution Neural Networks	Score level fusion — 89%
Explaining the Unique Behavioural Characteristics of Elderly and Adults Based on Deep Learning	Elderly people behavioural video	Skeleton heatmap and RGB video matching	Conducted over the average age of 77 years who performed 55 behaviours — 93% of normal behaviours were classified correctly

Deep Learning Approach for Screening Autism Spectrum Disorder in Children with Facial Images and Analysis of Ethnoracial Factors in Model Development and Application	Kaggle ASD facial image dataset	Deep learning model of Visual Geometry Group VGG 16	The accuracy and the F1 score achieved are 95% and 0.95 respectively

Datasets

The datasets for the human behavioural analysis to analyse the human behaviour on various scenarios can be obtained by using data repository of Mendeley which is a free open source data repository. It contains 11 million datasets that are indexed for easy accessibility and contains various research data such as raw or processed data, videos, images, audios etc. Also BARD dataset can be used which is a collection of videos majorly used for the human behavioural analysis and detection. The major advantage of BARD video is that the dataset videos contain open environment captured videos and are collected in uncontrolled scenarios which will be helpful to implement in any real time applications. Various other open source datasets are available to make the study on behavioural analysis both in the controlled or uncontrolled environment.

3   Conclusion and Future Work

The model studies the various algorithms and techniques involved in analysing human behaviour using deep learning algorithms. While performing the study over various research papers it is found that the common algorithm that were used over the human analysis or the behavioural identification includes Convolutional Neural Network. Upon which various models were built to improvise the performance or the accuracy of the classification. It is also identified that the data set used for the classification or the identification plays a significant role. When the data used for training or testing the model becomes higher the accuracy that the model produces is high. Thus the performance is directly proportional to the number of data used for training.

As a future work from this survey, a deep neural network can be implemented with an architecture for training the model to identify, analyse and classify the behaviour of the human is designed to produce high performance and accuracy.

References

1. Tufek, N., Yalcin, M., Altintas, M., Kalaoglu, F., Li, Y., Bahadir, S.K.: Human action recognition using deep learning methods on limited sensory data. IEEE Sens. J. **20**(6), 3101–3112 (2020)

2. Zhang, K., Ling, W.: Joint motion information extraction and human behavior recognition in video based on deep learning. IEEE Sens. J. **20**(20), 11919–11926 (2020)
3. Giritlioğlu, D., et al.: Multimodal analysis of personality traits on videos of self-presentation and induced behavior. J. Multimodal User Interfaces **15**(4), 337–358 (2020). https://doi.org/10.1007/s12193-020-00347-7
4. Mohan, K., Seal, A., Krejcar, O., Yazidi, A.: Facial expression recognition using local gravitational force descriptor-based deep convolution neural networks. IEEE Trans. Instrum. Measure. **70**, 1–12 (2021)
5. Byeon, Y.-H., Kim, D., Lee, J., Kwak, K.-C.: Explaining the unique behavioral characteristics of elderly and adults based on deep learning. Appl. Sci. **11**, 10979 (2021). https://doi.org/10.3390/app112210979
6. Angelina, L., Perkowski, M.: Deep learning approach for screening autism spectrum disorder in children with facial images and analysis of ethnoracial factors in model development and application. Brain Sci. **11**, 1446 (2021). https://doi.org/10.3390/brainsci11111446
7. Bala, B., Kadurka, R.S., Negasa, G.: Recognizing unusual activity with the deep learning perspective in crowd segment. In: Kumar, P., Obaid, A.J., Cengiz, K., Khanna, A., Balas, V.E. (eds.) A Fusion of Artificial Intelligence and Internet of Things for Emerging Cyber Systems. ISRL, vol. 210, pp. 171–181. Springer, Cham (2022). https://doi.org/10.1007/978-3-030-766 53-5_9
8. Ricard, B.J., Hassanpour, S.: Deep learning for identification of alcohol-related content on social media (Reddit and Twitter): exploratory analysis of alcohol-related outcomes. J. Med. Internet Res. **23**(9), e27314 (2021)
9. Alkinani, M.H., Khan, W.Z., Arshad, Q.: Detecting human driver inattentive and aggressive driving behavior using deep learning: recent advances, requirements and open challenges. IEEE Access **8**, 105008–105030 (2020). https://doi.org/10.1109/ACCESS.2020.2999829
10. Prabhu, K., Sathish Kumar, S., Sivachitra, M., Dinesh Kumar, S., Sathiyabama, P.: Facial expression recognition using enhanced convolution neural network with attention mechanism. Comput. Syst. Sci. Eng. **41**(1), 415–426 (2022).https://doi.org/10.32604/csse.2022.019749

An Improved Ensemble Extreme Learning Machine Classifier for Detecting Diabetic Retinopathy in Fundus Images

V. Desika Vinayaki[1][✉] and R. Kalaiselvi[2]

[1] Noorul Islam Centre For Higher Education, Kumaracoil, Tamil Nadu, India
vdesika.id@gmail.com
[2] Department of Computer Science and Engineering, RMK College of Engineering and Technology, Puduvoyal, Gummidipoondi, India

Abstract. This paper presents an automatic diabetes Retinopathy (DR) detection system using fundus images. The proposed automatic DR screening model saves the time of the ophthalmologist in disease diagnosis. In this approach, the segmentation is conducted using an improved watershed algorithm and Gray Level Co-occurrence Matrix (GLCM) is used for feature extraction. An improved Ensemble Extreme Learning Machine (EELM) is used for classification and its weights are tuned using the Crystal Structure Algorithm (CRYSTAL) algorithm which also optimizes the loss function of the EELM classifier. The experiments are conducted using two datasets namely DRIVE and MESSIDOR by comparing the proposed approach against different state-of-art techniques such as Support Vector Machine, VGG19, Ensemble classifier, and Synergic Deep Learning model. When compared to existing methodologies, the proposed approach has sensitivity, specificity, and accuracy scores of 97%, 97.3%, and 98%, respectively.

Keywords: Fundus image · Diabetic retinopathy · DRIVE · CRYSTAL algorithm · Improved watershed algorithm · Ensemble Extreme Learning Machine

1 Introduction

DR [1] is a major cause of blindness that normally occurs in diabetic patients. Diabetes is caused by a high blood glucose level in the blood, which affects the patient's internal organs as well as their eyes. The different diseases caused in the eye by diabetes are cataracts, glaucoma, and DR. The main reason for DR is the fluid leak in the small blood vessels which mainly results in visual loss. Based on the reports published by WHO, DR is the main reason for the 2.6% of deaths that occur worldwide. DR can also occur due to hereditary reasons. It is classified into three types namely: Proliferative, non-proliferative, and diabetic maculopathy. In this paper, we are mainly focusing on the non-proliferative DR known as exudates which are mostly known as hard exudates and various authors presented different techniques to identify the exudates in the retina [2].

© IFIP International Federation for Information Processing 2022
Published by Springer Nature Switzerland AG 2022
L. Kalinathan et al. (Eds.): ICCIDS 2022, IFIP AICT 654, pp. 332–344, 2022.
https://doi.org/10.1007/978-3-031-16364-7_26

Different vision-related issues arise when the tiny blood vessels in the retina are damaged. If early identification and treatment are not provided, the patient will lose all of his or her eyesight. In the earlier stages, this disease does not cause any symptoms and is only detected in different tests such as pupil dilation, visual acuity, etc. [3]. To prevent vision loss in diabetic patients, early DR diagnosis is necessary. The existing systems are both processing intensive and consume higher time [4]. The ophthalmologists mainly identify DR via lesions and vascular abnormalities, even though this is highly effective it is resource-intensive in nature. Since the diabetes rate in the local population is high, experienced professionals and equipment are scarce in many areas. This arises the need for an automated DR detection system to prevent more people from blindness. This can be achieved via image processing, pattern recognition, and machine learning techniques invented which have been gained quite a high popularity [5].

The first issue is identifying a high-quality dataset with accurate labels [6, 7]. Since there are more high-quality datasets available for the DR problem, several researchers face the challenge of unbalanced data. The next challenge is the problem associated with deep learning model design. The deep learning models are often prone to overfitting and underfitting issues and they also have to be trained via different trial and error methods. The main point to be noted here is the loss value obtained should be low and the learning capability should be high. The hyperparameter tuning process of the deep learning classifiers is another big issue that needs to be tackled. The limited processing resources, high time consumption, high false positives, and low accuracy is the problems faced by most of the existing techniques. To tackle these issues, in this paper we propose a Crystal Structure Algorithm (CRYSTAL) algorithm optimized Ensemble Extreme Learning Machine (EELM) architecture for DR classification. The major contributions of this paper are delineated as follows:

- This paper uses GLCM for feature extraction and an improved watershed algorithm for segmentation.
- The CRYSTAL optimized EELM classifier classifies the input obtained after segmentation and feature extraction into three types namely exudates, Micro Aneurysms, and Hemorrhages. The model proposed minimizes the time complexity and intricate processing issues associated with the state-of-art classifiers.
- The efficiency of this technique is evaluated using two datasets (DRIVE and MESSIDOR) in terms of accuracy, sensitivity, specificity, ROC curve, and confusion matrix.

The rest of this paper is arranged accordingly. Section 2 presents the related works and Sect. 3 elaborates the different steps incorporated in the proposed methodology in detail. Section 4 summaries the findings of the proposed methodology acquired by testing the approach using two datasets in terms of different performance measures, and Sect. 5 concludes the paper.

2   Review of Related Works

The different works conducted by various authors in this domain are represented as follows: Saranya P et al. [8] utilized Convolutional Neural Network (CNN) for automatically identifying the non-proliferative diabetic retinopathy present in retinal fundus images. They evaluated their model on two popular datasets namely MESSIDOR and IDRiD. This work mainly reads the data from the image information and identifies the severity of DR. This technique has been mainly designed to be run in small and compact systems. This method offers a maximum accuracy of 90.89% in the MESSIDOR dataset.

Kanimozhi J et al. [9] presented an automatic lesion detection model for identifying the severity of DR via screening. Luminosity, contrast enhancement, blood vessel, and optic disc removal, and lesion detection and classification are the main steps. Contrast limited adaptive histogram equalization (CLAHE) and gamma correction are utilized for contrast and luminosity enhancement. Using morphological operations and classification, the lesions are identified after background removal. The efficiency of this technique is evaluated via the publicly available.

Dutta A et al. [10] used individual retinal image features to identify the grades (binary or multiclass classification) of DR. They are identifying the different DR grades by integrating different machine learning classifiers by applying both single and multiple features observed. A fine-tuned VGG-19 model powered using transfer learning is used to evaluate the efficiency of this technique on both single and multiclass classifications of digital fundus images The results show that the accuracy is improved up to 37% when compared to the existing models. This work helps the ophthalmologist to identify the different grades of DR from a single feature alone.

Melo T et al. [11] detected microaneurysm in color eye fundus images using a sliding band filter and ensemble classifier. The initial microaneurysm candidates are obtained via the sliding band filter technique and the final classification is done by a set of ensemble classifiers. Based on the confidence value assigned for each sample, a score is computed. This technique yields a sensitivity value of 64% and 81% for the e-ophtha MA and SCREEN-DR dataset respectively. Pachiyappan A et al. [14] utilized morphological operators for identifying the DR from fundus images. The model was trained using 85 fundus images where at least 85 images have a presence of mild retinopathy. The accuracy of this technique is estimated to be 97.75% for optical disk detection.

Shankar K et al. [12] utilized a Synergic Deep Learning (SDL) model for detecting DR from the fundus images. Their model mainly classifies the RD cases based on their severity. The different stages involved in the proposed work are pre-processing, segmentation, and classification. In the preprocessing stage, the unnecessary noises are removed such as the edges in the images, and the region of interest is extracted using the histogram-based segmentation technique. The SDL model classifies the output based on the different levels of severity using the Messidor DR dataset. However, the authors have not taken steps to overcome the complexity associated with the SDL model.

Katada Y et al. [13] utilized artificial intelligence techniques for automatically screening the DR from fundus images. They utilized both Japanese and American datasets for these experiments and classified the input images based on the severity level. A Deep Convolutional Neural Network (DCNN) and Support Vector Machine (SVM) classifier are trained using the 35,1256 and 200 fundus images acquired from both the American and Japanese datasets. For the Japanese dataset, their model provided a sensitivity and specificity score of 81.5% and 71.9% respectively. For the American dataset, their model yielded a sensitivity and specificity score of 81.5% and 71.9%.

3 Proposed Methodology

In this work, we mainly aim to classify the DR fundus images with a maximum detection rate. The overall architecture of the proposed work is depicted in Fig. 1. The steps used in the proposed work are presented as follows:

3.1 Image Preprocessing

Initially, the input image from the DRIVE and MESSIDOR dataset is preprocessed using the adaptive histogram equalization technique, image enhancement and resizing techniques [18]. In this step, the background is normalized and illumination is equalized.

3.2 Image Segmentation

The segmentation process is mainly applied to the optical disk region of the fundus image. The optical disk is the brightest part in the fundus image and is based on the intensity value it is extracted. The segmentation process is mainly conducted using the improved watershed algorithm [17]. The reason for selecting this algorithm is its

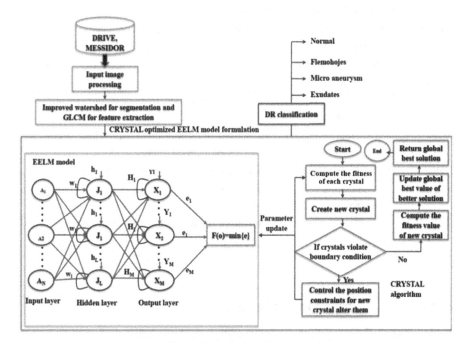

Fig. 1. Proposed architecture

capability to identify the bright areas and solve the low color discrimination problem. In this way, this algorithm helps in improving the accuracy of the classifier even more. The threshold value is mainly selected by the assumption that it should not exceed the maximum intensity value of the fundus image. The region in the optical disk is identified by segmenting the region around the final optimal disk candidate. The boundary of the optical disk can be identified via the thresholded region.

3.3   Feature Extraction Using GLCM

The preprocessed image is given as an input to the GLCM [16] for retina removal and blood vessel segmentation. The GLCM is mainly applied to the preprocessed image. The heterogeneous surfaces can be identified via the GLCM indices and the texture is the term used in GLCM to identify the grey level variations and tones present in the input fundus image. Every GLCM index highlights a certain property such as irregularity, smoothness, etc. The GLCM structure is formulated as follows: Initially, a grayscale matrix of the image is transformed into an integer matrix by dividing the continuous pixel value range to M bins. The bins are equally valued and are mapped into a single grey level. The pixel adjacency can be represented using four different angles: 0°, 45°, 90°, and 135°. The generic element of the matrix is represented as g(x, y). The details of the GLCM indices selected are presented in Table 1.

Table 1. GLCM indices description

Name	Description
Contrast	The change in the gey level pixels present in between continuous pixels is noted and exponential weightage is given
Homogeneity	The similarity of the grey level pixels to their neighboring pixels is measured
Dissimilarity	The elements are weighted linearly
Entropy	The disorder of the image is measured here which is negatively interrelated with the energy
Energy	The pixel pair repetition and texture uniformity are measured here. A constant grey level distribution possesses a high energy value
Mean	The average value of the grey level pixels in the image is computed
Variance	When there is a variation of the grey level values when compared to the mean, variance increases which is mainly a measure of homogeneity
Correlation	The linear dependency of the image is measured in relation to the adjacent and grey level pixels

3.4   CRYSTAL Optimized EELM Classifier

GS Algorithm: The crystalline solids structure is the basic concept of crystal structure (CS) algorithm [15]. Various geometrical structures is defined as the Lattice and the

Bravais model (A) describes the mathematical representation of lattice.

$$A = \sum m_i B_i \tag{1}$$

where m_i is the integer and B_i represent the smallest path with respect to the directions of principal crystallographic. From this, i denote the number of corners in the crystals.

Mathematical Model: In the lattice space, the single crystal deems optimized candidate solutions. For the iterative purposes, determines the number of crystals.

$$D = \begin{bmatrix} D_1 \\ D_2 \\ \vdots \\ D_3 \\ \vdots \\ D_n \end{bmatrix} = \begin{bmatrix} z_1^1 & z_1^2 & \cdots & z_i^j & \cdots & z_1^d \\ z_2^1 & z_2^2 & \cdots & z_2^1 & \cdots & z_2^d \\ \vdots & \vdots & \vdots & \vdots & \vdots & \vdots \\ z_i^1 & z_i^2 & \cdots & z_i^j & \cdots & z_i^d \\ \vdots & \vdots & \vdots & \vdots & \vdots & \vdots \\ z_n^1 & z_n^2 & \cdots & z_n^j & \cdots & z_n^d \end{bmatrix}, \quad \begin{cases} i = 1, 2, \ldots, m \\ j = 1, 2, \ldots, c \end{cases} \tag{2}$$

where c describes the problem of dimensionality and m is the total number of crystals. In the search space of the network, the nodes are placed randomly.

$$z_i^j(0) = z_{i,\min}^j + Random\left(z_{i,\max}^j - z_{i,\min}^j\right) \tag{3}$$

From the above equation, $z_i^j(0)$ is the nodes initial population as well as $z_{i,\min}^j$ and $z_i^j(0)$ are the minimal and maximal values. For the ith candidate solution, the jth decision variable with the random number tends to the interval $[0, 1]$.

The base nodes B_N and the best configuration with the randomly selected mean values and the best node configuration is M_D and D_O. The below-mentioned step describes the best candidate node from the network search space.

(a) **Cubicle based on best crystals:**

$$D_{new} = D_{old} + A_1 B_N + A_2 D_O \tag{4}$$

(b) **Simple Cubicle:**

$$D_{new} = D_{old} + A B_N \tag{5}$$

(c) **Cubicle depends on best and mean nodes:**

$$D_{new} = D_{old} + A_1 B_N + A_2 M_D + A_3 D_O \tag{6}$$

(d) **Cubile based on mean nodes:**

$$D_{new} = D_{old} + A_1 B_N + A_2 M_D \tag{7}$$

From this, D_{new} and D_{old} describes the new and old updated positions, where, A_1, A_2, and A_3 are the random numbers.

3.5   EELM for Classification

EELM is the modified version of the Single-layer Feed Forward Neural Network (SLFNN) [20]. The SLFNN comprises two different layers namely the input layer and the output layer. No computation is performed in the input layer and the output layer is obtained when diverse weights are applied on the input node. The EELM classifier is trained with the appropriate input images from the dataset to obtain the optimal outcome for the input. The reason the EELM architecture is selected in this work is for its improved generalization performance, faster learning speed, minimal training error, easier implementation, independence of user input data, and offers significant accuracy.

The GLCM features extracted is represented as M with an already specified target (A_i, Y_i) where $A_i = [A_{i1}, A_{i2}, .., A_{iM}]^T \in R^m$ and $Y_i = [Y_{i1}, Y_{i2}, .., Y_{iM}]^T \in R^n$. . The total number of output classess is represented as m which is 4 in our paper. Each hidden node (H) in the SLFNN network comprises of bias (β_i) and random weights (ω_i). Using the output weight matrix O_i and activation function Z, the output (X_i) of the SLNN is obtained and this process is explained as follows:

$$\sum_{k=1}^{H} O_i Z(\omega_i, \beta_i, A_i) = X_i ; \; k = 1, 2, ..., M \tag{8}$$

The zero mean error of the $\sum_{k=1}^{H} O_i Z(\omega_i, \beta_i, A_i) = \|X_i - Y_i\| ; \; k = 1, 2, ..., M$ value is approximated using $\omega_i, \beta_i, and \; 0_i$ and it is represented as follows:

$$\sum_{k=1}^{H} O_i Z(\omega_i, \beta_i, A_i) = Y_i ; \; k = 1, 2, ..., M \tag{9}$$

The above M equations can be compressed using a hidden layer output matrix (J) as follows:

$$JO = Y \tag{10}$$

By utilizing the smallest norm least square equation, the output weight matrix $\hat{O}$ is computed as follows:

$$\hat{O} = J^{\div} Y \tag{11}$$

The $J^{\div}$ value in the above equation mainly represents the Moore Penrose generalized inverse of J. The orthogonal projection technique is applied to prevent the singularity of the matrixes and the output weight matrix derived is presented as follows:

$$\hat{o} = \begin{cases} J^T \left(\frac{1}{\lambda} + JJ^T\right)^{-1} Y & ; \; If \; M < O \\ \left(\frac{1}{\lambda} + JJ^T\right)^{-1} YJ^T & ; \; If \; M < O \end{cases} \tag{12}$$

For each image sample, the input feature (Ai) is trained repeatedly to obtain the $\hat{O}$ value of the EELM classifier. The weight assigned to each member of the ensemble is optimized using the CRYSTAL algorithm and the efficiency of the CRYSTAL algorithm is evaluated using the loss function. The predicted samples of root mean square error (RMSE) is the selected loss function. The below equation explains the RMSE value.

$$RMSE = \sqrt{\frac{1}{E} \sum_{g=1}^{G} \left(\hat{Z}_g - Z_g \right)^2} \tag{13}$$

The predicted model to the input is $Z_g^{\wedge}$ and the real value is Z_g. The total amount of forecasted samples is G.

The proposed model classified three types of DR abnormalities namely Hemorrhages, Micro Aneurysms, and Exudates. The micro aneurysms are the initial stage of DR where red spots start appearing in the retina with a size of fewer than 125 μm. The lesions have sharp margins and mainly occur due to the weakness of the vessel walls. The hemorrhages have an irregular margin and are large in size (> 125 μm). There are two types of exudates namely hard and soft. The hard exudates are mainly caused due to the leakage of plasma and they are in the form of bright yellow spots. The soft exudates are mainly caused due to the swelling in the nerve fiber and they are usually represented as visual spots. The three types of abnormalities classified by the proposed methodology are demonstrated in Table 2 and it is explained as follows:

Table 2. Classifications of DR abnormalities

Image	Corresponding classes
	Exudates
	Micro Aneurysms
	Hemorrhages

4 Experimental Results and Analysis

The proposed model is implemented in Matlab in a DELL Inspiron One 27 Ryzen 7 PC equipped with a 16 GB DDR4 RAM, 1 TB HD capacity, 256 GB SSD, and Windows 10 OS. The details of the dataset, performance metrics used, and the results observed are presented in this section.

4.1   Dataset Description

Digital Retinal Images for Vessel Extraction Dataset (DRIVE): The DRIVE [19] is a retina vessel segmentation dataset that comprises a total of 40 JPEG images. A total of 7 abnormal cases is found in this dataset. The image has a resolution of 584 * 565 pixels with eight bits per three color channel. The 40 images are partitioned in a 5:5 ratio where 20 images are used for testing and the remaining 20 are used for training.

Messidor Dataset: This dataset comprises a total of 1200 eye fundus color images from which 800 images are obtained with pupil dilation and another 800 are obtained without pupil dilation [8]. The 1200 images in the dataset are partitioned into three sets which are then compressed into four subsets. In our proposed work, we partition this dataset in an 8:2 ratio, where 80% of images are used for training while the remaining 20% is used for testing. The retinopathy grade is mainly given based on the number of microaneurysms, hemorrhages, and neovascularization.

4.2   Performance Evaluation Metrics

The different performance evaluation metrics used in this paper is presented as follows:

Sensitivity: It is the capability that the classifier can recognize the DR abnormalities correctly and it is computed as follows:

$$Sensitivity = \frac{A^+}{A^+ + B^-} \tag{14}$$

Specificity: The ability of the classifier to accurately reject the DR abnormalities correctly and it is computed as follows:

$$Specificity = \frac{A^-}{A^- + B^+} \tag{15}$$

Accuracy: It determines the capability of the classifier to accurately discriminate between the actual and predicted samples and it is computed as follows:

$$Accuracy = \frac{A^+ + A^-}{A^+ + A^- + B^+ + B^-} \tag{16}$$

Here the true positive, true negative, false positive, and false negative values are represented as A+, A−, B+, and B−.

ROC Curve: The ROC curve plots two parameters namely the True positive rate and false positive rate and identifies the performance of the classifier at varying thresholds.

Confusion Matrix: This matrix mainly identifies the confusion of the classifier in predicting two classes. It is mainly computed by evaluating the model on the testing data whose ground-truth value is actually known.

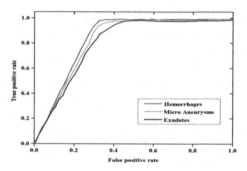

Fig. 2. ROC classification results for three classess

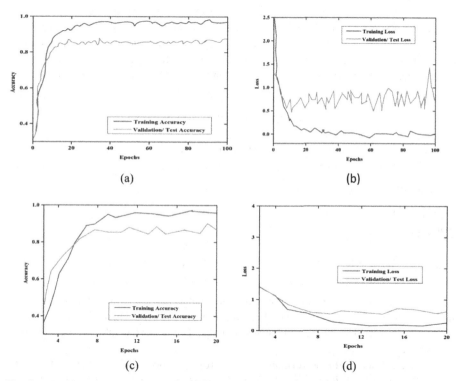

Fig. 3. (a) Training and testing accuracy of DRIVE dataset (b) Training and testing loss of DRIVE dataset, (c) Training and testing accuracy of MESSIDOR dataset (d) Training and testing loss of MESSIDOR dataset

4.3 Results and Discussion

The ROC curve obtained for the three classes is presented in Fig. 2. The ROC curve shows improved classification accuracy for the three classes. The accuracy and the loss curves of the proposed model on the two datasets (DRIVE and MESSIDOR) are presented in Fig. 3 (a)–(d). For both the training and testing datasets, we can notice an exponential rise in the accuracy with an increase in the number of epochs. The loss percentage also decreases with an increase in the number of epochs. These results show the capability of the model in predicting the false positives and false negatives accurately.

The efficiency of the proposed model is evaluated using different performance metrics such as accuracy, sensitivity, specificity, and computational time. The results obtained are shown in Table 3 and it is self-explanatory. The proposed methodology takes a total of 15.36 s to identify the abnormalities present in the input dataset. The time efficiency of the proposed model shows its efficiency in reducing the ophthalmologist time for disease diagnosis. The efficiency of the proposed methodology is also compared with existing techniques such as CNN [8], VGG19 [10], Ensemble [11], SDL [12], and SVM [13]. The experimental results demonstrate the proposed model's capability in identifying the DR accurately. Even though the deep learning models offer higher performance in terms of accuracy, sensitivity, and specificity, the time and computational complexity associated with it are high. The existing methodology mainly takes more time to diagnose due to the usage of intricate feature extraction techniques which individually take 10 min for processing. The segmentation and the feature extraction techniques used in our paper are also less time-consuming when compared to the existing techniques.

Table 3. Comparative analysis

Techniques	Sensitivity (%)	Specificity (%)	Accuracy	Computational time (s)
CNN [8]	91	91	91	85.36
VGG19 [10]	93	91	92	47.65
Ensemble [11]	87	85	84	19.36
SDL [12]	88	89	90	25.69
SVM [13]	86	85	84	17.56
Proposed method	97	97.3	98	15.36

Figure 4 expresses the results of the confusion matrix and the figure is plotted between the actual and predicted classes of diabetic retinopathy detections namely Hemorrhages, Micro Aneurysms, Exudates, and Normal classes. The accuracy value of each class is represented using green color.

Table 4 delineates the performance of the proposed model with respect to four different classes namely Hemorrhages, Micro Aneurysms, Exudates, and Normal classes. The accuracy, specificity, and sensitivity measures analyze the performance of these diabetic retinopathy classes. The accuracy, specificity, and sensitivity for Hemorrhages are 98.36%, 96.34%, and 95.56% as well as for Micro Aneurysms are 98.24%, 98.43%, and 96.34%. Based on the Exudates, we have obtained 98.42% accuracy, 94.47% specificity, and 97.03% sensitivity results. For normal classes, the proposed method offered 98.31%, 97.78%, and 98.45% performance values are achieved with respect to the accuracy, specificity, and sensitivity results.

Actual class		Predicted class			
		Hemorrhages	Micro Aneurysms	Exudates	Normal
	Hemorrhages	98.36%	1.25%	0.39%	0%
	Micro Aneurysms	0%	98.24%	0.78%	0.98%
	Exudates	0.96%	0%	98.42%	0.62%
	Normal	0.67%	1.02	0%	98.31%

Fig. 4. Confusion matrix

Table 4. Performance analysis of proposed method based on different classes

Name of the classes	Performance metrics		
	Accuracy	Specificity	Sensitivity
Hemorrhages	98.36%	96.34%	95.56%
Micro Aneurysms	98.24%	98.43%	96.34%
Exudates	98.42%	94.47%	97.03%
Normal	98.31%	97.78%	98.45%

5 Conclusion

This paper presents an efficient DR diagnosis system that improves the decision-making capability of ophthalmologists. A CRYSTAL optimized EELM architecture is proposed in this paper to identify the three different types of DR classes namely exudates, Micro Aneurysms, and Hemorrhages. The efficiency of this methodology is verified using two different datasets namely DRIVE and MESSIDOR. The GLCM technique is used for feature extraction and the improved watershed algorithm is used for segmentation. The CRYSTAL optimized EELM architecture is used for classification. The efficiency of the proposed technique is evaluated in terms of sensitivity, specificity, ROC curve, and confusion matrices. The proposed approach offers sensitivity, specificity, and accuracy score of 97%, 97.3%, and 98% when compared to the existing techniques. The computational time of the proposed methodology is 15.36 s which is relatively low than the other techniques. In the future, we plan to identify the different types of glaucoma in diabetic patients.

References

1. Kadan, A.B., Subbian, P.S.: Diabetic retinopathy detection from fundus images using machine learning techniques: a review. Wirel. Pers. Commun. **121**(3), 2199–2212 (2021)
2. Akram, M.U., Akbar, S., Hassan, T., Khawaja, S.G., Yasin, U., Basit, I.: Data on fundus images for vessels segmentation, detection of hypertensive retinopathy, diabetic retinopathy and papilledema. Data Brief **29**, 105282 (2020)

3. Tsiknakis, N., et al.: Deep learning for diabetic retinopathy detection and classification based on fundus images: a review. Comput. Biol. Med. **135**, 104599 (2021)
4. Ravishankar, S., Jain, A., Mittal, A.: Automated feature extraction for early detection of diabetic retinopathy in fundus images. In: 2009 IEEE Conference on Computer Vision and Pattern Recognition, pp. 210–217. IEEE, June 2009
5. Bonaccorso, G.: Machine Learning Algorithms: Popular Algorithms for Data Science and Machine Learning. Packt Publishing Ltd. (2018)
6. Dai, L., et al.: Clinical report guided retinal microaneurysm detection with multi-sieving deep learning. IEEE Trans. Med. Imaging **37**(5), 1149–1161 (2018)
7. Fu, H., et al.: Evaluation of retinal image quality assessment networks in different color-spaces. In: Shen, D., et al. (eds.) MICCAI 2019. LNCS, vol. 11764, pp. 48–56. Springer, Cham (2019). https://doi.org/10.1007/978-3-030-32239-7_6
8. Saranya, P., Prabakaran, S.: Automatic detection of non-proliferative diabetic retinopathy in retinal fundus images using convolution neural network. J. Ambient Intell. Human Comput. (2020). https://doi.org/10.1007/s12652-020-02518-6
9. Kanimozhi, J., Vasuki, P., Roomi, S.M.M.: Fundus image lesion detection algorithm for diabetic retinopathy screening. J. Ambient Intell. Humaniz. Comput. **12**(7), 7407–7416 (2020). https://doi.org/10.1007/s12652-020-02417-w
10. Dutta, A., Agarwal, P., Mittal, A., Khandelwal, S.: Detecting grades of diabetic retinopathy by extraction of retinal lesions using digital fundus images. Res. Biomed. Eng. **37**(4), 641–656 (2021)
11. Melo, T., Mendonça, A.M., Campilho, A.: Microaneurysm detection in color eye fundus images for diabetic retinopathy screening. Comput. Biol. Med. **126**, 103995 (2020)
12. Shankar, K., Sait, A.R.W., Gupta, D., Lakshmanaprabu, S.K., Khanna, A., Pandey, H.M.: Automated detection and classification of fundus diabetic retinopathy images using synergic deep learning model. Pattern Recogn. Lett. **133**, 210–216 (2020)
13. Katada, Y., Ozawa, N., Masayoshi, K., Ofuji, Y., Tsubota, K., Kurihara, T.: Automatic screening for diabetic retinopathy in interracial fundus images using artificial intelligence. Intell. Based Med. **3**, 100024 (2020)
14. Pachiyappan, A., Das, U.N., Murthy, T.V., Tatavarti, R.: Automated diagnosis of diabetic retinopathy and glaucoma using fundus and OCT images. Lipids Health Dis. **11**(1), 1–10 (2012)
15. Talatahari, S., Azizi, M., Tolouei, M., Talatahari, B., Sareh, P.: Crystal structure algorithm (CryStAl): a metaheuristic optimization method. IEEE Access **9**, 71244–71261 (2021)
16. Park, Y., Guldmann, J.M.: Measuring continuous landscape patterns with gray-level co-occurrence matrix (GLCM) indices: an alternative to patch metrics? Ecol. Ind. **109**, 105802 (2020)
17. Zhang, L., Zou, L., Wu, C., Jia, J., Chen, J.: Method of famous tea sprout identification and segmentation based on improved watershed algorithm. Comput. Electron. Agric. **184**, 106108 (2021)
18. Dubey, V., Katarya, R.: Adaptive histogram equalization based approach for SAR image enhancement: a comparative analysis. In: 2021 5th International Conference on Intelligent Computing and Control Systems (ICICCS), pp. 878–883. IEEE, May 2021
19. Niemeijer, J.S., Ginneken, B., Loog, M., Abramoff, M.: Digital retinal images for vessel extraction (2007)
20. Sahani, M., Swain, B.K., Dash, P.K.: FPGA-based favourite skin colour restoration using improved histogram equalization with variable enhancement degree and ensemble extreme learning machine. IET Image Process. **15**, 1247–1259 (2021)

Text Mining Amazon Mobile Phone Reviews: Insight from an Exploratory Data Analysis

S. Suhasini[✉] and Vallidevi Krishnamurthy

Department of Computer Science and Engineering, Sri Sivasubramaniya Nadar College of Engineering, Chennai, India
{suhasinis,vallidevik}@ssn.edu.in

Abstract. Online shopping is used widely by all the people across the world. Only due to that the sites of e-commerce is receiving more review about the products which includes positive, negative and neutral reviews about the product purchased by the people. Those reviews would help the peer customer to know about the quality of the product and to suggest the best one to purchase. Similarly, this paper is focused on analyzing the reviews of the mobile phone purchased on the amazon website and gives a graphical analysis of all reviews given after each purchase of the mobile phone and this analysis could help the customers to easily identify the best mobile phone. In future such analysis would be helpful for the amazon mobile phone buyers to make a better purchase. The exploratory data analysis is made with amazon review dataset which visualizes the review data and analysis of star rating, features rating of each mobile phones. The overall review of the product is also mapped using EDA.

Keywords: Exploratory data analysis (EDA) · Online reviews · Sentiment analysis · Star rating

1 Introduction

Online reviews have become useful for people to make their purchase in online. As people are showing more interest in online shopping, to make a useful online shopping the reviews are collected from every customers for each product which would be make the purchase increase or decrease depending upon the review given by the previous customer. In fact, clients frequently simply demand a modest number of positive ratings. Some internet retailers have systems in place to detect which evaluations are the most beneficial to customers.

Most of the online reviews have aided buyers in determining benefits as well as the drawbacks of various products, parallelly it helps in assisting the selection of the quality product according to everyone needs, they also gave an provocation in terms of analyzing this massive amount of data. The review data is rapidly growing in size day by day. Some users have started to include videos and photos of the product in the website in order to make them more appealing and user-friendly. As a result, the online purchase review

© IFIP International Federation for Information Processing 2022
Published by Springer Nature Switzerland AG 2022
L. Kalinathan et al. (Eds.): ICCIDS 2022, IFIP AICT 654, pp. 345–355, 2022.
https://doi.org/10.1007/978-3-031-16364-7_27

comments dataset could be viewed as a big data analytics provocation. Businesses can be interested in digging into the review data to learn more about their product.

To handle huge amount of review data, the prediction analysis techniques is used which would directly fetch the over analysis of the reviews given by the prior customers [9, 11]. The analysis which gives the overall positivity and negativity of the product instead of reading each review. This prediction analysis method will save the computation time for finding the review of the product.

Customers can grade or score their customer experience and happiness with a product on most ecommerce platforms. One of the key aspects that determines online customers' purchasing behavior is product ratings and reviews. According to research, not all online consumer evaluations are helpful and have an impact on purchasing decisions. Consumers who perceive online reviews to be useful have a bigger influence on purchasing decisions than those who do not. Previous research on feedback provided in ecommerce purchase had concentrated on quantitative aspects like frequent purchase and helpfulness of each brand [2, 8, 11].

First and foremost, consumers receive firsthand information from their peers about a product, allowing them to make an informed judgement about the product or service. However, due to the large number of reviews, they are overburdened with knowledge on the reviews. Any customer would be unable to look through all of the reviews before making a decision.

Reading all the reviews points is not possible and it may take more time and also we should consider all the review commands to select the best product. To overcome this complexity, an data analysis technique can be done with the amazon dataset which will process all the review commands and gives an overall statistical representation of product, with that representation the best product can be easily identified. Based on the different features of the mobile phones the analysis can be made. This paper focuses on the Exploratory Data Analysis(EDA) for the statistical prediction.

2   Literature Review

Few works has conducted research into both customer and seller feedback. Customers research a variety of factors before making a purchase, including customer reviews, seller feedback, pricing comparisons with other marketplaces, and so on. All of these factors were employed in this study to estimate customers' willingness to pay using a conceptual model based on consumer uncertainty [1]. Unlike previous researchers who have studied online shopping reviews, and few people conducted study on online travel reviews. Reviews produced by the customer were found to have a greater influence on travel decisions such as place of visit, restaurants for food and stay which are offered by the travel and tourism team [3, 4]. They looked into the relationship between the review text's qualitative qualities and the influence of review usefulness on review score. They utilized a dataset which contains around 37,000 reviews acquired from Amazon to validate their model. They also used intergroup comparisons to look into the connections between exceptionally helpful and unhelpful reviews, as well as overwhelmingly positive and negative reviews [3, 4]. Instead of making these comparison we can have a pictorial representation which includes comparison as well as it gives an highly purchased product.

Similarly, the content-based and source-based review elements that were examined have a direct impact on the usefulness of product reviews. Customer written reviews with less content abstractness and good comprehension were also shown to be more useful [2]. According to the conclusions of some studies that focused on only a small number of evaluations, a company's reputation may alter too frequently in its early years. There may be a minor shift in a company's reputation over a longer period of time. A business with a steady start, as opposed to one with an unsteady start, is more likely to maintain and grow its reputation later on [6]. According to a comparison analysis, the effect of the emotional intensity index on the final review score is not significant. Depending on the business context, the emotional intensity index might be chosen. The parameters used for analysis reveals that the loss is not only the attenuation coefficient, it also ensure that the parameters are correct, but it can verify that the parameters are correct loss avoidance psychology of the decision maker. But ensure that the review ordering is somewhat stable in a when the parameters vary, a specific range [7].

They addressed the problem on handling large amount review which confuses the frequent buyers who tend to look at the previous experience. This is handled by ranking the reviews provided in websites in order to improve the experience of the customers. The process is done by creating a CNN-GRU model [8]. Handling of large amount of review can be made easier by using EDA, which shows all the reviews in a graphical format.

3 Dataset Description

The purpose of this research is to find out what factors determine the usefulness of online reviews.

The dataset was collected from the amazon website using web mining technique. During August 2020, Amazon.in collected 73,151 online reviews for various smart phone brands such as Oneplus, Apple, Vivo, Mi, Redmi, Honor, Samsung and Oppo, with only 12,389 reviews being valid. It is because of the large number of zero counts in the whole sample, the analysis was limited to 12,389 online reviews.

The collected dataset was cleaned through the pre-processing tasks using the python programming library Pandas. The reviews of the respective mobile phone brands were processed and stored in a worksheet. The length of the review is also calculated with help of WordNet which count the number of words used in the review post.

Using the dataset, the sentiment analysis can also be made with three categories of sentiment as positive, negative and neutral.

4 Methodology

A statistical method is identified to measure the overall review of the product. Using the exploratory data analytics and data visualization method. EDA (Exploratory Data Analysis) is a data analysis approach that uses visual methods to summarize data. The primary focus of EDA is on examining what data can tell us in addition to formal models or the hypotheses are being tested. Initial Data is not the same as EDA. IDA is a type of analysis that focuses on confirming assumptions. It is crucial for the model and hypothesis evaluation, as well as resolving missing values and changing variables in the suitable manner where, IDA is included in EDA. The Bag-of-Words model is a technique for extracting textual features. It's a textual representation of the situation. The occurrence of words in a corpus of text There are two of things: a lexicon of well-known terms and the measuring of the presence of words that are well-known. It is called a "bag" because it contains everything. There is no information on the arrangement of the words or their meaning ignored.

5   Implementation

By applying the EDA, amazon review dataset was analyzed using data visualization libraries in python. The various visualizations for the amazon review dataset are shown below.

5.1   Trustworthiness of Online Purchase of Mobile Phones

Initially, the overall rating of the online purchase of a mobile phone is visualized. Figure 1 illustrates the distribution of rating given by the different customers who have purchased mobile phones through Amazon. It is inferred from the Fig. 1 that the online purchase of the mobile phone is recommended, as, more than 6000 customers have given the star rating as 5 with their highest satisfaction of online purchase. Further through the sentiment analysis the positivity level of the review comments given for the online purchase of mobile phones is identified as 66.73%. Similarly, the negativity level and the neutral levels for the online purchase of mobile phones is identified as 8.08% and 25.07% respectively.

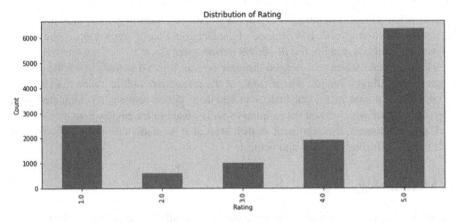

Fig. 1.   Overall distribution of star rating for online mobile phone purchase

5.2 Reviews Available for the Different Brands of Mobile Phones

In Fig. 2, it visualizes the number of reviews given to each mobile phone brands by the customers irrespective of positive, negative, and neutral reviews. Based on that it is observed that the frequently purchased mobile is Oneplus with more than 3500 reviews by the customers. Table 1, represents the ranking for each mobile phone brand based on the no of review comments given by customers.

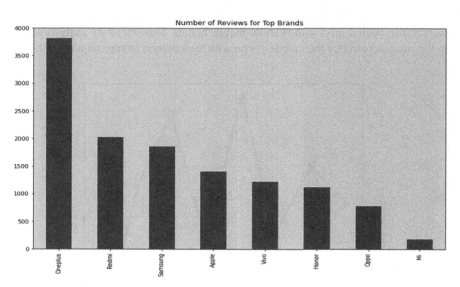

Fig. 2. Distribution of frequent purchase of mobile phone brand

Table 1. Ranking for each mobile phone brands

Brands	Ranking
Oneplus	Rank 1 (reviews were given by more than 3500 customers)
Redmi	Rank 2 (reviews were given by more than 1800 customers)
Samsung	Rank 3 (reviews were given by more than 1500 customers)
Apple	Rank 4 (reviews were given by more than 1300 customers)
Vivo	Rank 5 (reviews were given by more than 1200 customers)
Honor	Rank 6 (reviews were given by more than 1000 customers)
Oppo	Rank 7 (reviews were given by more than 500 customers)
Mi	Rank 8 (reviews were given by more than 50 customers)

5.3 Comparison of Different Features of Various Brands

Apart from the above analysis, few customers were interested in good camera quality, and few may concentrate only on battery life, etc. To satisfy those customers, the below graphs were visualized by comparing with different features of various smart phone brands. This visual analysis would help the customer for easy purchase based on unique requirement. Figure 3 represents the analysis of battery life for each mobile phone brand and the analysis is made based on the star rating given by the previous buyers. Similarly, Fig. 4, 5 and 6 represents the star rating analysis of picture quality, sound quality and camera quality with respect to the various mobile brands. This star rating analysis would help the customers to buy the mobile phone with their unique feature requirement.

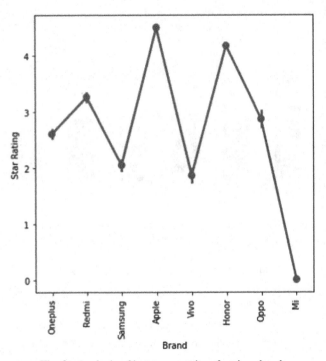

Fig. 3. Analysis of battery capacity of various bands

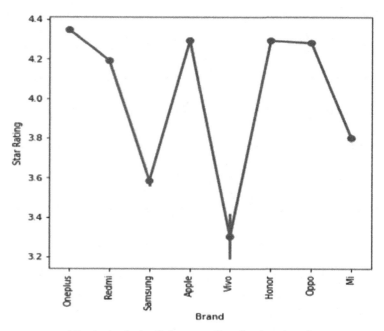

Fig. 4. Analysis of picture quality of various brands

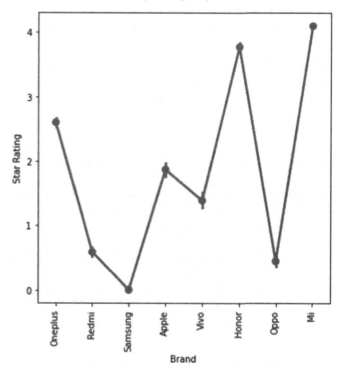

Fig. 5. Analysis of sound quality of various brands

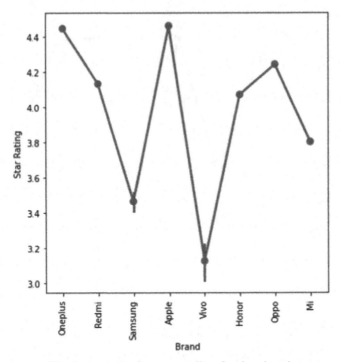

Fig. 6. Analysis of camera quality of various brands

Figure 7, illustrates the distribution of rating by various customers after purchase of the respective brands of their interest. It visualizes the number of customers who have given 1 star rating, 2 star rating, for each brand, and so on.

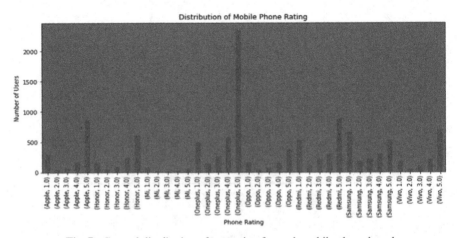

Fig. 7. General distribution of star rating for each mobile phone brand

From the Fig. 8, the overall distribution of rating score for each brand ranging from 1 to 5 is visualized. Oneplus mobile phone brand is having average of 4.1 star rating out of 5. Similarly, the average star rating for other mobile phone brands is also visualized and the same is summarized in Table 2.

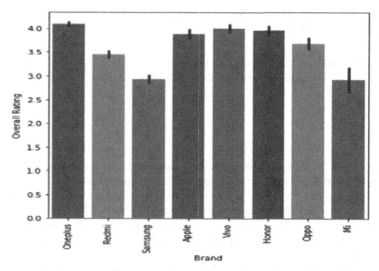

Fig. 8. Distribution of rating score for each mobile phone brand

Table 2. Rating score for each mobile phone brand

Mobile phone brand	Rating score
Oneplus	4.1
Redmi	3.4
Samsung	2.9
Apple	3.8
Vivo	4.0
Honor	3.9
Oppo	3.6
Mi	2.9

6 Conclusion and Future Work

Data visualization for the customer reviews of the following are explained in this paper:

1. Trustworthiness of online purchase of mobile phones: From the analysis it is clearly observed that online purchase of smart phones has positive review.

2. Reviews available for the different brands of mobile phones: From the analysis it is inferred that the filtering of mobile phone brand for online purchase could be made easier than checking the review comments of the various brands.
3. Comparison of various features of the different brands of the mobile phones were analysed in this paper with the help of the amazon review dataset. This comparative analysis would help the future buyers to find the best phone which is more frequently brought through online.

This study gathered information on only few mobile phone brands. Future research could include other brands, and additional features of the various brands for comparative analysis.

References

Wu, J., Wu, Y., Sun, J., Yang, Z.: User reviews and uncertainty assessment: a two stage model of consumers' willingness-to-pay in online markets. Decis. Support Syst. **55**(1), 175–185 (2013)

Li, M., Huang, L., Tan, C.H., Wei, K.K.: Helpfulness of online product reviews as seen by consumers: Source and content features. Int. J. Electron. Commer. **17**(4), 101–136 (2013)

Sparks, B.A., Perkins, H.E., Buckley, R.: Online travel reviews as persuasive communication: the effects of content type, source, and certification logos on consumer behavior. Tour. Manag. **39**, 1–9 (2013)

Spool, J.: The Magic behind Amazon's 2.7 Billion Dollar Question (2009). http://www.uie.com/articles/magicbehindamazon/2009

Korfiatis, N., García-Bariocanal, E., Sánchez-Alonso, S.: Evaluating content quality and helpfulness of online product reviews: the interplay of review helpfulness vs. review content. Electron. Commer. Res. Appl. **11**(3), 205–217 (2012)

Bilal, M., et al.: Profiling users' behavior, and identifying important features of review "helpfulness." IEEE Access **8**, 77227–77244 (2020)

Dong, H., Hou, Y., Hao, M., Wang, J., Li, S.: Method for ranking the helpfulness of online reviews based on SO-ILES TODIM. IEEE Access **9**, 1723–1736 (2020)

Basiri, M.E., Habibi, S.: Review helpfulness prediction using convolutional neural networks and gated recurrent units. In: 2020 6th International Conference on Web Research (ICWR), pp. 191–196. IEEE, April 2020

Kong, L., Li, C., Ge, J., Ng, V., Luo, B.: Predicting product review helpfulness a hybrid method. IEEE Trans. Serv. Comput. (2020)

Benlahbib, A.: Aggregating customer review attributes for online reputation generation. IEEE Access **8**, 96550–96564 (2020)

Fouladfar, F., Dehkordi, M.N., Basiri, M.E.: Predicting the helpfulness score of product reviews using an evidential score fusion method. IEEE Access **8**, 82662–82687 (2020)

Urologin, S.: Sentiment analysis, visualization and classification of summarized news articles: a novel approach. Int. J. Adv. Comput. Sci. Appl. **9**(8), 616–625 (2018)

Khan, S.A., Velan, S.S.: Application of exploratory data analysis to generate inferences on the occurrence of breast cancer using a sample dataset. In: 2020 International Conference on Intelligent Engineering and Management (ICIEM), pp. 449–454. IEEE, June 2020

Alamoudi, E.S., Al Azwari, S.:. Exploratory data analysis and data mining on yelp restaurant review. In: 2021 National Computing Colleges Conference (NCCC), pp. 1–6. IEEE, March 2021

Alamoudi, E.S., Alghamdi, N.S.: Sentiment classification and aspect-based sentiment analysis on yelp reviews using deep learning and word embeddings. J. Decis. Syst. **00**(00), 1–23 (2021)

Gupta, R., Sameer, S., Muppavarapu, H., Enduri, M.K., Anamalamudi, S.: Sentiment analysis on Zomato reviews. In: 2021 13th International Conference on Computational Intelligence and Communication Networks (CICN), pp. 34–38. IEEE, September 2021

Supervised Learning of Procedures from Tutorial Videos

S. Arunima⬤, Amlan Sengupta, Aparna Krishnan, D. Venkata Vara Prasad$^{(\boxtimes)}$, and Lokeswari Y. Venkataramana

Department of Computer Science and Engineering, Sri Sivasubramaniya Nadar College of Engineering, Kalavakkam, Chennai 603110, India
{arunima17016,amlan17008,aparna17012}@cse.ssn.edu.in,
{dvvprasad,lokeswariyv}@ssn.edu.in

Abstract. Online educational platforms and MOOCs (Massive Open Online Courses) have made it so that learning can happen from anywhere across the globe. While this is extremely beneficial, videos are not accessible by everyone, due to time constraints and lower bandwidths. Textual step-by-step instructions will serve as a good alternative, being less time-consuming to follow than videos, and also requiring lesser bandwidth. In this work, the authors aim to create a system that extracts and presents a step-by-step tutorial from a tutorial video, which makes it much easier to follow as per the user's convenience. This can be mainly accomplished by using Text Recognition for academic videos and Action Recognition for exercise videos. Text Recognition is accomplished using the Pytesseract package from Python which performs OCR. Action Recognition is performed with the help of a pre-trained OpenPose model. Additional information is extracted from both exercise and academic videos with the help of speech recognition. The information extracted from the videos are then segmented into procedural instructions which is easy to comprehend. For exercise videos, a frame-wise accuracy of 79.27% is obtained. Additionally, for academic videos, an accuracy of 78.47% is obtained.

Keywords: Openpose · OCR · Exercise videos · Academic videos · Speech recognition · Action recognition · Text recognition

1 Introduction

Knowledge and the dissemination of knowledge has become truly global with the advent of the Internet. Therefore, tutorial videos have become one of the primary media through which knowledge is imparted. However, video streaming is resource-intensive and not always possible in remote areas with poor connectivity or due to time constraints. In such situations, having a system that creates a procedural set of instructions from the tutorial videos, which can be read and reread as per the user's convenience. Text Recognition and Action Recognition can be used to interpret the information relayed in tutorial videos and segmented into a procedural set of instructions. With longer, more detailed tutorial

© IFIP International Federation for Information Processing 2022
Published by Springer Nature Switzerland AG 2022
L. Kalinathan et al. (Eds.): ICCIDS 2022, IFIP AICT 654, pp. 356–366, 2022.
https://doi.org/10.1007/978-3-031-16364-7_28

videos this system is the most impactful as it can turn a complicated video into an easy-to-follow textual tutorial.

The recognition and interpretation of human or robot-induced actions and activities have gained considerable interest in computer vision, robotics, and AI communities and as more and more videos are being made, and more students refer to them for their studies, there is an increase in the need to properly recognize the text content in the video. This is the base of text recognition. Text recognition is mainly carried out using Optical Character Recognition (OCR) techniques. The modeling and learning of the extracted features are a critical part of the action recognition, in improving its accuracy. Some popular techniques include optical motion detection, 3-D volume representation, temporal modeling, Hidden Markov Model (HMM) training, Dynamic Time Warping (DTW), multi-view subspace learning, and Pose estimation.

The motivation behind building such a system is to make the dissemination of knowledge over the Internet, especially in these times of online learning, easier and more accessible. The objective of this system is to ensure that students are not deterred by bandwidth restrictions or videos that are too long to be comprehensible. This is the first step towards building systems that make knowledge truly universal.

2   Related Work

Several research work has been reported in the field of procedure extraction from demonstration videos. The authors [1] proposed the method of temporal clustering and latent variable modeling, and developed a general pipeline for procedure extraction. Temporal clustering consists of aggregating similar frames or images and removing duplicates from the pipeline. This method is further evaluated using improved metrics like temporal purity and temporal completeness are proposed.

Another approach called Video Instance Embedding (VIE) framework [2] which trains deep nonlinear embedding on video sequence inputs is introduced. The authors feel that a two-pathway model with both static and dynamic processing pathways is optimal as it provides analyses indicating how the model works.

DPC [3] is used to predict a slowly varying semantic representation based on the recent past. A video clip is partitioned into multiple non-overlapping blocks, with each block containing an equal number of frames. An aggregation function is then applied to the blocks and this is used for creating a context representation which helps identify the actions.

The concept of using CompILE [4], is a framework for learning reusable, variable-length segments of hierarchically-structured behavior from demonstration data is also introduced. CompILE utilizes a novel unsupervised, fully-differentiable sequence segmentation module to learn hidden encodings of sequential data that can be re-composed and executed to perform new tasks. Once trained, the model generalizes to sequences of longer length and from environment instances not seen during training. CompILE has been tested in 2D multi-task environments successfully.

There are many approaches to action recognition. A key aspect of using machines to have an understanding of people in images and videos lies on Realtime multi-person 2D pose estimation [5]. This work is used to detect the poses of multiple people in an image. This method uses Part Affinity Fields (PAFs), to learn to associate the values that represent body parts to the individual people identified in the image. Since it is a bottom-up approach, it obtains higher accuracy and real time performance, regardless of the number of people in the image.

3 Methodology

The dataset [7] for this project has been majorly taken from YouTube. For exercise videos, a compilation of different exercises along with various angles for each exercise has been given for training. For academic videos, simple tutorials have been used. The proposed methodology employs two different approaches for exercise and academic videos as shown in Fig. 1.

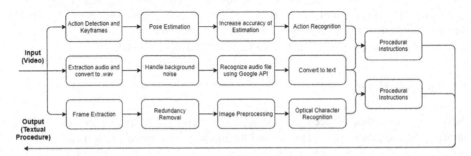

Fig. 1. Flow of exercise cycle

The overall process is divided into three separate flows of recognition: video, audio, text.

In the video recognition flow, the action shown is detected and keyframes are extracted. Then, pose estimation is performed using OpenPose on the detected actions. These estimates are further fine-tuned and the action (exercise) performed is identified.

In the audio recognition flow, the audio is converted to a usable format. The background noise is handled appropriately and lessened. Then, Google's Speech-to-Text is used to transcribe the audio file, and the text contained in the audio file is effectively obtained.

In the text recognition flow, video frames are extracted and redundant frames are removed. Each frame is preprocessed to make the text content clearer, and Optical Character Recognition is performed to identify the text.

The outputs from the three flows are combined to form the procedural instructions, which will be the final output of the system.

3.1 Exercise Videos

The actions performed in exercise videos are identified by using OpenPose estimation [8].

OpenPose Methodology. OpenPose is divided into three different blocks which are (a) body+foot detection, (b) hand detection, and (c) face detection. The core block is made up of the combined body+foot key poinAdditionally Convolutional Neural Networks (CNNs) i.e. 3 layers of convolutions of kernel 3 are used to obtain reliable local observations of body parts. DeepSort algorithm [6] is also used for tracking multiple people. Consequently, the deep neural networks are used to recognise the actions performed by each person based on single frame-wise joints detected from OpenPose.

t detector. The points recognised in the body and foot locations help the facial bounding box to be more accurate and also get more key areas like ears, eyes, nose and neck.

The single frame-wise joints are given as input to a baseline CNN network like VGG-19. The feature map thus obtained is processed through a multi-stage CNN pipeline which generates the Part Confidence Maps and Part Affinity Field (PAF) [5].

A Confidence Map consists of a 2D representation of the belief that any given pixel can be used to map a particular body part.

$$S = (S_1, S_2, ..., S_B) \ where \ S_b \in R^{w*h}, b \in 1...B \tag{1}$$

The above Eq. 1 denotes a confidence map where B is the number of body part locations and each map has a dimension of w * h. This confidence map is used to denote high probabilities where a body part might be in the input image.

The location and orientation of limbs of multiple people is encoded using a set of 2D vector fields known as Part Affinity Fields. The data is encoded in the form of pairwise connections between body parts i.e. it points the direction from one limb to another. The following Eq. 2 represents part affinity fields L, which consists of a set of "N" 2D matrices for each limb in the body. The dimension of L is w * h * n.

$$L = (L_1, L_2, ..., L_N) \ where \ L_n \in R^{w*h*n}, n \in 1...N \tag{2}$$

In case of multiple people detection due to the presence of multiple heads, hands etc., there is a need to map the correct body parts with the correct person. Equation 3 is used to calculate a predicted part affinity which is based on two of the locations of the body parts D_{b1} and D_{b2}. This predicted PAF value is used to determine E_n for the set of D_b values and maximize it.

$$max_{Z_n} E_n = max_{Z_n} \sum_{m \in D_{b_1}} \sum_{n \in D_{b_2}} E_{mn}.Z_{b_1 b_2}^{mn} \tag{3}$$

Action Detection Procedure. The input consists of a single exercise video from which frames are individually taken and the 36 points corresponding to distinct body parts are identified by the pose estimation algorithm. These points are stored as decimal values in a text file which is further converted into.csv format after adding an additional column i.e. class label. The actions are mapped to their respective class labels in Action function. The csv is then given as input for training the model. The number of cases for each class is given in Table 1.

Table 1. Training data for action recognition

Exercise	No. of lines	Class
Squat	445	0
Jumping jacks	715	1
Push ups	351	2
High knees	744	3
Total	**2255**	

A sequential keras model with relu activation function is used for training. A classifier trained for specific action is saved. The classifier is loaded along with a pre-trained VGG model [9] to perform action recognition.

Figure 2 is an example of how the points are mapped in an image. A text file is generated containing the sequence of distinct exercises performed throughout the video as shown in Fig. 3.

Fig. 2. Pushups

🗒 output - Notepad

File   Edit   Format   View   Help

Exercises to be performed are:
highknees
squats
pushups
jumpingjacks

Fig. 3. Output for activity recognition

3.2 Academic Videos

The input video is split into various frames at regular intervals using packages like OpenCV [10]. To remove similar and repetitive frames, similarity measures like Mean

Squared Error (MSE) and Structural Similarity Index (SSIM) are used. Structural Similarity Index (SSIM) is a metric used to measure the similarity between two images. The similarity between the two images is represented by a value between -1 and $+1$. A SSIM of $+1$ denotes that the two images are very similar. A value of -1 denotes the images have very less similarity. The system uses 3 aspects to measure similarity: Luminance, Contrast and Structure. The more similar images are, the more likely they are to be removed. The final collection of images has less similar images thereby making it easier for image pre-processing and recognition of text. For pre-processing, the colorspace of the images is changed to grey-scale using the cvtColor() method, and a Gaussian Blur is added using low-pass filters using the GaussianBlur() method. The text in the image is detected by covering the textual area with a rectangle and OCR is performed using Pytesseract [11] on that area.

A single academic tutorial video is given as input from which frames are extracted at 3 second intervals. Redundant frames are removed using SSIM [12]. Image preprocessing is performed on the images which includes converting to grayscale, blurring the background and thresholding. OCR using Pytesseract is performed to extract text from the preprocessed images. A text file is generated for recognized text as shown in Fig. 4.

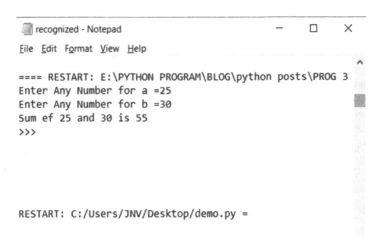

Fig. 4. Output for text recognition

3.3 Speech Recognition

Speech recognition is used to note down the additional information given in the videos. For example, instructions given by the trainer in exercise videos and important points discussed by the teacher in academic videos. Therefore, speech has been included in both types of videos. The audio from the input video is extracted and chunked to sub-clips using the extract- subclip() method from the moviepy python package. Each sub-clip is converted to .wav format and written back to the sub-clip files. Using each sub-clip as the source, the adjust-for-ambient-noise() method is used to remove noise from the sub-clip. Once each sub-clip has been pre-processed, the recognize-google() Speech Recognition method from Google is used to recognize the text from each clip. The chunked clips from which the text is recognized are later joined. The text obtained from speech recognition is punctuated using the python library punctuator and then added to the final output for both types of videos.

4   Results

For exercise videos, an output file (Fig. 5) is generated which contains the sequence of exercises performed throughout the video along with the instructions spoken by the trainer.

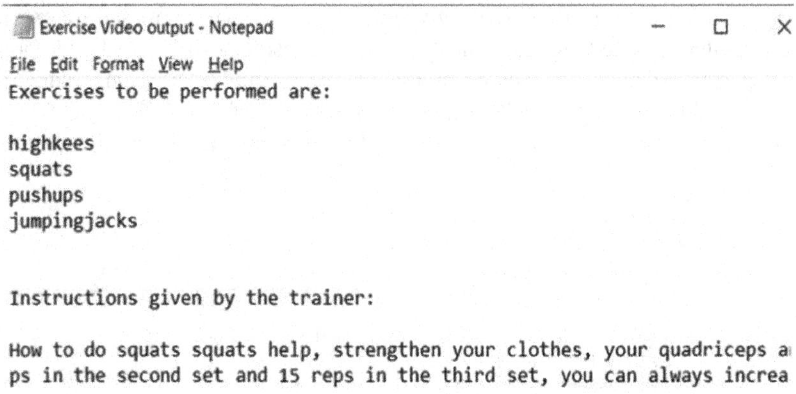

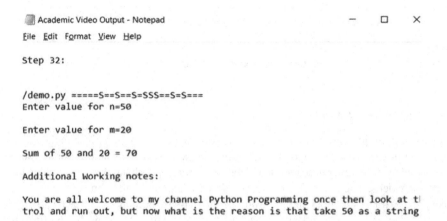

Fig. 5. Output for exercise videos

For academic videos, an output file (Fig. 6) containing the procedural information from the video along with the additional information spoken by the tutor is generated.

Fig. 6. Output for academic videos

4.1 System Performance

The performance and accuracy of the first part of the system is measured in terms of the percentage of exercise instances classified correctly in the case of Exercise Videos. The following graph in Fig. 7 shows the accuracy of OpenPose when run over our data.

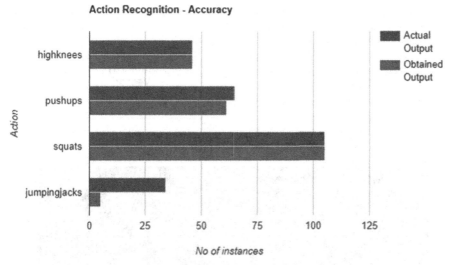

Fig. 7. OpenPose accuracy

The test video consisted of 304 frames, out of which 292 keyframes were detected and action was performed on these keyframes. Out of the 292 keyframes, 51 were classified incorrectly thereby leading to 241 frames classified accurately. This provided the model with a frame-wise accuracy of 79.27%.

The model involves using OpenPose specifically for exercise videos when compared to other models which include a multitude of actions. It is able to accommodate as many exercises as needed. The training is short when compared to other deep learning methods as the pose estimation points are extracted during the length of the video. The classifier containing the pose estimation points for each exercise is trained using a keras sequential model with relu activation. The training time is optimal as only points are given as inputs rather than images.

For Text Recognition, the accuracy is measured using Document Similarity as shown in Fig. 8 and 9. Document similarity is calculated by calculating document distance. Document distance is a concept where words (documents) are treated as vectors and is calculated as the angle between two given document vectors. Document vectors are the frequency of occurrences of words in a given document.

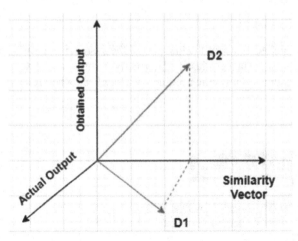

Fig. 8. Document similarity - accuracy of text detection

```
PS C:\Users\Aparna Krishnan\Desktop> python simi.py
File actual.txt :
840 lines,
187 words,
41 distinct words
File output.txt :
6149 lines,
1187 words,
261 distinct words
The distance between the documents is:  0.678415 (radians)
PS C:\Users\Aparna Krishnan\Desktop> []
```

Fig. 9. Document similarity - output

In the case of Academic Videos, the percentage similarity between the gold label document and the output document determines the system performance as shown in Table 2.

Table 2. Accuracy for text recognition

File 1	File 2	Percentage similarity
AT	OF	69.42
AT	OF-RR	78.47
OF	OF-RR	89.72

Key for Table 2:

1. *AT: Actual text that is manually extracted from the video*

2. *OF: Output obtained without performing redundancy removal*
3. *OF-RR: Output obtained by using redundancy removal on the frames*
4. *File 1: The first file taken for calculating similarity*
5. *File 2: The second file taken for calculating similarity*

The actual.txt file is manually created to act as the gold labels for checking the accuracy of the files that are obtained as output from the OCR function. The output without redundant frames being removed when compared to actual.txt is comparatively lesser than the accuracy obtained from comparing actual.txt and the output obtained from the frames where redundancy removal has been performed. The similarity between the two generated output files is quite high since only redundant and repetitive frames have been removed.

5 Source Code and Computational Resources

The source code for implementing this system has been uploaded in this GitHub repository [13]. The resources used for implementing the system were a 2.60 GHz Intel core i7 CPU and Intel UHD Graphics 630 4 GB GPU.

6 Conclusion

This work is the first step in making education accessible to all. With the world going increasingly online, education is now more global than ever before in history. The hope and end goal is to contribute to this success with this system.

This system creates a set of readable instructions that are much more accessible and less resource intensive than videos. The many advantages of such a system will be most felt in places with low internet connectivity, and for longer videos with many steps and sub steps. Based on the results of the first iteration, this system can be extended to other classes of tutorial videos. The accuracy of recognition can also be exponentially improved with each iteration.

As a part of future work, the authors intend to apply this system to all kinds of tutorial videos. The authors also aspire to make this system multilingual, thereby providing unfettered access to tutorial videos globally. Using advanced computing resources such as a dedicated Graphical Processing Unit (GPU), the accuracy of real-time text and action recognition can be increased manifold. This is due to the fact that the model can be trained on a larger data set and for many more epochs using a GPU. With higher computing power and a more diverse data set, the accuracy and efficiency of this model, and the ease of extension to other classes of videos will increase exponentially.

References

1. Goel, K., Brunskill, E.: Unsupervised learning of procedures from demonstration videos (2018)

2. Zhuang, C., She, T., Andonian, A., Mark, M.S., Yamins, D.: Unsupervised learning from video with deep neural embeddings. In: Proceedings of the IEEE/CVF Conference on Computer Vision and Pattern Recognition, pp. 9563–9572 (2020)
3. Han, T., Xie, W., Zisserman, A.: Video representation learning by dense predictive coding. In: Proceedings of the IEEE International Conference on Computer Vision Workshops (2019)
4. Kipf, T., et al.: Compile: Compositional imitation learning and execution. In: International Conference on Machine Learning, pp. 3418–3428. PMLR (2019)
5. Cao, Z., Hidalgo, G., Simon, T., Wei, S.-E., Sheikh, Y.: OpenPose: realtime multi-person 2D pose estimation using part affinity fields (2019)
6. Wojke, N., Bewley, A., Paulus, D.: Simple online and realtime tracking with a deep association metric (2017)
7. https://github.com/arunimasundar/Supervised-Learning-of-Procedures/tree/master/dataset
8. https://github.com/CMU-Perceptual-Computing-Lab/openpose
9. Simonyan, K., Zisserman, A.: Very deep convolutional networks for large-scale image recognition (2015)
10. https://pypi.org/project/opencv-python/
11. https://pypi.org/project/pytesseract/
12. https://www.mathworks.com/help/images/ref/ssim.html
13. https://github.com/arunimasundar/Supervised-Learning-of-Procedures

Author Index

Printed in the United States
by Baker & Taylor Publisher Services